CONTAGION

CONTAGION

HOW COMMERCE HAS SPREAD DISEASE

MARK HARRISON

YALE UNIVERSITY PRESS
NEW HAVEN AND LONDON

Copyright © 2012 Mark Harrison

All rights reserved. This book may not be reproduced in whole or in part, in any form (beyond that copying permitted by Sections 107 and 108 of the U.S. Copyright Law and except by reviewers for the public press) without written permission from the publishers.

For information about this and other Yale University Press publications, please contact:
U.S. Office: sales:press@yale.edu   www.yalebooks.com
Europe Office: sales @yaleup.co.uk   www.yalebooks.co.uk

Set in Minion Pro by IDSUK (DataConnection) Ltd
Printed in Great Britain by TJ International Ltd Padstow, Cornwall

Library of Congress Cataloging-in-Publication Data

Harrison, Mark.
  Contagion: how commerce has spread disease/Mark Harrison.
    p. cm.
  Includes bibliographical references and index.
  ISBN 978-0-300-12357-9 (cl:alk. paper)
1. Epidemics.   2. Communicable diseases.   3. International trade.   I. Title.
  RA651.H37 2013
  614.4—dc23
                                                                        2012017219

A catalogue record for this book is available from the British Library.

10 9 8 7 6 5 4 3 2 1

Contents

	Abbreviations and acronyms	*vi*
	List of illustrations	*ix*
	Preface and acknowledgements	*xii*
1	Merchants of death	1
2	War by other means	24
3	The evils of quarantine	50
4	Quarantine and the empire of free trade	80
5	Yellow fever resurgent	107
6	A stranglehold on the East	139
7	Plague and the global economy	174
8	Protection or protectionism?	211
9	Disease and globalization	247
	Conclusion: Sanitary pasts, sanitary futures	276
	Notes	*282*
	Bibliography	*332*
	Index	*364*

Abbreviations and acronyms

AIDS	Acquired Immune Deficiency Syndrome
AHC	Archivo Histórico Colonial, Lisbon
AHU	Archivo Histórico Ultramarino, Lisbon
APAC	Asian, Pacific and African Collections
APEC	Asia-Pacific Economic Cooperation
APS	Library of the American Philosophical Society, Philadelphia
BCA	Baltimore City Archives
BL	British Library, London
BOIE	*Bulletin de l'Office International des Épizooties*
BOIHP	*Bulletin de l'Office International d'Hygiène Publique*
BRO	Bristol Record Office
BSE	Bovine Spongiform Encephalopathy
CCA	Chester and Cheshire Archives
CEO	Chief Executive Officer
CER	Chinese Eastern Railway
CFBF	California Farm Bureau Federation
CWF	Compassion in World Farming
DCRO	Derbyshire Country Record Office
DGNMS	Director-General of the Naval Medical Service
DNB	*Dictionary of National Biography* (United Kingdom)
DSB	Dispute Settlement Board
EU	European Union
FAO	Food and Agriculture Organization

FO	Foreign Office
GATT	General Agreement on Tariff and Trade
GCRO	Gloucestershire County Record Office
GOI	Government of India
HIV	Human Immunodeficiency Virus
HMSO	Her/His Majesty's Stationary Office
ICC	Isthmian Canal Commission
IHR	International Health Regulations
IMS	Indian Medical Service
IPPC	International Plant Protection Convention
KCTU	Korean Federation of Trade Unions
LOC	Library of Congress, Washington, DC
MHS	Marine Hospital Service
MPPS	Manchurian Plague Prevention Service
MSA	Maharashtra State Archives, Mumbai
NAFTA	North American Free Trade Association
NAI	National Archives of India, New Delhi
NARA	National Archives and Records Administration, College Park, MD
NGO	Non-Governmental Organization
NMM	National Maritime Museum, London
OIE	Office International des Épizooties
OIHP	Office International d'Hygiène Publique
Parl. Debates	Hansard, Parliamentary Debates
PH	Public Health
PP	Parliamentary Papers (United Kingdom)
PRC	Peoples' Republic of China
PVCS	Procès Verbaux du Conseil Superieur (Constantinople)
RCPE	Library of the Royal College of Physicians of Edinburgh
RCPL	Library of the Royal College of Physicians of London
SARS	Severe Acute Respiratory Syndrome
SPS	Sanitary and Phytosanitary
TNA	The National Archives, London
vCJD	variant Creutzfeld-Jakob Disease
WBSA	West Bengal State Archives, Kolkata

WHO	World Health Organization
WL	Wellcome Library, London
WOAH	World Organization for Animal Health
WSRO	West Sussex Record Office
WTO	World Trade Organization

Illustrations

Unless otherwise stated all images are courtesy of the Wellcome Library, London.
1. Monro Scott Orr, map showing the history and distribution of the Black Death around the world, c.1920.
2. Pierre Mignard after F. de Poilly and Henri Arnauld, Saint Carlo Borromeo administering communion to victims of the plague in Milan of 1576, c.1645.
3. Small poster or leaflet from Ferrara dated 11 December 1679 proclaiming restrictions on trade, from G.M. Estense Tassoni, *Prohibitione di comercio di tutto il Trentino e de Siginori Griggioni* ... (Ferrara, 1679).
4. 'Orders conceived and published by the Lord Major and Aldermen of the City of London, concerning the Infection of the Plague', 1665.
5. Augustus Earle after Edward Francis Finden and Maria Callcott, European men examining slaves at the slave market of Rio de Janeiro, 1824.
6. Slave quarters on a sugar plantation in Martinique, from *Voyage pittoresque dans les deux Amériques. Résumé général de tous les voyages de Colomb, Las-Casas, Oviedo ... Humboldt ... Franklin ... etc* (Paris, 1836).
7. 'Hospital for the Sick' and 'The Hall for the Sale of Stuffs and Cloths ready made' in Batavia, Jakarta, from Awnsham Churchill, John Churchill, John Locke and John Nieuhoff, *A collection of voyages and*

travels, some now first printed from original manuscripts, others now first published in English (London, 1744–6).

8. Health pass ('fede di sanità'), Officio della sanità di Venetia, Venice, 1713.
9. James Barry, R.A., the Thames as a gateway to foreign trade routes, 1791.
10. Sea view of the lazaretto at Genoa, from John Howard, *An account of the principal Lazarettos in Europe* (Warrington, 1789).
11. The quarantine station ('Offizia della Sanita') at Naples, from John Howard, *An account of the principal Lazarettos in Europe* (London, 1791).
12. Sketch of a model lazaretto, from John Howard, *An account of the principal Lazarettos in Europe* (London, 1791).
13. W. Ward after J. Jackson, portrait of Sir James McGrigor, *c*.1815.
14. Nicolas-Eustache Maurin, the plague in Barcelona, 1821.
15. Tiberius Cavallo, portrait of Clot Bey, from Giovanni Birch, *Teoria e pratica dell'elettricità medica; e della forza dell'elettricità nella cura della suppressione de' mestrui, del chirurgo* (Naples, 1784).
16. Henry Cousins after Martin Archer Shee, portrait of Sir William Burnett, *c*.1840.
17. Unknown artist, portrait of James Ormiston McWilliam, *c*.1845.
18. Front cover of James Ormiston McWilliam, *Medical History of the Expedition to the Niger during the years 1841–2: comprising an account of the fever which led to its abrupt termination* (London, 1843).
19. 'The Royal West India Mail Steamer "Tyne" on Shore at St. Alban's Head', from the *Illustrated London News*, 30, 1857.
20. Juan Manuel Blanes, yellow fever in Buenos Aires, 1871.
21. Front cover of the passenger list of the White Star Line steamship *Britannic* for its journey from New York to Liverpool, 27th October 1883. Charles Brodie Sewell (1817–1900).
22. Cutting through the Culebra Mountain, general view looking west, towards Panama, from *Illustrated London News*, 92, 1888.
23. Man with Clayton disinfector, from Oswaldo Cruz, *Os serviços de saúde publica no Brasil: especialmente na cidade do Rio de Janeiro: de 1808 a 1907* (Rio de Janeiro, 1909).
24. Quarantine Station, Culebra Island, Panama, 1909.

25. 'Blue stage of the spasmodic cholera', November 1831, from the *Lancet*, MDCCCXXXI-XXXIII, 4 February 1832.
26. 'Map of some of the Principal Places in Asia & Europe, visited by the Contagious Cholera', from James Kennedy, *The history of the contagious cholera; with facts explanatory of its origin and laws, and of a rational method of cure* (London, 1832).
27. 'Mistaking cause for effect', from *Punch*, 1849.
28. John Peters, 'Actual & Supposed Routes of Cholera from Hindoostan to Europe and to North and South America in 1832, 1848, 1854, 1867 and 1873', from Edmund Charles Wendt, *A Treatise on Asiatic Cholera* (New York, 1885).
29. 'The Cholera in Egypt: Quarantine Examination at Brindisi', 1883.
30. Device for fumigating letters, c.1870, from Dr Cabanes, *Moeurs intimes du Passé*, c.1923.
31. The Staffordshire Regiment cleaning plague houses, Hong Kong, 1894.
32. Unknown artist, Bombay at the time of the plague, 1896 or 1897.
33. Albert Lloyd Tarter, *The plague spreads to America*, 1940s.
34. Albert Lloyd Tarter, *A rat stowing away on a ship carrying the plague further afield*, 1940s.
35. Albert Lloyd Tarter, *The global spread of plague, carried by rats*, 1940s.
36. H. S. Melville, 'Inspection of foreign cattle at the Metropolitan Cattle Market', from the *Illustrated London News*, 47, 1865.
37. Text of a bishop's prayer for protection from cattle plague, 1866.
38. Broadsheet issued by the County Police Office, Dunse, 19 November 1867, in the County of Berwick, Scotland.
39. Satirical drawing, 'The demon butcher, or the real rinderpest', from *Punch*, 1865.
40. Jung Yeon-Je, protests against beef imports, South Korea, June 2008. Courtesy of Getty Images.
41. Travel Ink, SARS inspection of passengers leaving plane, 2003. Courtesy of Getty Images.
42. AFP, quarantine checkpoint at infected chicken farm, South Korea, April 2008. Courtesy of Getty Images.
43. Romeo Gacad, crop duster plane spraying fungicide on bananas, Mindanao, Philippines, 2008. Courtesy of Getty Images.

Preface and acknowledgements

In the 1860s, the English physician William Budd warned of two 'great tyrants' – war and commerce – which had left pestilence in their wake and threatened to do so again.[1] The world was then facing the twin menace of cholera and cattle plague and in the coming decades it would experience two terrifying pandemics of influenza and one of plague. These catastrophes were all, in some measure, the product of conflict and trade. Budd had also many historical precedents on which to draw. In his own century, the march of armies, the dislocation of civilians and the destruction of civic infrastructure had been implicated in numerous epidemics. Typhus had devastated many parts of Europe during the French wars, for example, not to mention the army of Napoleon on its retreat from Russia. Commerce, too, was widely held to be the vehicle of cholera and yellow fever – two of the greatest epidemic diseases to blight the world in the first half of the century. As well as being a consistent factor in the spread of disease, commerce had transformed the societies it had bound together, reshaping the landscape of production as horizons shifted to distant markets. Rapid industrialization, urbanization and the reorientation of agriculture towards world markets had dramatically changed disease environments, for good and ill.

The nineteenth century was a period of rapid change and saw an unprecedented redistribution of infections, but commerce had long shaped the epidemiological fate of humanity. Indeed, of Budd's two great tyrants, commerce is arguably the greater, especially when its effects are measured over a long duration. From the fourteenth century, when the Black Death spread westward with merchandise from Asia, pestilence and trade became intimately associated in many cultures. And, because of these long-standing associations, disease began to represent concerns about commerce – about its propensity to transform societies and about the relationship between private profit and the common good. The notion of contagion also became a metaphor for cultural contamination and for the corruption of public life: anxieties which are as acute today – in an age of globalization – as at any time in the past. Recent outbreaks of BSE, *E. coli* and influenza have reminded us of the propensity of commerce to bring misery as well as wealth, and have raised awkward questions about the markets and regulators on which we depend. Is it possible to have freedom of movement and commerce without the globalization of disease? Just what is the right balance between sanitary protection and commercial freedom? How far should an individual state submit to the norms and regulations of the international community?

The history of commerce and contagion may not provide definitive answers to these questions but it can help us to understand the challenges we face. It is imperative that we do because the sanitary control of trade currently offers neither adequate protection nor economic stability. While specific aspects of trade-related disease are dealt with by bodies such as the World Trade Organization and the World Health Organization, preventive measures remain fragmented and crisis-driven. Each major outbreak of disease elicits a rash of hasty and opportunistic interventions intended to gain economic advantage as much as to protect health. Even when the validity of such measures is challenged – as it often is – the aggrieved parties cannot be sure of satisfaction, and compensation is generally tardy and inadequate. Quarantines have become tariffs by another name and the disputes they are apt to generate constitute a persistent and serious threat to global trade.

There is nothing new in the abuse of sanitary controls for economic and political gain but it is perhaps surprising that the situation persists despite international agreements and institutions of global governance. There are two reasons for this. The first is the competitive pressure exerted by globalization. The second is that disease prevention has come to be seen largely in terms of security, whether it is 'biosecurity' within individual states or sanitary barriers between them. This process has not gone far enough for some,[2] but the protection afforded by such measures is exaggerated. From a humanitarian perspective, too, the 'securitization' of health tends to serve the interests of the richer nations as against the poorer, and of producers over consumers. The obsession with security fosters an atmosphere in which stringent quarantines and trade embargoes can be more easily justified and this affects developing countries disproportionately because they depend more on agricultural exports. However, the ill effects of sanitary protectionism are by no means confined to poorer nations. The mistrust and retaliation it engenders presents a threat to the global economy every bit as serious as disease itself.

One purpose of this book is to understand the situation that has come to pass but it is also an historical study in a more traditional sense, in that it places the spotlight on an aspect of human history which is too often neglected. Trade is generally recognized as an important factor in most histories of epidemic disease but it is seldom the principal object of study, whereas there are numerous works on war and epidemics.[3] Commerce and its maladies are better served in histories of quarantine but these tend to focus on national experiences, rather than on comparisons or connections between different countries. Histories of commerce mention epidemics from time to time but in most cases they do not figure prominently and there is seldom serious consideration of trade and its relationship to public health. This book seeks to fill this gap by exploring two interrelated themes: the ways in which disease spreads as a result of long-distance trade and the measures which have been taken to prevent it. Its starting point is the Black Death, not because this was the first pandemic in which trade played a major role – that might also be said of the Plague of Justinian centuries before – but it was from this point that the two came to be closely connected in folk memory and public policy. The mid fourteenth century brought the first measures designed to control the spread of trade-borne infection and these have shaped our responses to disease ever since. While much has undoubtedly changed since then, these continuities suggest the relevance of history to an understanding of the present.

Writing a history that spans nearly seven centuries and six continents presents formidable difficulties. Anyone who attempts such a thing must rely to some extent on the scholarship of others, yet there are remarkably few specialized studies of commerce and disease from which to construct a general narrative. This has meant adopting a rather different approach: synthesizing existing histories where possible but relying overwhelmingly on original sources and case studies to illuminate general themes. The book's coverage is not always global in a strict, geographical sense but it traces interactions over a long time-scale and a substantial geographical range. It is in this sense – in the exploration of connections beyond traditional frames of reference – that this book may be regarded as a 'global' history.[4] Its case studies have been chosen to illustrate major changes in disease distribution, as well as the various ways in which diseases have been tackled. The earlier chapters are devoted to particular regions or trading networks, while the later ones examine global integration. I have resisted the temptation to project the concept of globalization on to the distant past because it is not strictly appropriate.[5] Neither disease nor commerce became truly global until the late nineteenth century and, even then, the first global economy was a very different creature to the globalized world of today.

The most basic task of this book is to trace the spread of disease along the arteries of commerce. Without such knowledge it is impossible to understand either the impact of disease upon business or the dangers which trade posed to public health. However, some historians will object to any attempt to identify diseases in the past, especially the distant past.[6] To them, I can only say that if we rule out informed speculation on this subject we have no way of explaining why diseases appeared, how they spread, or why they disappeared. Despite all the problems inherent in such a task, these questions are surely worth asking, especially in a study such as this. Provided we are prepared to acknowledge a degree of uncertainty, I see no reason why this approach cannot be used alongside an attempt to recover the views of contemporaries. It is vital to understand how they perceived the threat of trade-borne disease but to do that effectively it is helpful to know what they may have been responding to.[7]

When explaining variations in sanitary regulation over time and across nations, the historian encounters problems of a different kind. Public health measures are conventionally portrayed as advances in enlightened government, during which vested interests are swept away in the pursuit of progress.[8] This is clearly true at one level but it is also rather naïve for it ignores the ways in which public health was and is politicized.[9] At the opposite extreme is the view that public health measures were intended primarily to enhance the security of the state and to manage its population.[10] This is a view shared by writers from a variety of ideological standpoints but it tends towards oversimplification and cannot account for significant differences in the approaches taken by different states. No single factor – be it geographical location or political ideology – can explain these differences, either. Public health measures are best understood as unstable compromises between disparate and sometimes conflicting interests. Governments have always balanced the prospect of infection against the losses that may result from curtailment of commerce, for example, and that balance has shifted as economies have become more intertwined. But countries have rather different stakes in the international economy and their decisions about how to regulate trade have been shaped by both public opinion and the diplomatic context in which they are forced to operate.

The sanitary regulation of trade thus resembles a game of chess with many players, each of which attempts to second-guess the actions of the others. This has resulted in some rather perverse outcomes, with defences sometimes being lowered in the face of impending threats or, more commonly, the use of sanitary measures for economic protection and political leverage. Institutions such as the World Health Organization and the World Trade Organization have brought a degree of order to this chaos but the regulations which such bodies attempt to enforce are flawed and we face a future in which both commerce and health are unnecessarily endangered.

The idea for this book developed as I was researching other topics in which the relationship between disease and trade was in some way important. My interest was first aroused while I was studying the sanitary policies of the British government in India but over the last dozen years I have had the opportunity to extend the scope of my research to other periods and places. Over that time I have incurred more debts than I can adequately acknowledge but I should like to thank those who offered comments and criticism at formative stages, as well as providing useful advice on sources. I am particularly grateful to Katherine Arner, Josep Barona-Vilar, Karen Brown, Pratik Chakrabarti, Peter Christensen, Rhian Crompton, Sanchari Dutta, Simon Finger, Madeline Fowler, Katherine Foxhall, Karl-Erik Frandsen, John Henderson, Ryan Johnson, Catherine Kelly, Amna Khalid, Jeong-Ran Kim, Krista Maglen, Rita Mevelo, the late Harry Marks, Saurabh Mishra, Randall Packard, Margaret Pelling, Bob

Perrins, Loreen Salleh, Paul Slack and Michael Worboys. Heather McCallum at Yale University Press read two drafts of this book and made many valuable suggestions. I am profoundly grateful to her and the anonymous referees who read the first draft. I should also like to thank Lucy Isenberg and Tami Halliday at Yale University Press for their extremely careful reading of the proofs and editorial advice.

Thanks are also due to those who offered comments on earlier versions of this work at the following seminars and conferences: the International Conference of Economic History, Buenos Aires (2002); the American Association for the History of Medicine meeting at Boston (2003); Academia Sinica, Taipei (2003 and 2011); the 'Empires and Networks' conference, University of Kyoto (2004); the Historical Geography seminar, University of Cambridge (2004); the conference on 'Globalization, Nature, Health', University of Geneva (2005); the History of Medicine seminar, Johns Hopkins University (2005); the History of Medicine seminar, University of Valencia (2005); the annual lecture, London School of Hygiene and Tropical Medicine (2005); the 'International Workshop on Epidemic Diseases, Environmental Change and Global Governance', Osaka (2005); the James Martin Institute, University of Oxford (2006); the 'Explorers and Explored' conference, National Maritime Museum, London (2006); 'The Importance of Medical History' conference, Mumbai (2007); the Saxo Institute, Copenhagen; the History of Medicine seminar, Wellcome Centre for the History of Medicine at UCL (2009); annual History and Geography lecture, the University of Hull (2010); Korean Maritime University, Busan (2011); Port Cities Research Centre, University of Kobe (2011). Not least, I should like to thank those who have loyally attended the research seminars of the Wellcome Unit for the History of Medicine at Oxford, where I have been able to present on several topics from this book over the years.

I am deeply indebted, too, to many librarians and archivists around the world for their patience and their assistance in navigating collections which were sometimes very unfamiliar. In particular, I should like to thank the staff of the British Library's Asia, Africa and Pacific Collection, the Bodleian Library, the Gloucestershire Record Office, the National Archives, London, the Wellcome Library, London, the American Philosophical Society, the Baltimore City Archives, the Library of Congress, Philadelphia, the National Archives and Records Administration, College

Park, the National Archives of India, the Maharashtra State Archives, and West Bengal State Archives, Kolkata. I also owe a special debt of gratitude to the Wellcome Trust, which provided funding for much of this project, to my colleagues – Sloan Mahone and Erica Charters – for managing the Wellcome Unit so skilfully during my absence on research leave, as well as to the Unit's secretary Belinda Michaelides for help with so many things. Chapter 3 of this book has drawn upon my article 'Disease, Diplomacy and International Commerce: The Origins of International Sanitary Regulation in the Nineteenth Century', *Journal of Global History*, volume 1(2), pp. 197–217 (2006), © London School of Economics and Political Science, published by Cambridge University Press. I should like to thank Cambridge University Press for their permission to reproduce material from this article in revised form. Last but not least, I should like to thank my family for their love and forbearance.

CHAPTER 1

Merchants of death

Prior to the fourteenth century, much of Eurasia enjoyed centuries of freedom from epidemic disease. There had been no major outbreak of pestilence since the Plague of Justinian (AD 541–762), or First Plague Pandemic, which had fatally weakened many of the empires around the Mediterranean.[1] After the plague died out, the populations of many parts of Europe, Asia and North Africa began to recover and by the thirteenth century Europe and Asia were enjoying great prosperity and an expansion of trans-continental trade. By the middle of the fourteenth century, however, many parts of Europe had become over-populated and food was increasingly scarce. Cattle murrains and famines brought many societies to the brink of ruin and when plague arrived from Central Asia in the 1340s, it pushed them over the edge. Around a third of Europe's population – and that of affected parts of Asia and Africa – perished within years or even months. In Europe, population collapse resulted in severe scarcity of labour; this increased its price and, in the longer term, dissolved feudal bonds, allowing capitalism to flourish. Coping with plague also stimulated state-building and administrative centralization, giving rulers the means to raise revenue for armed forces and commercial expeditions.[2] Together, these changes paved the way for centuries of European expansion. But what, if anything, do these momentous events have to do with trade? To answer this question we need to consider the origins of the Black Death and the ways in which this and subsequent epidemics may have spread.

The geographical location and dissemination of the Black Death are questions which have generated rather more heat than light. As most interpretations are rather speculative, there is little to be gained by considering all of them but one of the most influential (and controversial) has been William H. McNeill's suggestion that the plague originated in or near China. In his famous work *Plagues and Peoples* (1976), McNeill claimed that the Second Plague Pandemic began in the khanate of Mongolia, where a terrible epidemic was recorded in 1331–2. He supposed that the plague bacterium (*Yersinia pestis*) might have been carried there after it was contracted from one of two hypothetical 'reservoirs' among wild rodents: the most likely one being on the Yunnan–Burma border, the other possibly on the Manchurian–Mongolian plateau. From China, the disease seemed to make its way west along caravan routes controlled by the Mongols to reach the northern shore of the Caspian Sea, infesting other rodent populations on the Steppe. The plague then appeared to move south to Azerbaijan and west to the Black Sea, again along caravan routes, from where it was conveyed by merchant ships to Europe and Africa.[3]

This version of events ascribes an important role to commercial activity, yet there is no firm evidence to support it and other geographical origins have been proposed, ranging from Kurdistan to southern Russia.[4] None of these alternatives can be completely ruled out but it seems likely that McNeill's original hypothesis was correct. It is generally acknowledged that China experienced major epidemics from the 1330s to the 1350s, resulting in a massive fall in population.[5] This demographic collapse was comparable to that in Europe after the first visitation of plague in the 1340s and it is difficult to imagine any disease which could have caused similar mortality over such a long duration.[6] Moreover, we now have good reason to suppose that a wild reservoir of plague did exist in the vicinity of China. The DNA signature of *Yersinia pestis* points to the bacterium evolving between 26,000 and 2,600 years ago, in or near China.[7] After China was united under the Mongolian Yüan Dynasty (1260–1368), it would have been relatively easy for plague to spread from these remote locations to major centres of population. Indeed, if plague was present in China prior to the fourteenth century, it is likely that many infected rodents would have been displaced by the floods and earthquakes which were widely reported immediately before the Black Death. As a result of these calamities, marmots might have come into closer contact with domestic rodents, their fleas subsequently carrying the plague bacterium to humans.[8]

These theories are plausible but we have no way of knowing for certain whether the great mortality which occurred in China was caused by the disease which gave rise to the Black Death. Once plague arrived at the Black Sea, however, there is substantial evidence from which to form an impression of how it may have been conveyed to other parts of the world. Most contemporary accounts start with graphic descriptions of the plague which raged at the Crimean port of Kaffa in 1346. The city was then besieged by the army of the Muslim Turkic ruler Kipchak Khan Janibeg, who had pursued Christian merchants from the city of Tana where they had been granted licence to trade. Kaffa was founded by the Genoese in 1277 as a trading post and it was regarded by them as a safe haven where they could purchase and store sought-after commodities such as silks from China, before selling them on to buyers in Europe.[9] As long as the silk routes across Central Asia were protected by the Mongols this arrangement was viable but as the Empire began to disintegrate, and some of the khanates converted to Islam, Christian merchants found that they were no longer welcome. The slightest disagreement between Christians and Muslims had the potential to escalate into a major incident, as the Genoese found to their cost at Tana. But the plague set them free: its ravages among the khan's army broke the siege at Kaffa and provided the merchants with an opportunity to escape to Constantinople.[10]

During 1347, the ancient Byzantine capital suffered severely from plague and many people fled from it, taking the disease with them.[11] The Genoese merchants who had escaped from Kaffa were among them and many died of plague before they reached their intended destinations in the Mediterranean. They first set foot in the Sicilian port of Messina, where the merchants hoped to take on provisions before returning to their native city. This was the first European port at which plague was recorded, in October 1347. The rest of the Genoese ships continued westwards along the Mediterranean coast, the merchants hoping to sell the goods they had brought with them from Kaffa. One ship reached Marseilles in 1348 and the others made it as far as Spain.[12] At this point, however, it becomes difficult to distinguish fact from fiction. There were tall tales of a single fleet of 'death ships' spreading the disease along the Mediterranean and far up the Atlantic coast. Many people refused to purchase the new season's spices from the Levant, fearing they might contract plague from them.[13] Whilst they seem fanciful today, these rumours show the anxiety which then prevailed. Moreover, they tell us that plague and commerce had become closely intertwined.

Within two years of the initial outbreaks in Europe, merchant vessels had carried the plague along the Atlantic coast and into the Baltic and North Seas, reaching as far north as the Faroe Islands.[14] In all these places, the disease found a hospitable environment. Most people lived in dwellings that were crawling with rats, whose fleas were subsequently to convey the disease to humans. The black rats which lived in close proximity to humans – unlike the less domesticated, indigenous brown rats – were themselves a legacy of trade with the Orient. They had arrived in Europe from the Levant, having probably originated even further east, in the rainforests of Burma and India.[15] Infection and re-infection of local rodents by ship-borne rats and fleas were to be the major source of plague epidemics in most European ports and, although these connections were unknown at the time, merchant vessels were often held responsible for the arrival of plague. It was also known that plague travelled inland along the routes regularly traversed by merchants, such as major rivers like the Rhône, Loire, Rhine and Po, or via overland tracks like those used by Italian traders to cross the Alps into Austria and Central Europe. Pestilence connected the commercial centres of Europe like dots in a child's puzzle.[16] A few major trading cities, such as Milan, managed to escape,[17] but the association between disease and commerce was underscored repeatedly by the vulnerability of places which regularly received goods from the East.[18]

Many of the goods purchased by traders in the Levant came from the interior of Asia. Although the disintegration of the Mongol Empire temporarily disrupted overland trade, it was revived during the fifteenth century by the Turco-Mongol rulers who controlled most of the region. As a result, silks, metals and other manufactured goods continued to be exported from China to South-West Asia. Europeans were marginal to this trade – especially after the conquest of Constantinople by the Ottomans in 1453 – but they were able to purchase Asian goods from enclaves in the Ottoman dominions.[19] Plague-bearing fleas encased in cloth, grain and other produce could easily have been carried over long distances from Central Asia to the Crimea and from there, on rats, by merchant vessels to the Middle East, North Africa and Europe.[20] In time, the Levant and Egypt came to be regarded as 'seats' of pestilence in their own right. If the disease was not actually endemic there, it occurred with sufficient frequency to identify the region as a major source of outbreaks in Europe. European physicians and natural historians enshrined this belief in their writings and it informed the precautions which some of the Mediterranean states subsequently took against infection.[21]

It is less easy to determine whether trade had any role in spreading plague to other parts of the world. Even after the fourteenth century, the epidemiological history of China remains an enigma. There were further reports of mass deaths during the fifteenth century, for example, but they were ascribed to a variety of causes, including wars, floods and famines, as well as to disease.[22] In all probability, China was afflicted by mixed mortality crises such as those which ravaged many parts of Europe in the three centuries after the Black Death, conflict and natural disasters paving the way for plague and other epidemics. Trade may have played some role in the distribution of disease at this time but there is no firm evidence that this was the case. However, there were certainly parts of Asia whose distance from the principal overland trade routes probably protected them. There are no records of plague reaching Korea or Japan in the fourteenth or fifteenth centuries, nor is there any evidence that it was transmitted to South East Asia.

The case of the Indian subcontinent is more intriguing, for India was once supposed to be the source of plague. In 1344, a Moroccan traveller, Ibn Battuta, witnessed a great epidemic in south-east India after leaving the service of the Muhammad bin Tughluq, Sultan of Delhi (r.1325–51). Some later writers took it that Ibn Battuta was referring to plague, and speculated that the disease later reached Central Asia from India.[23] However, there is no good reason to suppose that the disease described by Ibn Battuta was plague, nor do there appear to be other accounts of plague-like diseases in India at this time. In some ways, it is surprising that India was spared the ravages of plague, especially in view of the many commercial ties which then existed between the subcontinent and Central Asia. Yet it is possible that the great cold of the mountain passes which separated India from the rest of Asia blocked the transmission of infected hosts and vectors.[24] Of course, much would have depended on when plague was present on the northern and western side of the mountains, for if it had appeared there in summer there would have been no such impediment.

Over the next two centuries, there were further reports of epidemics in India, some of which might well have been plague. In 1443 there was an epidemic in which Sultan Ahmed I lost much of his army to *ta'un*, the Arabic term for a disease which closely resembles plague. In the 1590s there are also accounts of pestilence causing great mortality after famines, although this was referred to by the more generic term *waba*, meaning simply 'epidemic', and was just as likely to have been anthrax or typhus as plague in the modern sense of the term.[25] But after the Mughal conquests and the establishment of Portuguese colonies during the sixteenth century, reports of plague-like diseases become more frequent. A deadly 'bubonic fever', known to the Portuguese as *carazzo*, was often reported along the northern coast of the Arabian Sea, afflicting Indian ports such as Bassein, Surat and Daman during the 1690s.[26] The ravages of *ta'un* were recorded in Goa in 1681, Gujarat between 1681 and 1690, the Deccan in 1689, and in Bombay in 1689–90 and 1702.[27] These epidemics probably originated in trade between India and Arabia or the Persian Gulf, for some later outbreaks in western India appear to have followed the same route, the plague returning with traders in cotton textiles.[28]

Such a scenario is plausible in view of the vigorous trade which then existed between western India and the Persian Gulf; a trade conducted largely, though not exclusively, by Indians.[29] Textiles came not only from Western India but also from further south. Cotton goods made in the Coromandel region, for example, were often taken overland from there to ports such as Surat, from where they were shipped across the Arabian Sea.[30] These complex trade routes may have allowed plague to extend into the interior of India, especially as there was an upsurge in plague in nearby Persia from the 1570s and further epidemics throughout the next century.[31] Another route by which plague may have entered India was overland. The memoirs of the Mughal Emperor Jahangir (r.1605–27), or Tuzuk-i-Jahangiri, refer to a great plague in northern India between 1616 and 1619, which raged in the vicinity of Lahore before moving south-eastwards to affect Agra and Delhi. Its symptoms included buboes in the armpits and groin and there were reports of the death of rodents preceding that of humans.[32] Jahangir was also told of an epidemic (*waba*) to the north, in the mountainous region of Kashmir.[33] Although the emperor dismissed some of the wilder opinions on the causes of these northern epidemics, the reports suggest a pathway of infection between the north Indian plains and trade routes with Central and Western Asia. If plague appeared at the right season, it is conceivable that it could have passed through the mountains with caravans from the north.

The connection between trade and the spread of plague is therefore plausible if not always incontrovertible. With this element of doubt in our minds it is important to take account of the view – expressed by some historians and biologists – that the Black Death and some later outbreaks could not have been caused by plague. One problem in making a match between plague then and now is that the Black Death travelled quickly: the French historian J.-N. Biraben calculated that it moved at an average of four kilometres per day,[34] faster than some more recent epidemics of plague. Other aspects of plague epidemiology also seem to be inconsistent, as do certain descriptions of symptoms.[35] It is distinctly possible that some of the plague epidemics reported in the past were caused by different diseases or by a combination of infections. And yet, it seems likely that the core of these epidemics – at least those which occurred in Europe and the Middle East – was plague as we know it. In Arabian and Persian sources, many of the epidemics were known by the specific term *ta'un*, rather than the generic *waba*. Similarly, in Europe, the words *pestis*, *peste* and 'plague' came to have a fairly precise meaning which was often used to differentiate the disease from other epidemic fevers. Physicians sometimes speculated that plague might be related to other maladies but they generally regarded it as relatively distinct, on account of its characteristic swellings or buboes.[36] The description of these and other plague symptoms became more detailed and consistent over time.[37] Compelling evidence for the presence of plague in medieval and early modern Europe is also provided by analysis of ancient DNA, which shows that many medieval and early modern 'plague' victims were infected by *Yersinia pestis*.[38] Nevertheless, plague has clearly behaved in different ways over time, perhaps as a result of genetic mutation or of the various social and economic conditions in which epidemics have developed.

The more important question, perhaps, is how these epidemics were understood by contemporaries. What did they think caused them and what action, if any, did they take to prevent them? As might be expected, many people interpreted the arrival of plague in religious terms, attributing it to a vengeful deity; others blamed itinerant groups and ethnic minorities, especially the Jews. Explanations were diverse and often ephemeral, yet some persisted over time.[39] There was a pervasive and enduring understanding of what we might term 'contagion', of the transmission of plague by people and merchandise.[40] Although such notions were instinctive and rudimentary, they were the basis on which governments acted to combat plague, isolating the sick and restricting the movement of merchants and their goods.[41] Such measures date back to the Ordinances of Pistoia in 1348, which forbade access to persons from infected areas and to anyone carrying linen or woollen cloth.[42] Similar dictates were issued in other Italian cities. In Venice, for example, a sanitary council was established in 1348 with powers to isolate infected ships, goods and people on an island in the lagoon. As the main point of entry for the profitable traffic with Asia, Venice was normally the first European port to receive vessels carrying goods from the Levantine coast. After the city experienced fresh visitations of plague, it developed a permanent infrastructure for public health, culminating in the formation of a Council of Health in 1485, which drew up detailed regulations governing maritime trade and the management of lazarettos to contain the sick and those suspected of infection.[43]

The detention of ships for sanitary reasons came to be known as 'quarantine', a term originating in regulations devised by the Republic of Ragussa (Dubrovnik) in 1397, in which year it extended its existing precautions (of 1377) to enable the detention of vessels for up to forty days.[44] Other states menaced by plague soon took similar precautions: for example, Marseilles in 1383, Pisa in 1464 and Genoa in 1467. The close attention given to maritime quarantine attests to the strength of the perceived connection between infection and commercial intercourse. Although Jews, gypsies and foreigners were still popular scapegoats,[45] it was merchants who received the majority of blame, either for ignoring the dangers of infection or for evading sanitary regulations. In London, for example, the many plague scares of the late sixteenth and early seventeenth centuries usually implicated merchandise from overseas, particularly from Holland, which was then a leading trading partner.[46] This perception lasted right through to the final visitation of plague in England, in 1665, the author Daniel Defoe being one of many who believed that the disease had arrived in 'Goods brought over from *Holland*, and brought thither from the *Levant*'.[47] There were even reports of plague being exported as far as Bermuda in bales of Egyptian cotton.[48]

The association between plague and commerce proved enduring not simply because the presumed source of plague lay in the East but because many people distrusted the business in which merchants were engaged. The Bishop of Rochester in England was not alone in seeing the reappearance of plague in 1375 as chastisement for the 'great falsehood' that was 'practised in measures, charging interest, weights, scales, adulteration, lies and false oaths'. 'Each studies to deceive the next man,' he lamented, whereas the apostle Paul had enjoined them to serve one another in charity.[49] Descriptions of plague victims in Renaissance literature as 'merchandises', 'flitches of bacon' and 'pickled game' suggest that comparisons continued to be made between the ravages of plague and the rapacious business of the urban world.[50] Many Elizabethan playwrights therefore depicted plague as the scourge of merchants and those who practised usury and extortion.[51]

Although the link between trade and disease seemed obvious to many people it did not easily fit the precepts of most physicians. The very notion of contagion presented a challenge to doctors who were schooled in the ancient Graeco-Roman tradition and, in particular, in the immense body of work attributed to the second-century physician Galen. Although Galen had speculated about the existence of 'seeds of plague', they were relatively unimportant to his understanding of disease.[52] Renaissance Galenists, too, were more concerned with how potential external causes – be they 'seeds' or noxious miasmas – manifested themselves in the body than with how these seeds might have spread. They saw them as one of many factors which might contribute to an imbalance of the four substances or 'humours' which comprised the human body (blood, black and yellow bile and phlegm, respectively). Theirs was an individualistic conception of disease, in which maladies appeared differently according to the peculiarities of each bodily constitution, whether formed by heredity or habit.[53]

Popular ideas about plague initially jarred with the opinions of physicians, but the idea of contagion was gradually assimilated into learned medicine. Medical writings on plague came to acknowledge its dissemination by persons, animals and goods and, in so doing, blurred the boundary between contagion in the strict sense of the term and the 'infectious' atmospheres said to arise from filth and rotting matter.[54] Although plague often seemed to originate in a corruption of the air, it was thought that infectious particles might then adhere to persons, merchandise or animals, which could carry the infection as they moved.⁵⁵ Throughout the medieval and early modern periods, these explanations were often intertwined with notions of divine retribution. God sometimes appeared to intervene directly to punish wrongdoers but it also seemed that he could act indirectly by poisoning airs and waters.⁵⁶ These ideas of divine causation were beginning to fall out of favour among physicians by the end of the sixteenth century,⁵⁷ but they did not disappear altogether and the plague treatises of the seventeenth and early eighteenth centuries reveal an intellectual patchwork in which notions of divine and astrological causation coexisted with naturalistic explanations of disease.⁵⁸

The net result of these changes was that European medical thought became largely indistinguishable from that of the informed laity, particularly officials such as magistrates who often found themselves responsible for dealing with epidemics. By the beginning of the seventeenth century most European doctors and public officials accepted that contagion – including contagion by infected merchandise – was one of the primary causes of plague, albeit conditional upon other factors such as squalor and inclement weather.⁵⁹ Plague was by no means unique in this respect but of all the diseases then considered contagious it had the closest associations with commerce.

Outside Europe, however, the connection was less than clear. Despite being afflicted by plague more often than Europe, the Islamic polities of the Middle East and North Africa showed little inclination to emulate the measures taken by their Christian neighbours. Europeans travelling there were struck by what they saw as the 'fatalism' of Muslim peoples and the apparent indifference of Oriental rulers to the plight of their subjects.⁶⁰ In reality, the situation was more complex than these superficial comments suggest. According to many Islamic scholars, death from plague was to be regarded as divine mercy or martyrdom rather than as punishment. This belief distinguished Muslim opinion from that of most Christians, with the possible exception of Protestant sects such as the Calvinists, who believed in predestination. Some *hadiths* (sayings attributed to the Prophet) prohibited people from leaving areas infected by plague, yet other sayings warned travellers not to enter places where plague was known to be present, suggesting that martyrdom from disease was neither desirable nor inevitable.⁶¹ Medical texts exhibit similar inconsistencies. Like European physic, Islamic medicine was based on the writings of ancient Greek and Roman physicians and the notion of contagion was problematic for much the same reasons. The spread of disease through contagion also implied that things could occur independently of God's will, which was anathema to many Muslims. However, a minority did accept that possibility and considered contagion to be an earthly instrument of divine consciousness.[62] For this reason, references to contagion are occasionally to be found alongside atmospheric explanations of disease.[63]

But quarantine was still conspicuous by its absence. Even if it had possessed a strong medical or theological rationale, it is doubtful that large terrestrial empires such as those of the Ottomans or Mughals would have had the resources to enforce it. If these states took action against plague, it was generally by providing alms for the sick and destitute. By the sixteenth century, the Ottoman Empire was beginning to improve urban hygiene and to issue orders regulating the burial of plague victims during epidemics, but still stopped short of imposing quarantine. In China, however, rulers discharged their responsibilities solely through the relief of hardship during epidemics and the distribution of medicine. There was no conception of quarantine, for epidemics and similar catastrophes were commonly attributed to cosmological disharmony, sometimes precipitated by the failure of emperors to rule wisely.[64] The notion of some kind of polluting agent was not entirely absent from Chinese medical literature, however. There was a sense that certain substances could cause disease, but usually within a broader explanatory framework which included cosmological and environmental conditions. Indeed, the peculiarities of place were increasingly emphasized in Chinese medical texts from the seventeenth century. These notions were not unlike the classical learned medicine of Europe but there was one subtle difference: the Chinese concept of *qi* – the presumed agent of most epidemic fevers – encompassed energy as well as breath and air.[65]

If there were important differences between the ways in which Asian and European cultures understood epidemic disease, the decisive factors in shaping intervention were most likely political. Put simply, more was expected of European rulers and some, at least, had the capacity to meet such expectations.

Initially, some European states took action against plague for theological reasons, in the belief that visitations of pestilence were a form of punishment for the defilement of sacred spaces. In the cities of Venice and Lyons, such convictions resulted in the removal of beggars from the vicinity of churches and the cleansing of cathedral squares. But the propitiation of an angry deity gradually became less important than the removal of the immediate causes of infection or blocking the movement of suspect individuals.[66] This process of secularization – or more specifically the naturalization of disease and its placement within the affairs of state – set Europe apart from Islamic and Confucian Asia. And yet there were great differences in the energy with which European states tackled epidemics. Generally speaking, northern countries were slower to establish quarantines and lazarettos than those with Mediterranean coastlines, which were closer to founts of infection in the Levant and Central Asia. But the long reach of plague – which extended not infrequently into Scandinavia – suggests that one should not equate distance with complacency.[67] In the long run, the most important factor determining the strength of a state's sanitary defences was the emergence of ideas which associated freedom from epidemics with good governance.

This connection developed first in the Italian city states, where notions of the 'common good' became integral to civic culture. In these fiercely independent republics, community feeling was more highly developed than in countries in which people were bound rigidly by feudal obligations. Unity was the very foundation of republican government, and factionalism its greatest foe.[68] The introduction of quarantine and other sanitary precautions implicitly recognized that the interests of merchants needed to be curtailed for the public good. For their part, merchants went to great pains to demonstrate their public spirit by helping to found hospitals and charitable institutions for the poor.[69] This civic-mindedness became the basis of a political doctrine which equated protection against disease with virtue but also, increasingly, with power. The greatness of rulers was henceforth to be measured by whether their lands were populous and healthy.[70]

These ideas galvanized many Italian states into responding to the great plague epidemic of 1575–8, or the Plague of San Carlo as it became known. This was the first epidemic to embrace all of Italy since the 1340s and it brought a major shift in the way in which plague was understood and tackled. Physicians now paid more attention to the social causes of plague and explored the connection between its movement and forms of human interaction such as commerce. A flurry of tracts offered advice to rulers on how best to prevent the disease, stressing that the solutions to the problem were essentially political. In the course of the epidemic most Italian cities improved their provision of public health, establishing sanitary boards where these did not formerly exist, introducing tighter controls on the movement of goods and people, constructing new hospitals and lazarettos, and cleaning streets and public places.[71] Although the costs of doing so could be crippling, governments prided themselves on being able to tackle epidemics and the deeds of heroic or virtuous individuals were widely celebrated.[72] This, indeed, is how the Plague of San Carlo got its name. Carlo Borromeo (1538–84) was the Cardinal Archbishop of Milan who led efforts to help the sick when plague struck the city in 1576. He was subsequently canonized for his bravery on this occasion and also, no doubt, for his role in the Counter-Reformation.

In the years which followed the plague of San Carlo, most Italian states remained vigilant towards the prospect of infection from without and special attention was paid to its importation through commerce. In the 1580s, for example, ninety-nine of the 282 plague decrees issued by the sanitary board of Verona pertained to trade or to the quarantine of goods imported from infected places. The Italian states also improved their intelligence regarding the presence of plague, making use of a network of correspondents across Europe. This enabled Verona's sanitary board to target quarantines against certain countries, rather than to insist on a blanket embargo.[73] The growing sophistication of public health measures in the Italian states widened the gap between them and the rest of Europe but in the course of the next century the rulers of other countries began to emulate those of Italy. Followers of the Dutch humanist Erasmus imported notions of civic virtue from the Italian states,[74] as well as the idea that sanitary measures could enhance the power and wealth of nations.[75] Most states began to devise measures against the importation of disease, which usually took the form of embargoes on imports from infected countries or the quarantine of suspected persons and merchandise. In 1602, for example, the Lord Mayor of London forbade intercourse with the Dutch port of Amsterdam and the English port of Yarmouth, having received intelligence that infected goods from the former had reached the latter.[76]

Quarantine was not maintained simply as a defence against disease. By the late sixteenth century the internal problems of countries afflicted with plague were severely aggravated by the embargoes imposed by other states. The merest hint of plague in a foreign port could be enough to trigger such bans and in some cases rumours were deliberately put about to damage the commerce of rivals. In such circumstances it was doubly important to prevent the spread of plague and, above all, to be seen to do so. States with rudimentary provisions against plague were especially vulnerable to disruption of trade and in England, for example, the plague scares of the early 1600s stalled the country's recovery from a fifteen-year depression.[77] The plague orders issued by the new monarch James I therefore aimed to reassure neighbouring countries and trading partners that the necessary safeguards had been taken. Such orders were also used to deal with the growing problems of poverty and vagrancy. In fifteenth-century Italy it was noticed that plague occurred disproportionately among the poor and among persons classed as 'ruffians' and 'outlaws'. These shadowy figures, who represented an incipient threat to public order, were made the target of regulations ostensibly directed at combating plague.[78] New hospitals were created to incarcerate the sick and those suspected of infection; in quieter times they doubled as refuges for the infirm and poverty-stricken. Quarantines likewise acted as social filters, excluding undesirable characters while allowing important persons free passage, even if they did sometimes come from infected areas.[79]

While northern countries were passing their first plague legislation, many southern states were strengthening and extending their quarantine establishments in an effort to compete with one another. They were motivated partly by civic pride and partly by an attempt to gain a larger share of the lucrative business of quarantining ships sailing through the Mediterranean. There was good money to be had from the temporary storage and fumigation of goods, and by providing room and board to passengers. With this in mind, a new council of health was created in Pisa between 1606 and 1609; the Magistrato di Sanita in Venice was reorganized in 1630; and a board of health constituted in Marseilles between 1640 and 1654. Other ports such as Livorno, Dubrovnik, Palermo and Naples built additional lazarettos over the same period. These ports competed to attract passing ships which needed to perform quarantine in the Mediterranean before they could dock in countries to the north and west. In the case of Livorno, the physical structures in which quarantine was performed were also intended to afford protection from the marauding corsairs of the Barbary Coast.[80] Northern states had nothing to compare with these rather grand establishments and most made do with temporary pest-houses. There were no permanent lazarettos in northern Europe until one was constructed in Amsterdam in 1655.[81] Indeed, until the eighteenth century, most northern countries had neither lazarettos nor legislation setting out the duration of quarantine and the conditions under which it should be imposed. Their response to the threat of plague was ad hoc in nature and varied considerably from crisis to crisis.[82]

But if the infrastructure of quarantine was stronger in Mediterranean countries, its enforcement was irregular, to say the least. Complaints about exorbitant and arbitrary charges were common and there was little coordination or parity between institutions. Despite a long history of regional cooperation in France, for example,[83] it proved difficult to harmonize arrangements in the country's two lazarettos (at Marseilles and Toulon), let alone with cities inland. This resulted in the absurdity that maritime quarantine sometimes remained in force while plague rampaged through the interior. In 1683 a quarantine statute was passed in order to remove such anomalies but it had little effect.[84] It was relatively easy to control public health measures in a small state like Genoa or Venice but it was vastly more difficult in a country the size of France, where distance frustrated central control.

In many European ports, sanitary precautions were also reduced because of the political power wielded by merchants. For example, when the Board of Health in Seville took the decision in 1582 to prohibit commerce with cities infected by plague, irate merchants petitioned the authorities to relax restrictions, claiming 'intolerable harm and injury'. They protested that many of the goods they wished to import were safe from contagion but while some articles, such as shipments of wine, fruit and vinegar, were seldom considered dangerous in themselves, unloading them was a different matter. Removing merchandise from an infected vessel was said to expose dock workers to infection lurking in the ship and there was always the risk that the crew might pass the disease to locals. Under mounting pressure from merchants, and aware that the city's tax revenues would be adversely affected if the ban continued, the board of health relaxed quarantine to permit the landing of wine and vinegar under strict conditions. Other commodities, including shipments of seasonal goods like cherries, remained subject to a complete ban, much to the annoyance of merchants. Some nearby cities such as Carmona permitted trade to flow more freely and subsequently paid the price, as plague broke out and other ports issued embargoes against them.[85] Even in Italy, where maritime quarantines were long established and often popular, merchants were sometimes powerful enough to override them. In 1629, for example, cloth merchants in Venice were able to prevent the imposition of quarantine upon vessels from the Levant.[86] The situation was much the same in France. A royal ordinance of 1702 compelled ships' captains to declare the health status of their vessels before entering the harbour but it was often ignored because the authorities were in thrall to merchants and their allies.[87]

Sanitary practices evidently varied widely and could seldom be relied upon. Moreover, some port cities were beginning to relax quarantine in an attempt to gain advantage over their rivals. It was with a view to preventing this that some Italian states banded together and coordinated their actions, sharing intelligence about epidemics in their territories and in plague hotspots such as North Africa and the Levant. Florence, Genoa and Tuscany also sent diplomatic missions to inspect the health of their neighbours and scrutinize quarantine arrangements so that none gained an unfair advantage. These delegations normally consisted of one or two surgeons and physicians travelling together with officials such as magistrates. But although this *concerto* was taken seriously by Florence and Genoa, other states, such as Rome and Naples, continued to act independently. Nor did the Florentines and Genoese always see eye to eye. When plague broke out in Genoa in 1656, the concordat rapidly dissolved when the Florentines placed Genoa under quarantine.[88] This chaotic state of affairs persisted for years to come.

From the Old World to the New

In the three centuries following the Black Death, much of Eurasia and a substantial part of northern and eastern Africa became epidemiologically intertwined. The plagues of the later Middle Ages were the latest episodes in a process which had been under way, albeit intermittently, for millennia. Up until that time, there had been some mixing between the disease pools of Africa and Eurasia but other parts of the world were now sucked in. Most significantly, Columbus's voyage to the Americas in 1492 set in train an 'exchange' in which most diseases of the Old World arrived in the Americas and a single New World disease – syphilis – made its way east with explorers returning to Europe.[89] Controversies rage over the extent to which Old World diseases such as smallpox, measles, typhus and, possibly, plague assisted the European conquest of the Americas,[90] as well as over whether syphilis was an American disease or the mutant offspring of disease germs already present in Europe.[91] To some extent these are still open questions, but it is obvious that the 'Colombian Exchange' was an unequal one. Although syphilis was initially a virulent and horrific disease, the impact it made on Europe can scarcely be compared to the depopulation of the Americas which followed the Columbian contact. Lacking immunity to Old World germs, the peoples of South and Central America, and later North America, perished in extraordinary numbers, often shortly after coming into contact with Europeans.

The demographic collapse that followed was almost certainly comparable to that resulting from the Black Death and probably a good deal worse in some areas. It came about as the result of several distinct waves of infection, beginning in 1518 with an epidemic of smallpox in the Caribbean islands which soon spread to the mainland of Central America. Within a decade it seems to have travelled as far south as what is now Bolivia and as far north as what is now the United States. The disease penetrated the interior with the Spanish conquistadores and was spread unknowingly by native peoples in the course of their trade along the coast and through the Panamanian isthmus and other corridors.[92] Subsequent waves of infection followed in the 1530s, including what may have been pneumonic plague (the most deadly and contagious form of the disease), influenza and measles. In the 1540s there were further epidemics in Central America, which included typhus and possibly plague, both diseases reaching as far north as Florida through coastal trade and/or Spanish expeditions. A decade later, the misery was compounded by further outbreaks of smallpox and measles; and so it continued, with fresh waves of disease brought in from Europe and, later, from regional foci of infection.[93]

Although the impact of disease varied widely from region to region,[94] the death of millions of indigenous Americans provided a stimulus to European involvement in the African slave trade, which had hitherto operated on a small scale under Arab control. The Spanish and Portuguese, and later the British, Dutch and French, required African slaves to toil in plantations and mines in the Americas and the Caribbean.[95] Their labour in these colonial peripheries produced vast wealth which fuelled commercial, agricultural and industrial development in the homelands of the colonial powers.[96] The slaves themselves were taken mostly from tropical West Africa, where they were procured by Europeans from African chiefs and traders who established themselves on the coast, keeping their human cargo in densely packed 'barracoons'. These prisons became melting pots for a variety of germs from the interior, among them dysentery, malaria and smallpox.[97] Until this point, malaria had been absent from the New World but malaria parasites and malaria-bearing mosquitoes arrived in the Americas with slaves and also with migrants from low-lying parts of Europe where the disease was endemic. In the long term, however, it was the African connection that would prove the more significant. Many Africans carried with them the parasite *Plasmodium falciparum*, which was far deadlier than the *vivax* parasite which predominated in Europe. As malaria became entrenched, European mortality began to increase and forced European colonists to rely more heavily upon Africans, who appeared to enjoy some immunity to the disease.

Unlike plague in medieval and early modern Europe, none of the diseases which had so far altered the history of the Americas were especially associated with commerce. Periodic epidemics of smallpox often followed the arrival of slaving vessels from Africa but such experiences were so frequent that they quickly became normalized. Quarantine was sometimes used in the Iberian colonies to prevent smallpox from arriving on slaving vessels – in the 1620s, at Buenos Aires and Pernambuco, for example – but such measures were easily evaded and slaves were often permitted to land despite cases of smallpox having occurred among them. The demand for slaves was such that buyers were prepared to risk infection rather than purchase them at greater cost on account of the delays caused by quarantine.[98] By the eighteenth century, most maritime quarantines against smallpox had fallen into disuse and attempts to prevent the spread of contagious diseases inland were equally rare.[99] Some colonies, such as New Spain (later Mexico), did briefly reimpose quarantine to prevent infection by ships from South America, but a great deal of latitude was given to local officials and the quarantines were seldom effective.[100] The continued absence of quarantine in the South American and Caribbean colonies may also be due to the practice of inoculating slaves against smallpox on vessels sailing from Africa, as well as in some plantations, which reduced the need for disruptive quarantines.[101] But while this practice was relatively common in French and British colonies from the early eighteenth century, it was resorted to less frequently in other European territories such as Portuguese Brazil.[102]

In the middle of the seventeenth century a new disease made its appearance in the Americas. Yellow fever, as it came to be known, was one of many maladies shipped to the New World after the conquest but, as a relative latecomer, and a disease often fatal to Europeans, it attracted more attention than most. Its first documented appearance was on Barbados, in September 1647, when there were reports of a 'new distemper' characterized by black vomit. By the time the epidemic had ceased, in September 1649, some 6,000 people had died. After this outbreak, yellow fever spread to the nearby island of St Christopher, where it raged for eighteen months and killed one third of the inhabitants. In 1648, a ship that had sailed from St Christopher arrived at the French island of Martinique with many of its crew and passengers dead or dying. An epidemic then broke out on the island and lasted for twenty months. From there, the disease moved westwards through the Caribbean to Havana and the Yucatán, where it claimed many victims among peoples of all races.[103]

In most cases the descriptions given of this disease – its high fatality rate and its symptoms – tally with each other and with what is now known about yellow fever. We cannot be certain of its biological identity but contemporaries were adamant that it was a new disease, although some likened it to the plague because of its severity and apparent contagiousness. The Protestant pastor Charles de Rochefort, who compiled a natural history of the Antilles in 1658, wrote that this 'malignant fever' was the only blight on what were otherwise healthy islands, blessed with many natural advantages. In his opinion, and that of the physicians he had spoken to, the disease had been imported by ships whose holds contained 'bad air' originating on the African coast.[104] Although he could not have known the true cause of the disease, Rochefort was probably correct about the geographical origins of yellow fever. It most likely arrived on Dutch vessels carrying slaves for the Caribbean sugar plantations which were undergoing enlargement and consolidation. European indentured labourers who had formerly performed most of the arduous work on these plantations were replaced by Africans who were thought better adapted to humid conditions.[105] Some of these slaves may have been infected with the yellow fever virus, while the mosquito vector of the disease (*Aedes aegypti*) could have survived the long voyage across the Atlantic in barrels of drinking water.[106]

The next wave of yellow fever swept the Caribbean in 1685, this time extending as far south as Brazil, and within a few years the disease had become established in some of the mainland colonies and larger Caribbean islands, where imported African monkey populations served as permanent reservoirs of infection. New epidemics often arose from these local sources but importation from Africa remained significant.[107] Indeed, yellow fever did not become endemic on some smaller islands in the Caribbean; Barbados, for example, continued to enjoy a reputation for salubrity despite occasional imported epidemics.[108] From the 1690s, however, yellow fever began to spread north, appearing at many ports along the eastern seaboard of North America. The first epidemics occurred in Boston, Charleston and Philadelphia in 1693 and at Charleston and Philadelphia again in 1699. New York was affected for the first time in 1702 but there were no further outbreaks there until 1728, when the disease also returned to Charleston. On all these occasions, the outbreaks displayed many similar clinical features, including black vomit and yellow discoloration of the skin. After 1728, the disease appeared sporadically, typically producing several epidemics in two to three years and then none for decades, before it erupted again as deadly as ever. All told, there were probably twenty-five major epidemics of yellow fever in the North American colonies before the War of Independence.[109]

The region was prone to outbreaks of the disease because of the inexorable growth of sugar plantations from the middle of the seventeenth century. Slaving vessels crossed the Atlantic with greater frequency, often carrying slaves infected with the virus as well as the larvae of the *Aedes aegypti* mosquito. But an equally important factor was the environmental change which had taken place as the sugar economy developed: clearance of forests to make way for sugar cultivation reduced the habitats of the birds and other animals which might have feasted on the newly arrived insects, while water pots and storage tanks provided mosquitoes with numerous places to breed. And, most importantly, there was the sugar itself: sugar residue in the vicinity of plantations and, later, around the refineries established in coastal cities, boosted the nutrition of mosquito larvae by stimulating the growth of bacteria on which they could feed.[110]

Of course, none of this was known at the time and it was common to think of 'yellow fever' as a contagious malady which could be easily spread through human contact or infected merchandise. Like plague, it appeared in epidemics that were relatively infrequent, making it easy to conceive of the disease as imported. In a letter of 1740 to the London physician Richard Mead, the British doctor Henry Warren remarked that yellow fever had visited the island of Barbados twice in the past sixteen years, apparently arriving on vessels sailing from Martinique when that island was infected.[111] Warren was emphatically of the opinion that the malignant fever afflicting the island was not a miasmatic disease, for the island's air was 'remarkably fresh and pure' and probably more salubrious than that of any other sugar colony; the island was responsibly cultivated and devoid of lakes and marshes. 'Neither the Alteration of the Weather or Winds, nor the different seasons of the Year, have ever, of themselves, been able to produce this contagious Disease among us,' he insisted.[112] Moreover, the disease was entirely different from the common fevers of the island, affecting newcomers disproportionately, whereas Africans, whose 'rancid' diet supposedly predisposed their constitutions to putrefy, were little affected by it.[113] Warren believed that the epidemic was caused by poisonous particles that had a 'distinct specific power', producing much the same symptoms when passed from one person to another – the classic definition of contagion. In his view, the disease could be spread both by immediate contact with anyone who had contracted it and by their personal effects.[114]

This account was consistent with theories of that other great pestilential disease – plague; a disease to which yellow fever was often compared. Indeed, the recipient of Warren's letter, Dr Richard Mead, had been commissioned by the British government to write a report on plague after the disease had appeared in Marseilles in 1720.[115] Mead was a vigorous supporter of quarantine and took the view that plague was 'the Growth of the Eastern and Southern Parts of the World, and [was] transmitted from them into colder climates by the Way of Commerce'.[116] Warren's view of yellow fever was similar but he believed that it had been brought to the Caribbean not from Africa, as most people then thought, but from Asia,

French medical literature having referred to a disease known as 'la maladie de Siam'.[117] The French had encountered an unusually malignant fever in Indochina and Warren believed it to be identical with that which had raged in the Caribbean.

Regardless of the origins of yellow fever, most of those who wrote about the disease during the seventeenth and early eighteenth centuries saw a link between its spread and patterns of trade. As with plague, this perception was enshrined in the practice of quarantine, which, from the late seventeenth century, was occasionally imposed on vessels sailing to Europe from the West Indies when yellow fever was reported.[118] Quarantine was also sometimes applied to shipping from the West Indies in American ports, the first instance being in the Massachusetts Bay Colony in 1647, when it was established for ships arriving from Barbados.[119] From that time on, quarantine was occasionally imposed along the eastern seaboard of North America for yellow fever and other diseases deemed to be contagious. In 1760, for example, the assembly of the province of Georgia passed an Act to compel ships from places infected with 'epidemical distempers' to remain in quarantine for 'such time as shall be directed by the Governor or commander in chief'. All persons and merchandise arriving from such places were obliged to remain on incoming vessels until granted permission to go ashore.[120] In the same year, following reports of yellow fever in the West Indies, the Lieutenant-Governor of Virginia also issued a proclamation which called upon masters of vessels to declare where they had come from, and to give their word upon the health of the vessel. If their reply gave cause for concern, the superintendents in each port were empowered to place these vessels in quarantine.[121]

Such regulations conferred considerable power on local officials and their success depended on the integrity of captains and pilots. For this reason, sanitary restrictions on persons and goods entering American ports remained variable and the situation did not alter materially until after the American colonies won independence from Britain.[122] The efforts made by the North American colonies contrast sharply with the absence of precautions in the Caribbean and Latin America. Whether this was due to the greater frequency of epidemics in the Caribbean or to the lethargy of those in authority is hard to say, but in some respects the situation of the South American and Caribbean colonies resembles that of Middle Eastern countries affected by plague. The frequent outbreaks of yellow fever in the Caribbean and of plague in the Levant may have meant these diseases had less capacity to shock, while a degree of population immunity would have been established. In such circumstances, quarantine would have seemed almost superfluous, quite apart from the inconvenience and losses caused by regular disruptions to trade.

CHAPTER 2

War by other means

In the last chapter we saw how many European countries and some of the American colonies began to regulate trade in order to check the spread of disease. While such measures often enjoyed considerable support in port cities, they could be a major source of tension within and between nation states. Merchants regularly protested against the imposition of quarantine and did their best to evade it. Sometimes they also claimed that their business had been deliberately ruined by competitors who used sanitary measures to their advantage. Such issues had provoked argument among the Italian states for years but it was only in the 1660s that sanitary measures became a major bone of contention throughout Europe as a whole. In that decade, plague spread across much of the Continent causing great mortality in cities such as London and Amsterdam. Although this was to be the last time that the disease appeared in most parts of Central and Western Europe, the epidemics of the 1660s heralded the beginning of a new era in the history of commerce and contagion. Quarantines and sanitary embargoes came to be used consciously as instruments of statecraft and have figured prominently in international relations ever since.

'Obstructions upon our forygne trade'

The growing misuse of quarantine followed years of aggressive state-building and territorial expansion, culminating in major conflicts such as the Russo-Polish War of 1654–67 and the Great Northern War of 1700–21.[1] In the west, the United Provinces (the Netherlands) was competing with England and France for access to lucrative markets in the East and West Indies. Amsterdam, in particular, had grown rapidly as a result of its trade with the Indies and many exotic commodities – spices, sugar, china and cotton – were shipped from there to other parts of Europe. Rumours that plague had appeared in the great Dutch port in 1663 therefore caused panic throughout the region. The volume of trade with Amsterdam was such that most countries faced the prospect of infection if steps were not taken to prevent it. Neighbouring states responded to the threat of plague in the time-honoured fashion and imposed quarantines on all ships sailing from the city. But the duration of these quarantines was regarded by the government of the United Provinces as unduly severe and it suspected that plague was being used as a pretext by its rivals to destroy the Dutch trading empire, which had become the envy of the world.[2] The Dutch objected to a thirty-day quarantine imposed by England on the basis of what appeared to be mere rumours of a 'contagious malady' in Amsterdam. This disease (which the Dutch carefully chose not to term 'plague') had apparently abated and their government stressed the mutual advantage of 'uninterrupted commerce' between the two nations. Similar protests were made to the governments of Sweden and Denmark, which the Dutch also accused of using plague as an excuse to curtail their navigation.[3]

There may have been some truth in these allegations but those countries which traded with the Netherlands were right to be wary. Dutch navigators were notorious for flouting sanitary precautions and their government's reassurances about the outbreak could not be trusted.[4] The English, however, were initially reluctant to curtail their trade with the Netherlands. Many of their merchants depended on it, not least because of the Dutch connection with the Levant and Asia – the source of most of the silk and porcelain sold on the British market. Thus, while Amsterdam and its vicinity remained under quarantine, restrictions against ports such as Rotterdam and Dordrecht – at which no sickness had been reported – were relaxed.[5]

Like other governments, the English depended heavily on their consular officials and their informants (many of them merchants) when taking such decisions. But the question of whether to impose quarantine was now a political as much as a medical one. Over the previous decade, the English and the Dutch had clashed on a number of occasions and tension was mounting again over colonial trade. This meant that any attempt to regulate commerce using quarantines or sanitary embargoes was likely to be seen as a hostile act. This was certainly the interpretation placed upon the English quarantine by the Dutch, who immediately responded with an embargo on English vessels. The Dutch maintained this stance despite the fact that the original English quarantine had lapsed.[6] Having protested to no avail about these 'obstructions upon our forygne trade', the English retaliated in April 1664 and restored their quarantine against vessels sailing from Amsterdam.[7] At the same time as the new quarantine was introduced in England, efforts were doubled to ensure that it was enforced. After the first quarantine had been imposed in 1663, Dutch traders had connived with English merchants and harbour officials to smuggle goods into England, suspicion being aroused when a Dutch captain was discovered breaking the quarantine in London.[8] There were many similar reports from around the country.[9]

This 'tit for tat' diplomacy was symptomatic of a serious deterioration in relations between England and Holland, resulting from fierce competition for control of the West African slave trade and disputes over England's mercantilist Navigation Act. In October 1663 the English had raided some Dutch trading posts in West Africa. The following year, they captured the settlement of New Amsterdam (later to be named New York) and seized a number of Dutch merchant vessels in the Mediterranean.[10] These attacks brought reprisals, including a raid on Barbados which led to a formal declaration of war the following year. As conflict loomed, quarantine became useful as a way of augmenting the attacks on commerce which both rivals were pursuing vigorously. The great irony is that the English measures did nothing to prevent plague reaching England.[11] Indeed, the resulting epidemic weakened the English administration so much that it was forced to suspend naval operations and agree to a peace less favourable than expected.[12]

The death of over 35,000 people from plague in Amsterdam was not a matter to be taken lightly, and in 1664 most countries which traded regularly with the Netherlands extended or enhanced their sanitary precautions. In August of that year, the Dutch States-General was informed that the French had 'placed an embargo on all commerce and traffic coming from Holland and Zeeland, whether by sea or land, applicable to all merchandise without exception.'[13] Prohibitions on all forms of commerce were rare so this must have confirmed the Dutch in their opinion that Paris had an ulterior motive for curtailing their trade. The French embargo began as a precautionary measure but remained in force throughout 1664 and was extended in November of that year on the grounds that the epidemic in Amsterdam had not subsided.[14] There was no denying the risk from trade with Amsterdam but what rankled with the Dutch was the indiscriminate nature of the ban – which covered all merchandise – and the fact that it embraced Zeeland even though the disease had not been reported there.

The ways in which European countries responded to the epidemic in Amsterdam set the tone for the rest of the century. It was now clear that the imposition of sanitary measures was likely to be considered an act of aggression, even if it was not intended as such. In 1667, for example, Sir Robert Southwell, the English ambassador to Portugal suggested taking measures against Portuguese shipping in retaliation for what he regarded as 'unreasonable' restrictions enforced against English vessels entering Lisbon. In his view, the Portuguese quarantine was illegitimate because the danger of plague had abated. He also wrote of extortion by Portuguese quarantine officers. Southwell therefore recommended that Portuguese ships currently in English harbours be prevented from sailing, adding that 'untill something of that nature is done, we shall not be able to brake [sic] the work of this combination and constraint upon us'.[15]

It is not known whether the English government followed this suggestion but embargoes on trade and shipping were commonly used for such purposes when plague returned to Europe at the beginning of the eighteenth century. Epidemics of plague and other diseases added to the misery suffered by the people of Sweden, Russia, Poland and Denmark during the Great Northern War. Over 25,000 died in Danzig, 30,000 in Warsaw, and 33,700 in Vilnius; there were around 40,000 deaths in Sweden and 28,000 in Denmark.[16] News of the plague terrified the populations of countries which traded regularly with the Baltic, not least Britain, which was dependent upon the region for imports of hemp, flax, pitch and tar – all of which were vital to its navy. After the plague reached Danzig in 1709, orders were issued for a forty-day period of quarantine against ships from that port but it soon became apparent that these measures were being evaded. This realization prompted the passage, in 1710, of Britain's first quarantine Act. It did not remove the royal prerogative to impose quarantine but it showed the public that Parliament was backing the Crown.[17]

Legal provisions of this kind enabled more funds to be granted for quarantine establishments and for specific penalties to be imposed on those who broke quarantine.

The Act of 1710 placed sanitary measures in Britain on a similar footing to those in France and in theory allowed them to be used more effectively as tools of statecraft. However, in 1720 it became obvious that French precautions left much to be desired. When plague arrived at the Mediterranean port of Marseilles, it was the first time that the disease had been reported in the western Mediterranean since the 1660s. According to most contemporary accounts, the disease was imported by a merchant vessel which had left the Levantine port of Sidon, making its way to France via Tripoli and Livorno. At the latter port, the ship's master, Captain Chataud, reported the death of one of the Turkish passengers and eleven sailors. The officials at Livorno declared the men infected with a 'malignant and pestilential fever' rather than using the term plague, no doubt in order to prevent disruption to commerce.[18] On arriving in Marseilles on 25 May, instead of being sent to the quarantine station on Jarre Island, the captain was told to dock at the infirmary quay, where his ship's cargo was unloaded. A few days later, three other ships arrived from the Levant, their crew suffering from what was almost certainly plague. But according to some accounts, the inspecting surgeon was coy about declaring the disease and referred to it as a 'malignant and contagious fever'.[19] Others laid most of the blame on the 'Intendants of Quarantine' (a body of twelve merchants and landowners) for ignoring medical warnings that the disease was, in fact, plague.[20] Either way, the ships were ordered to remain in quarantine for forty days but at the dock rather than at the quarantine island. It was only after more deaths had occurred among the crew and some of the porters that the ships were eventually sent to the island.[21]

It appeared that quarantine officials had wilfully ignored the threat of plague, as the disease was widely known to be prevalent in Tripoli and the Levant.[22] As Marseilles conducted frequent trade with Levantine ports such as Smyrna, it seemed reasonable to suppose that there was a high risk of infection.[23] Indeed, suspicion that goods from the region might be contaminated was later confirmed by the fact that new cases occurred among the porters who opened bales of cotton unloaded from the infected vessels.[24] However, merchants involved in the export trade with the Levant had grown increasingly critical of these restrictions, which cost them dearly through delays, fees and the damage of goods by fumigation.[25] In Marseilles they had come to exercise a good deal of influence over the local administration and the port's formerly robust quarantine procedures had been permitted to lapse. When plague arrived, they attempted to evade responsibility by claiming that it had been imported in contraband carried by smugglers.[26] But the outbreak placed the merchants and their supporters on the defensive: there were repeated calls for repentance and denunciation of those who had corrupted public life.[27]

The arrival of plague realized the worst fears of many who harboured doubts about the opening of trade with the Levant. In 1669 the Minister of Finance, Jean-Baptiste Colbert, had issued an edict which stipulated that silks from Italy, Africa and the Levant could enter France only through the ports of Marseilles and Rouen. Furthermore, he announced that foreign merchants could not only reside in these ports but become naturalized Frenchmen if they married and purchased substantial property there. This edict brought greater prosperity to Marseilles but with it came a large influx of foreign merchants, many of them Ottoman Armenians and Jews. The increasingly cosmopolitan nature of Marseilles was not to the taste of all the city's inhabitants. Many feared that it was losing its Roman Catholic identity, while local merchants faced ever stiffer competition from newcomers.[28] These concerns were closely bound up with anxieties over the spread of disease from the Levant; anxieties which Colbert had hoped to assuage with the opening of a new lazaretto the year before his edict.[29] But fear of infection remained and became inseparable from the economic and cultural worries of the city's French inhabitants. The Chamber of Commerce – the first such institution in the world – therefore lobbied the king, with the support of some civic leaders, in an attempt to exclude foreign merchants, alleging conspiracies between the Jews and Barbary corsairs. By the early 1700s, many foreigners had been forced to flee the city but there was still a good deal of unease about its connections with the Orient. Merchants of all kinds – not just foreigners – were regarded as duplicitous and self-interested. The arrival of plague thus seemed to confirm suspicions about the nature of mercantile activity and the dangers emanating from the East.[30]

The initial response from the civic authorities in Marseilles also gave cause for concern. Belatedly, a land quarantine consisting of troops was placed around Marseilles to prevent the plague from spreading,[31] but the record of the city's leaders inspired little confidence and quarantines were imposed by most of the towns with which it was in regular communication.[32] The municipality of Aix, for example, ordered the city's gates to be shut and prevented its inhabitants from going to Marseilles on pain of death.[33] In October, the central government followed suit and the Duke Regent placed a guard on the rivers Durance and Rhône in order to prevent traffic between Marseilles and the interior. Only necessities and certain items deemed unlikely to harbour plague, such as fish, olives, citrus fruits, spices and perfumes, were permitted to pass.[34] The government's determination owed much to the disastrous effects that plague – and the restrictions imposed by other countries – were having on trade. By October, the disease had spread from Marseilles to other parts of Provence and Gévaudan, inducing neighbouring countries to place stringent restrictions upon vessels leaving *all* French ports.

Fearing retaliation from their larger neighbour, smaller states bordering France took pains to justify the precautions they were taking. The government of Geneva, for example, stressed that all nations had an interest in guarding their people from plague as well as in the freedom of commerce, and that the measures it proposed to take against the affected area of France had been drafted with such a balance of interests in mind.[35] But the British authorities were less than apologetic. The threat of plague spreading to Britain was taken sufficiently seriously for a new quarantine Act to be passed in 1721. This legislation strengthened existing provisions by creating a quarantine station in the River Medway and raising the maximum penalty for evasion to death.[36] These measures were later relaxed as the threat from plague diminished,[37] but at their height, the restrictions imposed on vessels sailing from French ports were a source of great irritation to merchants of all nationalities and more generally to the French. A number of dignitaries such as the statesman Cardinal Dubois were 'gravely inconvenienced' by these quarantines, which were regarded by the French as unfair because they encompassed all ports, including Atlantic ones like Cherbourg which were far from the seat of infection. French diplomats stressed that the king had taken every possible precaution against the spread of the disease. They also alleged that some states had used the outbreak of plague as a pretext to damage French trade in order to benefit their own. Although Venice was cited as an example, the clear implication was that Britain was doing the same.[38] In order to reassure international opinion, the French government stepped up measures to prevent plague spreading beyond the infected area.[39] But by September 1721 local doctors reported that the epidemic had ceased and the government began to threaten Britain with retaliatory measures if quarantines against its vessels were not relaxed.[40] On the 10th of that month, the British lifted their quarantine against all vessels from France.[41] It remains uncertain whether this was due to the reduction of the sanitary threat posed by France or to the prospect of retaliation, but in all likelihood the decision would have been taken with both in mind.

Outbreaks of plague in the Levant and in France caused concern in all countries which traded regularly with the eastern Mediterranean but they also presented an opportunity for some to gain advantage over their rivals. In the long run this was in no one's interests, and by the eighteenth century the boards of health in some ports of the western Mediterranean were again acting in concert. The Italian states began to share information with each other and with other quarantine stations, such as that at Marseilles. As one historian has put it, 'Any failure to toe the line meant retribution in the form of an arbitrary quarantine against their shipping.'[42] However, Spain and Portugal were not members of this 'health clique' and often acted in ways calculated to disadvantage their competitors. Britain, for example, was aggrieved by the measures imposed by Spain and Portugal (under duress from its neighbour) on its ships in the Mediterranean. Britain's trade with Portugal had increased following the signing of the Methuen Treaty in 1704, underscoring a strategic alliance against their mutual rival, Spain. Lisbon had since become home to a sizeable community of British merchants who played an important part in the trade between the two countries and in Portugal's trade with other countries, particularly those in the Mediterranean.[43] Restrictions against British vessels wishing to enter Spanish and Portuguese harbours were justified on the grounds that quarantine was not adequately enforced in British territories such as Gibraltar, which Spain had recently been forced to cede under the Treaty of Utrecht following the War of the Spanish Succession (1701–13).[44] Britain, in turn, put pressure on Portugal and secured exemptions for its men of war, although not seemingly for commercial vessels. As one delicately worded communiqué put it, this concession was intended 'to show His Majesty how much he [the king of Portugal] desires to favour His subjects notwithstanding those ships came from the Mediterranean, and had not performed the intire Quaranteen.'[45]

Portugal had gained independence from Spain in 1688 but there remained great animosity between the two nations, Portugal having sided with Britain in the recent conflict. Spain was still able to exert a great deal of influence over its affairs and could make life very difficult for the Portuguese commercially if it so chose. This became apparent in November 1721 when Henry Worsley, the British consul in Lisbon, was told by a Portuguese official that the viceroy of the Algarve was under pressure to introduce restrictions against British vessels or face punishment by Spain. The Spanish had issued an edict which prohibited all commerce with Gibraltar on the grounds that plague might be imported into the territory and thence to Spain. The viceroy was requested to do likewise and it seemed probable that all Portuguese harbours would be expected to follow.[46] It was a pattern that would become familiar over the next few years: pressed heavily by Spain, Portugal took measures against British shipping – targeting Gibraltar in particular – while the British, in turn, attempted to wrest concessions from Portugal.

These tactics became more obvious as tension rose between Britain and Spain over the latter's alliance with Austria (in 1725), which threatened British trading interests in the Mediterranean. By 1726, Britain was practically at war with Spain but frequent outbreaks of disease in ports such as Smyrna and Alexandria, with which Britain traded regularly, gave the Spanish measures a veneer of legitimacy.[47] In November 1726, the Portuguese Secretary of State, Diego de Mendonia, informed the Governor-General of the Senate of Lisbon that plague had been reported in Constantinople and Cairo, with thousands dying in those cities daily. The French had warned him that two vessels loaded with corn had arrived from the Levant at Marseilles. Their crews were suffering from a suspicious fever: some had died on the voyage and some in the city's lazaretto. The epidemic of 1720–1 fresh in their minds, the city's inhabitants became greatly alarmed and stringent quarantine was imposed in all French ports against vessels from the Levant. De Mendonia was determined that Lisbon should do likewise and requested the Senate to order that all ships from the Levant be 'strictly examined and not admitted without legal testimonies, of their being free from the said contagion, or performing quarantine according to the manner prescribed by the [Lisbon] Regiment [sic] of Health'.[48] There was nothing exceptional in this but it appears that the decision to impose quarantine had been taken under intense pressure from the Spanish court. A week prior to de Mendonia's letter, the Spanish consul in Lisbon, Don Jorge de Macazaga, had been told by a Spanish official that his government had 'absolutely prohibited' commerce with all vessels coming directly from the Levant and other Mediterranean ports further west which did not impose strict quarantine. Ships without a testimony to that effect were compelled to undergo a quarantine of forty days on reaching Spain.[49] It was clear that Spain expected Portugal to do likewise.

Two years later, the tension over quarantine came to a head. On 14 August the British envoy Lord Tyrawley informed Lord Newcastle, the Secretary of State that

> The Portuguese have of late taken it into their heads to oblige all our Ships, that come from the Mediterranean, to perform a Quarantaine, even tho' they have already performed it in other Ports, and tho' they bring certificates with them of their having done so, and others of our Ships that come from the Levant they wont admitt to pass a Quarantaine at all, but oblige them to goe out of the River againe. This making all our Merchants prodigious uneasie, they aplyed to the Consul for redress, who spoke several times to the [Portuguese] Secretary of State about it, but never could obtain any satisfactory answer....[50]

When he eventually replied, the Portuguese Secretary of State intimated that

> The Spanish Ambassador had given in a Memorial insisting upon the Portuguese, obliging all the English Ships to perform the same Quarantaine in the Ports of Portugal that they did in the Ports of Spain, and ending with a sort of Menace to the Portuguese, if they did not doe it.

The British consul thought this 'so extream bad' that he took it upon himself to do something about it. He called for a Mr Start, who was known to be friendly with the Portuguese Secretary of State, to see if he could prevail upon him to lift the quarantine, indicating Britain's displeasure that the king of Spain had been gratified at King George's expense, but assuring him of His Majesty's 'friendly' intentions towards Portugal. This veiled threat seems to have worked and the following day orders were sent to the health office in Lisbon to admit British ships.[51]

But the relaxation of quarantine was short lived. Caught between its powerful neighbour and its chief trading partner, Portugal swayed one way and then the other, as each exerted diplomatic pressure. Just two weeks after British ships had been admitted into Lisbon harbour, the Portuguese announced that, on account of plague in the Levant, a strict quarantine was to be placed on all ships sailing from there, their notion of the Levant encompassing the Mediterranean as far west as Italy. British merchants in Portugal complained loudly and Lord Tyrawley made their case strongly to the Portuguese Secretary of State.[52] However, Britain's position was undermined by a serious assault committed by the captain of one of its vessels upon a health official in Lisbon harbour. It was with great disappointment and some anxiety that Tyrawley informed Lord Newcastle:

> An English Merchant Man coming some days since into this river [the Tagus], the Officer of the Quarantaine went on board, which the Master was not willing to suffer, but the Officer insisting upon it, the Master took him on board and striped [sic], and whipped him, and set him on shoar again; and likewise committed several other irregularitys. The next morning, reflecting upon what he had done, and fearing to be punished as he well deserved, he got privately on board one of our Men of War, actually under sail for Gibraltar and made his escape.[53]

The Portuguese Secretary of State had already made an official complaint but he had known only about the refusal of the captain to permit the official on board; not about the assault. Tyrawley feared reprisals if the full story leaked out. Indeed, the Portuguese refused to lift their quarantine on British vessels and, by November 1728, there were reports that the lucrative wine trade out of Oporto had been severely damaged. It still seemed that Portuguese officials showed a preference to oblige the Spanish court rather than the British, but the British remained optimistic that these impediments to their trade would soon be lifted. Portugal was suffering from a severe shortage of grain and would soon be compelled by this to open its ports fully to trade with other countries.[54]

Similar incidents flared up periodically over the coming years with Spain continuing to pressurize Portugal into imposing restrictions on British shipping. As before, the Spanish focused much of their attention on the rock of Gibraltar. In February 1729 Lord Newcastle, who advocated a hard-line policy against Spain, complained to one of his officials:

> Contrary to our Treatys, they [the Spanish] have in effect cut off all communication between Gibraltar and the Ports of Spain and that on Account of our having a Correspondence with the Coast of Barbary, where there is not the least pretence of there being an infectious Distemper, at the same time that Vessels of other Nations are freely admitted from Barbary into the Spanish Ports; and you will see that even the Men of War have been refused admittance into the Port of Cadiz – without performing Quarantaine, a thing never before practised or insisted on.[55]

Newcastle feared that the Spanish measures were the harbinger of military action, their intention being to place Gibraltar under siege, as they had recently in 1727. His aggressive methods were ultimately successful and, in 1729, Spain was forced to abandon its harassment of Gibraltar, following the collapse of its alliance with Austria. But the British remained vigilant and suspected that Spain would attempt to disrupt communications with Gibraltar on the flimsiest of grounds. Such an opportunity arose in 1735, when outbreaks of plague at Smyrna and Alexandria led many states along the Mediterranean to place quarantine on shipping from the Levant. Some of the severest restrictions were imposed by Bashaw Hammett, the ruler of Tangier, who declared an embargo against all ships sailing from Turkey to the Barbary Coast, except for those which possessed a clean bill of health. This was a significant development, for Tangier was the first Muslim state to depart from the sanitary orthodoxy of the Ottoman Empire. The Ottomans had never imposed quarantine, but as ties with Constantinople weakened in these western outposts of the Empire, the rulers of states along the Barbary Coast began to emulate practices on the northern shore of the Mediterranean.

These countries sat between two 'world systems' – the Ottoman and the European – but obeyed neither. However, they were marginalized by European control of trade in the western Mediterranean and their attempts to make inroads were frustrated by the frequent and lengthy quarantines imposed on shipping from the Barbary Coast on account of its ties with the plague-ridden Levant. The activities of the notorious Barbary corsairs and the decision by some rulers to erect their own quarantines (partly in retaliation against states to the north) were attempts to chip away at European dominance.[56] The British, however, appeared pleased by the actions of Tangier because Gibraltar then conducted a busy trade with Barbary. Nevertheless, they were apprehensive that 'the Spaniards who are generally very ready to lay hold of any pretence that offers to burthern and embarrass our Trade' would use the plague outbreak again to prohibit all commerce with the peninsula.[57] Again, in 1740, British ships were denied entry to Atlantic ports because of plague on the Barbary Coast, although in this instance with some reason, for the disease had claimed thousands of lives in Tunis and Algiers.[58] The imposition of quarantine on Gibraltar remained one of the few bargaining chips available to Spain, which had no immediate prospect of recovering the territory militarily.

The mercantile interest

Despite its potential to generate serious disputes, quarantine remained firmly entrenched in Europe for the rest of the century. It was manifestly imperfect but it was an art of the possible and to abandon it was incompatible with contemporary theories of statecraft. Although public health had its origins in Renaissance conceptions of the common good, the enlightened regimes of the eighteenth century took a distinctly managerial interest in health.[59] Disease was widely regarded as a consequence of disorder and this was particularly true of plague, which often coincided with periods of dearth and war.[60] The absence of plague from the west and centre of Europe was therefore a matter of pride. But as Johann Peter Frank insisted in his multi-volume treatise, *A System of Complete Medical Police* (1786), it required constant vigilance to keep the disease at bay. An exponent of enlightened absolutism, this Viennese physician proposed a comprehensive system to improve the health of all persons through generous state provisions and sanitary regulations. In this system, quarantine played an important part by protecting enlightened states – like that of the Austro-Hungarian emperor, Joseph II – against incursions of disease from neighbouring territories. 'It is one of the foremost tasks of the state to prevent persons or animals, goods, and all objects to which or whom contagions cling, from entering the country,' he proclaimed, 'and there is no doubt that governments are entitled to use all suitable means that do not contravene international law in order to achieve this.'⁶¹ To this end, the Austro-Hungarian Empire maintained a 1,600 kilometre (1,000 mile) cordon along its borders with the Ottoman provinces of Moldavia and Wallachia, which were still occasionally ravaged by plague. The cordon was policed by watchtowers and roving bands of soldiers under orders to shoot on sight those who crossed the border without performing quarantine. Its sanitary functions developed gradually from 1710, having originated in the military cordon established against the Turks.⁶²

Although some historians have doubted the protection afforded by this cordon,⁶³ most contemporaries credited it with keeping plague at bay.⁶⁴ Nothing else seemed to account for the prevalence of the disease in the Ottoman provinces to the east of the cordon and its absence in the Hapsburg lands to the west. But the cordon had its disadvantages, too. Although he believed it had protected Austria from plague, the Viennese physician Paskal Joseph Ferro recognized the damage which quarantine could inflict upon trade and argued that such measures should be brought into conformity with enlightened principles, minimizing inconvenience and disruption.⁶⁵ He was aware that quarantines were rarely maintained solely for reasons of public health and that the Hapsburg cordon continued to have an important military role, providing an additional reserve of manpower which could be deployed elsewhere if necessary. These additional functions meant that the cordon could never be just a temporary expedient. The same was true of the cordon maintained by the Venetian Republic against the adjacent Ottoman provinces of Istria and Dalmatia, which was both a quarantine and a military defence.⁶⁶ Sanitary cordons could also be used offensively, to conceal impending acts of aggression. When plague ravaged Eastern Europe in 1770, for example, Prussia established a cordon that encroached upon Polish territory, its ostensibly defensive nature masking predatory intentions.⁶⁷

Sanitary disputes were also capable of provoking war, especially in the Mediterranean region, which was ravaged by major conflicts such as the Seven Years War and perpetual rivalry between European states and Islamic polities to the south. In 1789, for instance, the Tunisian ruler Hamuda Bey severed relations with Venice because one of its quarantine officers destroyed the cargo carried by a Tunisian vessel. Having been refused compensation, the Tunisian merchants complained to their ruler, who declared war against the Venetian Republic. Tunisian corsairs were dispatched to seize compensation from Venetian vessels and, in reprisal, a Venetian fleet bombarded the Tunisian cities of La Goulette, Sousse and Sfax. Eventually a ceasefire was agreed and commerce between the two countries resumed.[68]

The professionalization of diplomacy during the eighteenth century meant that decision-making on issues such as quarantine became more complex.[69] Those responsible for such decisions seldom took action without carefully considering the likely responses of other states. Thus, when plague appeared in western Russia in 1771, threatening the port of St Petersburg, quarantine was imposed upon all goods brought to the city for export in the hope that this would reassure other countries. But, despite these precautions, fear of plague led most northern European countries to impose quarantine against St Petersburg, much to the disappointment of the Russians and foreign merchants based in the city. Britain held out at first, but as Denmark, the Dutch Republic and Sweden had already imposed quarantines, it was forced to follow suit. Had it not done so, Britain would have faced embargoes similar to those imposed on Russia. Ultimately, the losses sustained by Anglo-Russian traders were not as great as some had feared but the damage to Russian merchants and their British creditors may have deepened the commercial crisis of 1772, which severely affected several British firms in St Petersburg.[70]

In most of the cases considered so far, the mercantile interest was seldom paramount and was often regarded with suspicion. As far as the Russian epidemics were concerned, most of the eminent practitioners who had been appointed to the Moscow Plague Commission (formed to guide plague policy in the capital) agreed that the disease had spread largely as a result of contact with infected persons and merchandise. As in the Marseilles outbreak, there was a broad spectrum of views on the relative importance of contagion and local environmental conditions, but rarely to the complete exclusion of the former. Those who leaned more in the direction of contagionism placed special emphasis on the textile trade as a vehicle for the spread of plague because clothing was said to be capable of retaining infection for a long period unless aired or fumigated. The frequent outbreaks of plague in the Ottoman Empire were attributed to this trade and to the failure of the Turks to take precautions against suspect goods and carriers.[71] One of the most forthright contagionists, the physician Dr Danilo Samoilovich (Samoilowitz), concluded that the epidemic had begun in Wallachia, where Russian troops had probably contracted plague brought into Bucharest by Turkish merchants.[72] But Samoilovich was acutely aware of the value of trade to the state and did not wish to curtail it. Rather, he advocated a system which took into account the exact nature of the merchandise which was imported, rags and wool being among the most susceptible as they were likely to accommodate the tiny 'animalcules' that he believed caused the disease.[73]

A considerable part of Samoilovich's 1783 memoir on plague was devoted to this problem and his recommendations, which were based on the methods devised by the Moscow Plague Commission, encompassed the disinfection and fumigation of goods purchased from potentially infected areas, using substances ranging from vinegar to specially prepared concoctions of nitre and myrrh. Under his proposals, disruption would be kept to a minimum by the use of certificates which attested to whether fumigation or disinfection had been performed; these would be awarded by inspectors of quarantine in each city (in liaison with designated physicians and surgeons) and given to the inspectors of city quarters into which merchandise was imported. In order to open for business at times when plague was reported, merchants and shopkeepers had also to present the inspectors with a certificate attesting to the fact that they had not been in contact with any infected person and were not concealing such persons or goods in their shops or homes. If these certificates were presented, then formal quarantine could be avoided; if not, then quarantines of between fifteen and twenty days would be necessary. Samoilovich insisted that such measures had been of great value in the past, enabling St Petersburg to escape plague during the 1709 and 1771 epidemics.[74]

One notable thing about these recommendations was that medical policing would be placed beyond the influence of merchants, whose self-interest and diverse ethnic origins rendered them suspect. In Russia, Turkish merchants were implicated in the initial spread of plague to Romania, while in later stages of the epidemic, Jews involved in the clothing and rag trades were sometimes expelled or refused entry into the czar's dominions. The fight against plague was widely seen as a patriotic struggle in which all the Empire's subjects ought to be actively engaged. Likewise, the failure of quarantine and other measures to contain the disease was often blamed on avarice and lack of national sentiment.[75] In other countries, too, there was a feeling that the pursuit of profit had been elevated above the public good. Looking back at the Marseilles plague epidemic, the town's librarian and historian J.P. Papon recommended that merchants be removed from public health bodies such as the port's quarantine board. Although they did not dominate the Marseilles board when it was first established in the 1640s, reforms in 1716 had allowed merchants to prevail by the time that plague arrived from the Levant. In Papon's opinion, this had led to regulations becoming lax, culminating in the epidemic of 1720.[76] By the same token, the containment of plague within a relatively small area of France was a victory for strong administrative authority.[77] It was for this reason, perhaps, that the Sanitätsmagistrat established in the Austro-Hungarian port of Trieste, in 1755, included no merchant representatives at all.[78]

Whatever lessons one chose to learn from the outbreaks of plague in Marseilles and Russia, opinion on quarantine was clearly becoming polarized. Supporters credited it with having secured the freedom of much of Europe from infection, while its opponents insisted that it was useless; if not theoretically, then practically. When plague arrived at Marseilles, for example, the Spanish authorities found it impossible to impose an effective embargo against ships from France, despite posting guards along the Mediterranean coast. Some ships had been able to avoid the guards altogether while others carried fraudulent bills of health, which claimed that they had sailed from non-infected ports. Cordons along land borders were even easier to evade and plague epidemics were often blamed on illicit traders who crossed undetected.[79] These difficulties were acknowledged by some supporters of quarantine, too, but they sought a remedy in more efficient systems of disease notification. The British physician William Brownrigg, for example, argued that quarantine measures could be targeted more effectively if the bills of health issued by plague-infected countries were more reliable.[80] Such bills were already provided by some of the Italian states and by European consuls in the Ottoman dominions. They normally declared the time and place from which they were granted, the names and numbers of crew and passengers, and indicated the status of vessels. Bills of health also recorded whether or not quarantine had been performed and the nature of any merchandise carried.[81]

Typically, there were two or three types of bill, and these had different names in different countries. The French nomenclature, which was widely used until the late nineteenth century, was 'patente nette' (a clean bill of health), 'patente soupçonnée' or 'touchée' (a suspect bill of health issued when there were unconfirmed rumours of plague) and 'patente brute' (a foul bill of health, issued when the existence of plague had been confirmed).[82] One of the problems with this system was that consulates had to depend on what were often unreliable sources of information and some merchants suspected that officials were deliberately fed false reports of disease by their commercial rivals. When a consulate received a report of plague – even a single case – it was supposed to issue a foul or at least a suspect bill of health, which would normally require shipping under its flag to perform quarantine somewhere in the Mediterranean or at its port of destination. Occasionally this allowed merchants flying different flags to steal their trade, as in the case of the Greek merchants who usurped much of the British trade from Smyrna – a port so frequently infected by plague that consulates were bound to take rumours seriously.[83]

This was just one of many complaints which merchants made about the iniquities of quarantine. By the early eighteenth century, the Lords of the Privy Council in Britain were receiving many petitions from merchants and ship owners protesting against the fact that their vessels had been detained on arrival in British ports, too. In most cases, the petitioners argued that their crews were healthy, that they had not touched at an infected port, and that they possessed affidavits attesting to the fact that they had already performed quarantine en route.[84] It was not only vessels from the Levant that were affected by quarantine but those involved in the Baltic trade, which was regularly disrupted by outbreaks of plague in Russia and neighbouring countries embroiled in the war with Sweden. In 1713, for example, one Henry Morris petitioned for the removal of his ships from quarantine in the Scottish port of Inverkeilling. His vessels had carried their cargo of flax from St Petersburg where no plague had yet been reported, nor had they touched at any infected port on their return journey. Furthermore, he claimed, his ships had already been detained in quarantine beyond the normal period.[85] He may well have been correct, for Scottish quarantine officials had a reputation for being zealous in interpreting such regulations.[86]

In the coming years, such petitions became more numerous and merchants began to represent their dissatisfaction collectively. Sometimes the objects of their indignation were domestic quarantines, but mercantile communities continued to complain about measures taken in foreign ports. In 1722, in the wake of the Marseilles plague, London merchants protested bitterly to the Privy Council about the restrictions which had been placed on the entry of British vessels into Spanish ports. In their view, the measures were absurd because vessels trading between Italian ports and Spain posed no conceivable threat, having no reason to put in at Marseilles. They were also moved to protest against the retaliatory measures which Britain had taken against vessels arriving from Spain, as these affected import houses in England as much as Spanish exporters.[87] Petitions were sent regularly by the Levant Company, protesting against quarantine in Britain and in the Mediterranean, as well as against the maltreatment of goods when aired or disinfected.[88] The Company claimed that merchandise was frequently damaged by being exposed to the elements and by the increasingly common practice of fumigation with chemical substances.[89]

These were difficult years for the Levant Company. It had steadily been losing ground to its rival, the English East India Company, which had been undercutting the price of imported silk.[90] Sanitary measures of the kind imposed in British and Mediterranean ports threatened to render the Levant Company uncompetitive. By the 1760s, however, there were demands to abolish all trading monopolies, including that of the East India Company, which by that time was becoming a territorial power. It was in these difficult circumstances that some merchants came to advocate the construction of a permanent lazaretto in Britain. They argued that such a building was necessary, not only to protect Britain as its commercial connections with the East proliferated, but because it would reassure other countries that Britain would not become a hub for disease.[91] Yet such requests were rare and the majority of merchants continued to regard quarantine as an unnecessary burden and expense.[92]

Complaints about the severity and duration of quarantines imposed in Britain against ships sailing from the Mediterranean were voiced by British merchants overseas as well. In 1766, traders at the British factory in Livorno protested that their business had been depressed by the quarantine imposed in Britain against ships coming from that and nearby ports. They were supported by merchants in London and by the government of Tuscany which submitted a memorial requesting the abolition of the quarantine imposed in 1759 against all vessels arriving in Britain from the Mediterranean. This occurred during the Seven Years War when fears about the spread of disease by naval as well as merchant vessels were heightened. An element of retaliation may have been involved, as British shipping had been subjected to quarantine by some of the Italian states. Venice had introduced measures against vessels from certain Mediterranean ports (including Livorno and Malta) which were used regularly by the British. The Venetians appear to have been particularly concerned about the activities of a British privateer which was operating clandestinely in the eastern Mediterranean.[93]

However, the quarantine imposed by Britain followed the disclosure of certain 'irregularities' practised by ships trading in the Mediterranean, which had evaded quarantine in an attempt to profit from the war. Britain was similarly concerned about the activities of its own privateers in the Mediterranean and subjected all of them to a forty-day quarantine when they returned.[94] The merchants of Livorno and the Tuscan government argued that sweeping restrictions were no longer necessary now that the war had ended and that trade might be safely resumed on a selective basis.[95] They also stressed that Livorno, of all Mediterranean ports, posed little risk 'inasmuch as there is no Place where the Rules for the Preservation of the publick Health are so strictly and exactly performed'.[96] The Privy Council believed that this claim was well founded and forwarded the petition to the king, noting that the matter merited special attention.

The recommendation was couched in terms which made it abundantly clear that public health ought not to be sacrificed to commercial interests. But the Privy Council also stressed the 'great Burthen and Expence' which the continuance of quarantine placed upon merchants, who claimed that the charges attending quarantine amounted to as much as one fifth of the value of their cargoes. Added to this was the loss of goods which perished or were damaged during quarantine, and the obstacles this placed in the way of 'Circulation and quick Returns'. On balance, the Privy Council concluded that the abolition of quarantine against Livorno would produce 'a great Saving and Relief to such of Your Majesty's Subjects ... as are concerned in carrying on Trade from that Place, as well as of general Use and Advantage to the Trade and Navigation of this Kingdom'.[97]

In this instance the government was prepared to make a concession which benefited mercantile interests, though purely in respect of Livorno. It may have been made because Britain was embarrassed by revelations that it had received ships from the Sardinian port of Villa Franca without their having to submit to quarantine. Writing on behalf of the Grand Duke of Tuscany, Marquis Botta Adorno informed the British ambassador that

'His Royal Highness flatters himself, that what the King of Sardinia has obtained, will not be refused to him; and so much the more as the precautions, that are taken at Leghorn in regard to the public Health, are much greater than what is practised in every other Port, and more particularly at Villa Franca.'[98] The concessions made to Livorno and Villa Franca opened the floodgates and the Privy Council was deluged with petitions from merchants who traded with other Italian ports. The following year, it received an appeal from a group of merchants in Liverpool pointing to the anomalous position of these two ports and the great damage that was being done to trade with Italy and, via Italian ports, with the Levant. They insisted that the other major ports which traded regularly with Britain, such as Messina, Palermo, Naples and Venice, had equally reliable quarantines and vessels proceeding from them posed little threat to public health. The merchants also pointed out that other European nations generally admitted ships from these ports without quarantine, providing they had a clean bill of health. Moreover, plague had retreated and was no more prevalent than before the regulations of 1759 had been imposed.[99]

There is no evidence of any further concessions to merchants, however, and stringent restrictions were re-imposed against all ships carrying cotton through the Mediterranean to Britain, excepting those arriving from Spanish ports and those of west Barbary (the reason for these exemptions is unclear). All such vessels were compelled to perform a quarantine of forty days on arrival in Britain and goods liable to quarantine were to be opened and aired for a week, apart from flax, hemp and certain other commodities deemed liable to harbour infection for longer, which were aired for a fortnight. These regulations were introduced in 1772 two years after similar measures had been imposed against ships arriving from Danzig and other Baltic ports following reports of plague in Poland. In 1780 an Act of Parliament was passed which placed existing provisions relating to ships from the Levant on a firmer basis.[100] French precautions were similarly robust. All ships from the Levant were subjected to a quarantine of between eighteen and thirty days, even if they were carrying cargoes deemed not susceptible to plague; susceptible goods such as cotton were quarantined for between twenty and thirty days. In some cases, the French authorities imposed quarantines of exceptional duration and when plague caused unusually high mortality in Levantine or North African ports, it was not uncommon for a quarantine of fifty days to be placed against vessels leaving them. This happened in 1784–5, in the case of ships sailing from Tunis for Marseilles; in 1785–6, in the case of vessels departing Bone, Calle and Collo; and in 1787 for those leaving Algiers. The quarantine board in Marseilles also thought it prudent to impose a regular quarantine of between ten and twelve days on vessels sailing from Gibraltar on account of the peninsula's frequent commerce with Barbary and Morocco.[101]

At this time, there was great alarm about the possible infection of North African ports because of a severe plague epidemic which had radiated out from the Levant to much of the Ottoman Empire.[102] There was a close trading relationship between Ottoman ports and those of the Barbary Coast and many Turkish merchants were based there. However, the measures taken against Gibraltar may well have had a political purpose, as France had assisted Spain in another siege of the British enclave between 1779 and 1783. There may also have been a retaliatory element in measures against the Barbary states, as their rulers had begun to impose quarantines which affected European shipping in the region. The development of quarantine institutions was driven by pecuniary motives, too. By 1781 a third lazaretto had been constructed at Livorno in response to the extension of quarantine facilities at Genoa. Both ports, it seems, were competing for income from vessels which were forced to undergo quarantine at some point in their journey west through the Mediterranean.[103] More attention was also paid to ships arriving from the West Indies, following a resurgence of yellow fever in the Caribbean. In 1770, for example, British officials considered quarantining ships carrying cotton wool from the Caribbean even though they doubted the legality of such a measure, the existing statute referring only to plague.[104] The readiness of even comparatively liberal countries like Britain to resort to quarantine suggests that the modest concessions which had been made to mercantile interests in the 1760s were very much the exception. Merchants thus continued to protest against the unfairness, as they saw it, of quarantine at home and in foreign ports, and at the damage to 'national' interests that came from relying on an 'outmoded' defence against disease.[105]

Contagion reconsidered

In the eighteenth century, the fragile consensus that seems to have existed over the necessity of quarantine was breaking down. Quarantine was still favoured by governments and received support from members of the medical elite but merchants and their supporters were growing more impatient with the practice and with the failure to win significant concessions. In some countries, merchants' influence over sanitary measures actually diminished. Opponents of quarantine began to condemn it as a medieval practice, mired in dogma and superstition, and in doing so they were able to draw upon a flurry of pamphlets, books and articles which claimed that quarantine was unnecessary. Medical practitioners were increasingly questioning the validity of the doctrine of contagion upon which quarantine was based. Even if they did not reject it altogether, they often sought to modify it in such a way as to show that contagion depended upon other factors, such as climate and ventilation. There had been relatively little debate about such matters during the seventeenth century but by the time of the Marseilles plague, professional opinion was sharply divided. Although sanitary cordons appeared to prevent plague from spreading beyond southern France, some physicians disputed the contagious nature of the disease. If plague were contagious, why did it appear so infrequently and usually at certain times of the year? Might not epidemics be related to other factors, such as seasonal changes and states of the atmosphere?

These climatic explanations had gained ground steadily following the revival of Hippocratic medicine in the Renaissance, and by the late seventeenth century they were articulated by the English physician Thomas Sydenham (1624–89), amongst others.[106] Many of the medical practitioners who commented on the epidemic in southern France employed such explanations as an alternative or supplement to theories of contagion. According to the French physician Jean Baptiste Sénac, the fact that the Levant was afflicted more often than Europe could be explained by its hot climate, the plague 'poison' arising from the rapid putrefaction of dead animals and plants. Likewise, plague tended to occur in Europe during the summer, when weather conditions approximated more closely to those in the East. The role of climate was acknowledged by some supporters of quarantine but its opponents placed much greater emphasis upon meteorological conditions and their role in limiting the distribution of disease. To such writers, quarantine seemed unnecessary, as well as injurious to trade.[107] In England, too, some practitioners bucked the official line of the Royal College of Physicians and stressed the necessity of extreme heat in the production of plague epidemics. One of Britain's foremost authorities on fevers, John Huxham, claimed that the spread of plague was arrested by a change of air from hot and moist climates to those which were cold and dry. Plague was therefore to be considered in much the same way as other fevers, which Huxham attributed to changes in the weather.[108] Having witnessed plague in the Levant, and having kept detailed meteorological tables, the Italian physician Dr Timoni also maintained that the 'seeds of contagion' were suppressed by the cold of winter but gathered strength in the summer and were at their height in autumn. In his view, the plague germs were most likely to be spread by winds at certain times of the year.[109]

Coinciding with growing dissatisfaction among merchants, medical accounts of plague and yellow fever were becoming increasingly politicized. Dale Ingram, who had practised as a surgeon and man-midwife on the island of Barbados for some years, adduced what he believed to be conclusive proof that plague and yellow fever were not contagious. 'Before the Plague in London,' he argued, 'we imported many thousand bags of cotton, and since that time some millions, even from infected places, and even the bale of cotton, so much talked of in 1665, as having conveyed the infection into London, is not uncontroverted.'[110] Like many contemporary writers, he attributed outbreaks of plague to heat acting upon populations reduced to eating unwholesome food and living in filth. In the latter respects, there was little to separate London and Cairo but the fact that London had a temperate climate meant that it was subjected to plague only rarely.[111] Ingram did not deny that plague or yellow fever was contagious but believed that they became so only in overcrowded and ill-ventilated conditions, and when the heat of the atmosphere had reached a critical point. In the case of yellow fever, newcomers to the tropics were always most at risk because their bodies had not yet acclimatized, he argued.[112] Since these pestilential diseases were, in Ingram's opinion, 'natural' to specific climates and localities, a blunt instrument like quarantine made no sense. He attributed the 'blind' attachment of many physicians and governments to quarantine to their uncritical faith in the 'absurd histories' of earlier periods and to a superstitious adherence to doctrines fabricated during the medieval period by the papacy.[113] Like some other English writers on quarantine, Ingram alleged that the theory of contagion had been devised as a pretext, on the basis of which certain delegates would be prevented from attending the Council of Trent.[114] The ascription of the doctrine of contagion – and of quarantine – to the papacy was a recurring theme in British (and some American) literature on plague through to the nineteenth century, reflecting prevailing distrust of Roman Catholic powers such as France and Spain.

Although comparatively few practitioners opposed quarantine outright, many were coming to believe that it was unnecessary in many instances. Thus Dr Mordach Mackenzie, who had witnessed plague at Constantinople, protested:

> To what purpose keep ships in Stangate-Creek [the quarantine station on the River Medway] for weeks, and even months, without landing and serening [airing] the goods? ... There is little to be feared from the bodies of men, who get in good health from Smyrna to England, which voyage is seldom performed in less than 7 or 8 weeks; which I presume will be thought too long for infection to remain in the blood without producing some effect.[115]

While there was no justification for quarantine against ships from the Levant in Britain, in his opinion, 'the case was different in Italy and in the South of France; to which countries a ship with a fair wind may perform a voyage in eight days from the Levant; during which time the person may have the plague about him'.[116] Views such as Mackenzie's could not be easily dismissed as special pleading on the part of physicians representing the mercantile lobby. His recommendations were based on a theory in which most governments implicitly based their trust. But those medical practitioners who denied the possibility of contagion, or who reduced its role to such an extent as to render quarantine unnecessary, were easy targets for opponents who claimed they were subservient to mercantile interests. One of the staunchest defenders of quarantine in Britain was the eminent physician Richard Mead, whose advice had led the government to strengthen its quarantine provisions in response to the epidemic in Marseilles. Mead had argued that 'The greater part of the French physicians denied it to be the effect of contagion communicated from abroad, and the interest of trade made it greatly to be wished, that their opinion was true . . .'.[117]

These remarks suggest a close affinity between medical doctrines and commercial interests but the relationship was more complex than Mead's simple formulation suggests. Critics of the theory of contagion were not simply stooges of merchants and their political allies but were often reformers who wanted to sweep away ancient doctrines and ground their practice in what they regarded as hard facts.[118] They were attracted to natural-historical explanations of disease not just because of their incompatibility with quarantine but because they liked the notion that diseases obeyed natural laws and were therefore in some measure predictable. For many practitioners, the allure of certainty was more attractive than the caprice of contagion. It pandered both to their professional vanity and to the spirit of optimism which characterized the Enlightenment. Nevertheless, the connections between commerce and medicine were very real and would grow stronger at the end of the century, as practitioners were increasingly called to account for the economic implications of their theories. In an era marked by war and revolution, debates over quarantine became ideologically charged and stark divisions began to open up between conservatives and those who extolled the liberal virtues of democracy and free trade.

CHAPTER 3

The evils of quarantine

Growing frustration with sanitary constraints upon trade boiled over towards the end of the eighteenth century as quarantine became a focus for radicals on both sides of the Atlantic. Many reformers believed that commercial and political freedoms were inseparable and demanded an end to a practice which seemed to violate both.[1] Having suffered from accusations of self-interest, merchants acquired fresh confidence as their complaints about quarantine began to be echoed by travellers and humanitarian reformers. In one way or another, the spirit of liberty dignified all of these protests and found its way into the medical writings of the period, especially those on epidemic disease. Heated debates broke out between those styling themselves 'anticontagionists' (mostly opponents of quarantine) and 'contagionists', who tended to defend the status quo. Anticontagionists were generally at the liberal end of the political spectrum whereas contagionists tended mostly towards conservatism.[2] But as we saw in the last chapter, neither of these positions was absolute. Even the most conservative monarchies could be flexible in their implementation of quarantine, while commercial nations like Britain made surprisingly few concessions to the demands of merchants. And yet politics, commerce and medicine were becoming closely intertwined. In the heady atmosphere of the 1780s and 1790s, many reformers – not least medical practitioners – began to portray quarantine as a relic of a less enlightened and brutal age.

'Reason and science'

In Great Britain, the merchants of the Levant Company had long complained about the injustices of quarantine and the practices of Mediterranean lazarettos. Their protests had usually fallen on deaf ears but in the 1780s they received unexpected support from the redoubtable reformer John Howard. Howard's probity and high-mindedness were widely respected and he was already well known for his efforts to reform prisons. His interest in bringing enlightened governance to institutions was not confined to Britain however, and he extended his inquiries to the lazarettos of the Mediterranean, now the subject of numerous complaints from travellers as well as merchants.[3] The issue of quarantine reform was slowly being transformed from a mercantile obsession into a matter of general concern. Indeed, when Howard inspected some of the lazarettos for himself, he found ample evidence to support the countless claims of cruelty and extortion which had been made in previous decades. Bribery and corruption were rife, while sanitary conditions were often as bad as those in the prisons he had lambasted at home. Both types of institution seemed calculated to produce misery and disease in equal measure.[4]

Although Howard's scathing report was welcomed by opponents of quarantine, he did not deny the utility of quarantine or the contagiousness of diseases such as plague. He merely urged that lazarettos be improved and their practices brought into line with enlightened principles. In the meantime, he advised the British government to allow ships leaving the Levant to perform quarantine at British ports, rather than undergo a 'long and tedious quarantine' in the Mediterranean whenever plague was suspected in the Levant. At that time, however, Britain had no permanent quarantine facilities and it would be necessary to build something like the Mediterranean lazarettos without duplicating their faults. Howard therefore sketched a model lazaretto which possessed facilities for the fumigation of goods, shops for the sale of provisions, rooms for visitors, and so forth.[5] As well as being more humane, he argued, such an arrangement would protect British shipping from commercial competitors who could hinder the passage home by making false reports of plague to the consul in Smyrna. The fact that British ships were compelled by the Quarantine Act of 1753 to perform a lengthy quarantine in the Mediterranean if granted a foul bill of health had enabled Greek merchants to seize much of their business.[6]

Howard's plan for a lazaretto in Britain was supported by one of the Levant Company's physicians, Patrick Russell, who had been stationed at the Company's factory in Aleppo.[7] Russell was a respected practitioner and a fellow of the Royal College of Physicians of London. His elder half-brother, Alexander, had also worked as a physician in Aleppo and had previously been engaged by the Prime Minister William Pitt the Elder as an advisor on quarantine.[8] This meant that the younger Russell's advice was likely to be valued by government but some of his proposals – especially that for an intermediate bill of health as used by the French – were unlikely to be welcomed by merchants. His suggested addition of a 'touched' or suspect bill of health, issued without cast-iron evidence of plague in foreign ports, would have exacerbated some of the problems of which the Levant merchants legitimately complained. It was therefore Howard, rather than Russell, who inspired the Levant Company to make new proposals for quarantine reform. There was evidently some sympathy in government circles for the merchants, for in 1793 the Prime Minister, Pitt the Younger, referred the matter to the Board of Health. The Board considered plans drawn up by the surveyor John Soane for a lazaretto on the River Medway but there were numerous delays and the plan did not receive parliamentary approval until 1800.[9]

These were modest reforms achieved by moderate reformers, but in the coming years opposition to quarantine became more strident. The first signs of this appeared across the Atlantic, where opponents of quarantine had clothed themselves in the mantle of liberty. In the colonial period, quarantine was often used to protect American ports from outbreaks of yellow fever in the Caribbean, but merchants and political radicals were beginning to question its efficacy. In 1762 the disease had managed to kill several thousand people in Philadelphia, despite the fact that quarantine was imposed. Critics claimed that such measures were useless because yellow fever was not imported and could occur anywhere when meteorological conditions were favourable. One of the most distinguished exponents of this view was the physician Benjamin Rush (1745–1813), a graduate of Edinburgh University who was known for his radical views. Rush opposed quarantine on medical and political grounds, insisting that yellow fever was not contagious except in cramped and ill-ventilated conditions. He also believed that quarantine was incompatible with human freedom, which he saw as essential for health and happiness.[10]

During the War of Independence from Britain, Rush's dislike of quarantine intensified; it seemed to him to exemplify the tyranny against which he and his fellow republicans were struggling. Yet his connections with friends in Britain remained strong and Rush reassured his former teacher, the Edinburgh professor William Cullen, that the members of the 'republic of science all belong to one family'. Indeed, several Edinburgh graduates were later to become supporters of quarantine reform and made common cause with its champions across the Atlantic.[11] Rush identified his campaign with philanthropists such as Howard, praising the 'immense services' that he had 'rendered to humanity and science' through his critical reports of prisons and lazarettos.[12] Rush realized that it was through such associations that quarantine reform acquired a moral imperative.

In 1793, the return of yellow fever to Philadelphia thrust quarantine to the fore of American politics. First reported in early August, the disease had claimed around 300 lives by the end of the month, by which time 55,000 people – 15 per cent of the city's inhabitants – had fled. Before the epidemic ceased, between 4,000 and 5,000 people had perished.[13] It was a devastating blow to the capital of a new nation and one felt particularly keenly by Rush, a signatory of the Declaration of Independence. After witnessing pitiful scenes of suffering, he wrote that his heart was 'torn to pieces' by the distress of the afflicted.[14] In the same year, yellow fever also broke out in New York and in Portsmouth, New Hampshire; a year later it returned, though less severely, to Philadelphia and New York, as well as to Charleston, Baltimore, Providence, and Norfolk, Virginia.[15] These epidemics occurred amidst political tumult, domestically and internationally. The infant republic was beginning to cleave into Republican and Federalist factions, the latter blaming the epidemic on the arrival of refugees from the French Caribbean colony of St Domingue following the slave revolt of 1793. In the coming years, yellow fever became one of the main bones of contention for the emerging parties and many – although by no means all – physicians held views which were compatible with their political sympathies. Most Federalist physicians appear to have supported the theory that the disease was imported, as did many who were non-aligned. Republican doctors like Rush, by contrast, were more likely to downplay the importance of contagion and to stress the importance of meteorological conditions and miasmas. Since the War of Independence, Rush and other medical practitioners had turned their attention to the improvement of the new nation's capital. They viewed disease as a sign of moral and physical decay and sought to cleanse the city, believing that filth and overcrowding were the source of the fevers suffered by many of its citizens.[16]

These different perspectives on the 1793 epidemic are best represented as tendencies rather than exclusive positions. The medical theory of the day revolved around the metaphor of seed and soil; the 'germ' of disease being the seed and the soil the environment or human body into which it was introduced. Debates over the contagiousness of the Philadelphia fever – and of the usefulness of measures such as quarantine and isolation of the sick – were chiefly about the relative importance of each, or, to put it another way, the conditions which were necessary for the imported 'seeds' to germinate.[17] Federalists tended to emphasize importation because it deepened suspicion of radicals in the French Caribbean and at home. Some Republicans, notably Thomas Jefferson, had strong French sympathies, and Federalists realized that they could use the issue of quarantine as a weapon to cast doubt on the patriotism of their opponents. This happened in 1794, when yellow fever returned to the city and an Act was passed to strengthen quarantine provisions for the port of Philadelphia, establishing at the same time the city's first Board of Health. Henceforth a physician was always to be stationed at the lazaretto, which was situated on an island in the Delaware River. Thereafter, quarantine remained the backbone of health policy in the city and fostered a sense of local and national identity, much as the Federalists had intended. In 1798, during the 'Quasi-War' with France, for example, it was used to block the immigration of suspected subversives from St Domingue.[18]

The importation theory of yellow fever had considerable resonance with the inhabitants of Philadelphia, for the alternative – home-grown pestilence – was offensive to patriotic sensibilities.[19] Some merchants also disliked the localism theory because it might lead other ports to stigmatize Philadelphia and cause its demise as a major port. But the Federalists' championing of the importation theory posed a dilemma to merchants who supported their cause. While they stood to benefit in the short term from the quarantine imposed against the French Caribbean islands, they might be damaged in the longer term by the increased rigour of quarantine around Philadelphia. Some feared that business might be diverted elsewhere to avoid these burdensome precautions. It is not surprising that many merchants and ships' captains did their best to circumvent quarantine, and the return of yellow fever to Philadelphia in 1798 was blamed on vessels which had apparently done so.[20]

The presumed evasion of Philadelphia's quarantine led, in 1799, to the passage of an Act which made it compulsory for any vessel that arrived from abroad between 1 April and 30 September to undergo quarantine at the mouth of the River Delaware. A sentence of up to five years' hard labour awaited anyone found evading or assisting the evasion of these regulations.[21] Similar legislation was passed in other Atlantic ports such as New York and Baltimore.[22] These measures met with the approval of correspondents for newspapers such as the *Federalist*, some of whom sought the suspension of trade during the 'sickly season', others conceding that goods might be landed forty or fifty miles from the city and placed in storage for inspection before they were allowed to proceed to Philadelphia.[23] But there were still bastions of implacable opposition to quarantine, not least Rush and his circle. In his correspondence with physicians around the world, Rush continued to insist that 'yellow fever' was nothing more than a severe form of the bilious remittent fever which arose locally from some inscrutable change in the atmosphere.[24] Many of the city's merchants also continued to complain about the quarantines,[25] and Rush protested to James Madison that

> The commerce of our country has suffered greatly by our absurd quarantine laws in the different states. These laws, which admit the contagious nature of our American yellow fever, have produced a reaction in the governments of Europe which has rendered our commerce with the cities of Europe extremely expensive and oppressive.[26]

The only solution, according to Rush, was the removal of all such legislation in American ports and he hoped that Madison would help convince his fellow politicians that quarantine afforded no protection at all. Rush confided to Thomas Jefferson, the most prominent member of the Republican faction:

> I wish this subject occupied more of the attention of legislators of all countries. The laws which are now in force in every part of the world to prevent the importation of malignant fevers are absurd, expensive,

vexatious, and oppressive to a degree.... We originally imported our opinions of the contagious nature of plague from the ignorant and degraded inhabitants of Egypt. It is high time to reject them from the countries where free inquiry is tolerated.[27]

Rush was probably wrong about opinion in Egypt but his view that some form of international agreement was necessary was echoed often in the years ahead. For the moment, however, outright opposition to quarantine was to be found only among physicians with distinctly Republican sympathies. Rush and twelve other members of the Philadelphia Academy of Medicine signed a statement to the effect that yellow fever was not contagious and that it could be prevented only by cleaning the city and carefully disposing of cargoes liable to putrefaction.[28] One of these doctors – Charles Caldwell – denounced quarantine as a medieval institution, founded on 'superstition and prejudice rather than on ... reason and science'. Echoing the sentiments of Dale Ingram earlier in the century, Caldwell claimed that it had been introduced at a time when the human mind had been led astray by 'the delusive wiles of priest-craft, and groaned under the heaviest papal tyranny'. The Church had propagated the false idea that plague had been inflicted upon the infidel in the East and had been communicated to the Christian West by 'the channels of commerce'. But there was not a scrap of evidence to support the spread of plague or yellow fever by commercial means, he insisted; nor had quarantine done anything to protect America from disease. Despite having stricter quarantine regulations than most other American ports, Philadelphia had suffered more from pestilential disease than its neighbours, with the single exception of New York.[29] Baltimore, by contrast, had remained relatively salubrious despite having a more liberal quarantine regime than either Philadelphia or New York, and a year after Caldwell wrote, it actually abolished quarantine.[30] The action taken by the Commissioners of Health in Baltimore set the tone for the future, and as Caldwell put pen to paper he observed that opinion in America and Europe was beginning to move against contagion and what he regarded as a despotic medical establishment.[31]

It would have been more accurate to state that the debate over contagion and quarantine was becoming more polarized. Elite institutions such as the English Royal College of Physicians and the French Academy of Medicine remained predominantly contagionist in outlook. But such opinions were also voiced loudly by those who had recently occupied positions of power and influence. These men were beneficiaries of the rough meritocracy of war. Most came from relatively humble backgrounds and lacked the means to study for a medical degree before they enlisted as military or naval surgeons. The long conflict between Britain and France provided the opportunity for some of those who survived to rise fast within their service, gaining important connections and patronage along the way.

One of those whose star shone brightly during the wars was the British military surgeon James (later Sir James) McGrigor. In the course of his military service, McGrigor encountered both plague and yellow fever and became convinced they could be easily transmitted from person to person. Initially, he steered clear of the controversy surrounding quarantine but his views hardened as he rapidly climbed the ranks. This would have endeared him to senior military men, such as Arthur Wellesley, the future Duke of Wellington, with whom he served in India before the wars. McGrigor was later appointed Director of the recently formed Army Medical Department and served in that capacity for many years, during which time he strenuously argued the case for quarantine. McGrigor was no reactionary but his politics were resolutely conservative.[32] At a time of war, when Britain faced the prospect of invasion, the maintenance of quarantine seemed to him – like many contagionists before – to be a patriotic duty. Whether it was the French or the germs of plague and yellow fever, the enemy had to be kept at bay.[33]

Another staunch supporter of quarantine was the physician Sir Gilbert Blane, head of the Navy's Medical Board. Hailing from a Scottish merchant family, Blane was able to afford to take a medical degree before entering the Navy and through his connections secured a position as a physician rather than a surgeon. However, as a graduate of Glasgow rather than Oxford or Cambridge, his entry into London society was by no means assured. Like McGrigor, Blane's rise as an authority on epidemic diseases was due largely to the experience he gained in the armed services. By 1801, Blane's advice was sought by the government as to whether it would be necessary to quarantine the Army on its return from Egypt, where the plague had broken out among French and British troops. Even the Prussian government requested his views on the prevention of yellow fever, when it looked as if the disease might cross the Atlantic. Like

McGrigor, Blane's politics were conservative and he counted many aristocrats and ultimately the king among his patients. He was vocal in his denunciation of those who sought to dismantle quarantine and remained so long after the wars with France.[34]

The most outspoken advocate of contagion and quarantine in this period was the British army doctor Colin Chisholm, who had experienced yellow fever in the Caribbean during the 1790s. Chisholm rose to the rank of Inspector-General of Hospitals in the West Indies and after leaving the Army he moved to Bristol, where he established a lucrative private practice. He was adamant that yellow fever was contagious and that it had spread from the coast of Africa to the West Indies and from there to the eastern seaboard of America. The books he wrote on the subject became the main counterpoint to Rush's influential account of yellow fever and were vigorously debated in relation to quarantine.[35] As we shall see in a moment, this made Chisholm the principal target for Rush's many champions.

Chisholm, Blane and McGrigor were men who had risen to the apex of their profession or close to it but many of their former colleagues were attempting to scratch a living following their demobilization from the services. Quite a few of these men made a virtue of intellectual dissent and vigorously attacked the doctrine of contagion as a theoretical position maintained by the elite. Many practitioners who had served overseas during the wars were also coming to see diseases such as plague and yellow fever as products of the atmosphere.[36] Service in warm climates made a profound impression upon those who went there and there was a kind of common-sense opinion, shared by local medical practitioners and laypersons, that climate determined many aspects of life, not least the production of disease.[37] The East India Company surgeon John Wade spoke for many when he declared that he had never encountered a 'single instance of contagion' during his service.[38]

In the French Revolutionary and Napoleonic Wars, the number of practitioners who gained experience of hot climates grew and the balance of medical opinion began to tilt in favour of anticontagionism. This may seem odd in view of the fact that the European armies suffered gravely from plague in Egypt and yellow fever in the Caribbean.[39] But these epidemics provided military and naval surgeons with an opportunity to study exotic diseases and to record the conditions in which they occurred. Plague and yellow fever were thereby demystified and more medical practitioners came to regard them, as Rush did, not as separate diseases but as varieties of 'epidemic fever'.[40] Most tended to emphasize the role of climatic and sanitary conditions in the production of epidemics, and the importance of environmental improvement – as opposed to quarantine – in preventing them.[41]

Remaining outside the corridors of power, British opponents of contagion were keen to make common cause with like-minded practitioners overseas. Some were in regular contact with Rush and his colleagues in America and aired their opinions in periodicals such as New York's *Medical Repository*,[42] the first medical journal to be published in the United States. It was founded in 1797 by three physicians – Samuel Latham Mitchill, Elihu Hubbard Smith and Edward Miller – following an aborted attempt by the lexicographer Noah Webster, who had been disturbed by the ravages of yellow fever on the Eastern Seaboard and its prejudicial impact on commerce. From the very beginning, the journal had a distinctly environmentalist bias, issuing a circular to physicians in America asking them to investigate the seasonality of disease and states of the atmosphere accompanying epidemics. As joint editors, Mitchill, Smith and Miller welcomed contributions on these themes from around the world. Their tone and subject matter were reformist and, among other things, there was a clear commitment to freedom of trade. It is no coincidence that 11 per cent of subscribers were merchants, which was probably unusual for a medical journal, even in this period.[43]

It is easy to see why the *Medical Repository* was angered by the publications of that arch-contagionist Colin Chisholm. Its editors gleefully pointed out that Chisholm had based his original hypothesis – that yellow fever had been introduced into the Caribbean from Africa – on an account which was later retracted. Chisholm had interviewed one of the survivors of the failed colony of Bulama on the West African coast after they had fled from there to Grenada – regarded as the origin of the Atlantic epidemic of 1793. This individual, one Mr Paiba, later produced a written account which called into question Chisholm's version of events, particularly the fact that the fever had been preserved between the filthy, ill-ventilated decks of the *Hankey* on its voyage from Bulam. Having undermined Chisholm's credibility, the editors of the *Repository* received a counterblast from Chisholm, who marshalled an impressive array of supporters from both sides of the Atlantic.[44]

Although Chisholm seems to have maintained his reputation, the *Repository* was unstinting in its attacks on quarantine and the doctrine of contagion. As might be expected, most articles on this subject were concerned with yellow fever but the journal's international outlook meant that it also took an interest in epidemics abroad, especially outbreaks of plague in the Mediterranean. It was a matter of great interest to one contributor that the port of Tripoli had enjoyed remarkable freedom from plague for some years despite having permitted its quarantine to lapse. The author, Jonathan Cowdrey, noted that the city had been thoroughly cleansed, that pools of stagnant water had been drained and that the dead were now buried at some distance from the residences of the living. As a result, Tripoli no longer suffered from the noxious exhalations which Cowdrey believed to be the cause of plague. For the *Repository* it was an object lesson in how to maintain public health without compromising either liberty or commerce.[45]

Yet, as the *Repository* knew only too well, Tripoli was very much the exception. Merchants and travellers in the Mediterranean complained bitterly about quarantine and the bewildering variety of regulations. Some lazarettos maintained forty-day quarantines against all vessels from the Levant, regardless of their bills of health, while others settled for shorter periods depending on the state of the bill. Irregularities in ships' manifests might result in the impounding of vessels when there was no disease on board, so travellers from the Levant often purchased bills of health separately from those of the crew.[46] Such impediments threatened to stall the revival of international trade after the end of the French wars. The dynamic force behind the recovery was Britain, now the predominant sea power, although non-European states such as Egypt also played their part in the birth of what some historians have referred to as a wave of globalization.[47] The term 'globalization' does not adequately convey the fractured nature of international trade at this time,[48] but the boom in commerce around the Mediterranean gave greater urgency to discussions of sanitary regulation. Some merchants, especially those involved in the cotton trade with Egypt, increased their demands for the relaxation of quarantine.[49] They again found allies in the medical profession, most prominently the former East India Company and British Army surgeon, Dr Charles Maclean (*fl.* 1788–1824).

A compulsive radical who railed against anything that smacked of tyranny, Maclean was an outspoken opponent of quarantine and denied that diseases such as plague and yellow fever were contagious in any sense.[50] After leaving (or, according to some critics, deserting) the Army, he worked for a time in the Levant where he claimed to have contracted plague, but his fate did nothing to alter his view that the disease arose from local conditions. Maclean was also vocal in criticizing his former employers, the East India Company, having been expelled from India for seditious activities in 1798.[51] He continued his attacks on the Governor-General Lord Wellesley after he returned to Britain, comparing his oppressive rule to that of Napoleon Bonaparte.[52] But Maclean reserved his greatest scorn for the Royal College of Physicians; in his eyes, an effete bastion of privilege which stood in the way of professional reform. Above all, Maclean disliked the physicians' unquestioning belief in the doctrine of contagion which, like some earlier writers, he regarded as Popish superstition.[53]

Curiously, Maclean wrote little about quarantine until relatively late in his career. His first direct assault upon the practice came in 1817, in his *Suggestions for the Prevention and Mitigation of Epidemic and Pestilential Diseases*, which was written shortly after he had studied plague in Constantinople. Over the next five years several more books and pamphlets issued from his pen, most of which advocated the abolition of Britain's 'antiquated' quarantine laws. His opposition to quarantine was consistent with his anticontagionism and, as he saw it, with his attack on other forms of tyranny including the reactionary administration of the Prime Minister, Lord Liverpool. But his avowal of free trade sat uneasily with some of his other pronouncements. After leaving the Army, Maclean appears to have been in dire straits financially and he courted the East India Company by defending its governance of India and its monopoly of commercial navigation to Indian ports.[54] But when the Company failed to keep its monopoly after its charter expired in 1813, Maclean was unambiguously on the side of free trade. In 1819 he was called twice as a witness to a government select committee which was convened to look into the question of the contagiousness of plague.[55] He appears to have been summoned because of his association with Lord Grenville, a Privy Councillor and Governor of the Levant Company.[56] Although the committee did not examine quarantine directly, its underlying purpose was to assess the likelihood of plague being imported from the Levant and whether or not existing regulations could be relaxed further. Twenty-one witnesses were questioned by the committee but only two denied that plague was in any sense contagious.

Most of the other witnesses had direct knowledge of plague or were involved in the Levantine trade. Although they did not deny the contagious nature of plague, some, like John Green, Treasurer of the Levant Company, believed that it was contagious only under certain atmospheric conditions. Still, the committee was inclined to agree with the majority of witnesses that plague was a contagious disease and Maclean's detractors in the Royal College of Physicians claimed that he had reduced the profession to a laughing stock.[57] The Levant Company was now keen to end its association with Maclean, whom it regarded as a liability. It was evident that only moderate proposals for reform would have any chance of success.[58]

Five years later, a row over quarantine and the Levantine trade flared up again. During the 1820s the Company lost much of its business to Dutch interests, partly because quarantine restrictions in Holland were less stringent than in Britain. Dutch merchants were able to import their cargoes from the Levant with the minimum of delay and to export them to other countries, including Britain. This led the commercial lobby to protest against the stringency of British quarantine laws, and another select committee was appointed, this time with the specific remit of investigating quarantine.[59] Maclean was excluded from giving evidence to the committee but he formed an alliance with some merchants in Liverpool and with a number of supporters in Parliament, including the Liberal MP John Smith and the Radical MP John Cam Hobhouse. These parliamentarians joined Maclean in calling for the repeal of the quarantine statute in its entirety.[60] Although they were unsuccessful in obtaining its abolition, an Act of 1825 brought some liberalization, permitting ships to be released from quarantine following a favourable report from a designated officer.[61]

These limited concessions were a disappointment to the likes of Smith and Hobhouse but failure was inevitable given their uncompromising demands. They had chosen a particularly bad time to press for the repeal of sanitary laws, as there was a heightened sense of the danger posed by infectious diseases originating outside Europe. Although some medical practitioners insisted that plague and yellow fever were not contagious, these diseases were once again causing great alarm among the lay public. Outbreaks of yellow fever in the West Indies had raised fears that troops sent back to Europe would carry the disease with them.[62] As a result, vessels sailing from the West Indies were frequently impounded on their arrival in European ports, much to the frustration of the naval authorities.[63] Epidemics of yellow fever in some Mediterranean ports in the 1800s and 1810s showed these fears to be justified,[64] while the appearance of plague on Corfu in 1816 and in Tunisia in 1818 served as a reminder that this ancient pestilence still had the power to escape the Levant.[65] In such circumstances, even a modest reform of quarantine was likely to be controversial. Dr A.B. Granville informed the President of the Board of Trade that the country's ships were already being placed under quarantine at Ottoman ports 'in consequence of the reported relaxation in the sanatory [sic] laws said to have been recommended by this government to the legislature'.[66] Britain was in the ignominious position of being denounced by a state synonymous with laxity and corruption. Granville pleaded with the government to think again, claiming that there was strong recent evidence that contagion had been spread by maritime trade, notably an outbreak of plague in Silesia in 1819, which appeared to have been conveyed from Smyrna in bales of cotton.[67] He went on to criticize the Levant traders who sought, as he saw it, to place their own interests above those of the nation.[68]

The resurgence of plague and yellow fever meant that most governments were reluctant to abandon their traditional defences despite increasing pressure from merchants and humanitarian reformers. This was true not only of Europe but also of ports in America. In Baltimore, which had formerly been envied for its liberal policies, quarantine was reintroduced in 1823, partly in response to the threat from yellow fever and partly because of the prospect of other diseases being imported with European immigrants.[69] In the coming years, this new statute was strengthened despite protests from the city's merchants. Referring to those involved in the coastal trade, the city's health officer, Samuel Martin, lamented that

> Many impediments have been thrown in my way ... and I have submitted, much against my inclination, to vexatious treatment for *peace-sake*. It has always been my most earnest endeavour, to keep in view the mercantile interests of this City impartially, taking care at the same time to keep a strict eye to the health of the City, as the principal object of my appointment. Many complain of the operation of our laws '*because they come from healthy ports*', forgetting that the contents of foul Holds of Vessels, when taken in the aggregate, being brought into the

shoal waters of our Harbour, may be highly instrumental in the generation of diseases.⁷⁰

As the health officer explained, 'The time taken up in examinations of, or in cleansing, ventilating, etc., at Quarantine, is very trifling when viewed in competition with the health of this much favoured and flourishing City, already proverbial for its salubrity and advantageous location. Our prosperity depends greatly on our reputation as to good health.'⁷¹

The health commissioners of Baltimore, like those of most American ports, had devised a set of regulations that aimed to balance mercantile interests with those of public health. This middle way – which took account of, but was not subordinated to, commercial considerations – was rapidly becoming the norm. Measures tended to discriminate between diseases, partly on medical grounds, and partly on account of the commercial disruption which sanitary measures would cause. Baltimore, for example, maintained strict surveillance of diseases regarded as easily contagious – such as smallpox – but permitted a more liberal regime for yellow fever. The official line was that the disease was not directly communicable from person to person but that it could be generated by rotting cargoes and unclean conditions on board ship. If weather conditions were right, these noxious miasmas could spread from incoming vessels or imported goods to the surrounding area. This was now sometimes referred to as 'infection' rather than 'contagion', although both terms covered a range of possibilities. The merit of declaring yellow fever to be an infectious rather than a contagious disease was that it permitted sanitary regulation with a light touch. As yellow fever could be spread only under certain conditions, there was no need to impose restrictions all year round, thereby reducing disruption to commerce. More importantly, this position allowed American merchants and sanitary authorities to claim that the chance of the disease spreading to cooler European ports was minimal, even in the summer months.

American trade was regularly disrupted by lengthy quarantines in Atlantic and Baltic ports, as well as in the Mediterranean, and merchants complained of the 'ruinous' expense of discharging their cargoes during quarantine. Although it was sometimes possible to avoid or evade such measures, it was often pointless to attempt trade with Europe at times when yellow fever was reported along the Atlantic coast.⁷² The Philadelphia businessman Stephan Girard (a Frenchman by birth) wrote regularly about this in his correspondence with one of the leading medical opponents of quarantine in France, Jean Devèze. Both men expressed annoyance with what they saw as the iniquities and absurdities of sanitary practices in European ports. Devèze denounced quarantine as a method whose origins lay in 'centuries of ignorance and barbarity' and told Girard that many chambers of commerce had written to the French government protesting about the imposition of quarantine for yellow fever – the disease which most concerned Girard as a merchant involved in the Atlantic trade.[73] The sanitary authorities in many American ports also made great efforts to convince European governments that yellow fever posed little threat to their countries. It was for this reason that Baltimore's consulting physician, Dr Horatio Jameson, visited Hamburg in 1830. There, by his own estimation, he performed an 'essential service to our commerce' by spreading the gospel of non-contagion among the city's doctors. He was delighted to find in Hamburg 'many non-contagionists' who felt 'equally solicitous with myself to rid commerce of the useless restrictions which have so long been imposed on our shipping in the north of Europe'. Jameson, however, did not hope to see quarantine abolished, even against yellow fever, but merely wished it to be modified.[74]

Moderate demands such as these were often successful but those who insisted that commerce be freed of sanitary restrictions continued to face indifference or downright hostility. By the 1820s, diehard opponents of quarantine were organizing internationally, picking causes which they believed offered some chance of success. If just one country fell under their spell and dismantled quarantine, they hoped, others might follow. In the 1820s the most likely candidate was Spain, which had recently adopted a liberal constitution and whose ships were then regularly placed in quarantine by other powers. Following a serious outbreak of yellow fever in Barcelona in 1821, British opponents of contagion such as Maclean and a former army doctor Thomas O'Halloran joined forces with fellow critics of quarantine in France such as Nicolas Chervin and Jean Devèze to denounce measures like the sanitary cordon imposed by France against Spain.[75] They disputed the claims of contagionists such as the French physician Étienne Pariset who saw Spain as a sanitary threat,[76] and insisted that the doctrine was merely a pretext for the exercise of 'tyranny' by the French. The French sanitary cordon was deeply resented not only because it disrupted commerce but because it was seen as hostile to the liberal government which had been formed in Spain in 1820. It was thought to harbour royalist spies and provide a refuge for traitors. The medical commission sent to Spain – led by Pariset – was also regarded with suspicion, although Pariset himself was no reactionary. A supporter of quarantine, he nevertheless favoured measures which struck a judicious balance between the protection of public health and the needs of commerce. And yet, he viewed arrangements in Spain as too liberal and suspected that ships' crews had disembarked with their cargoes at the northern port of San Sebastian to allow both to proceed secretly into France overland.[77]

Maclean thus found many deputies of the Spanish Cortes sympathetic to his views when he visited the country in 1822. Although the Cortes stopped short of accepting his uncompromising stance against all quarantine, he flattered himself that he had managed to delay the enactment of a code which aimed to establish permanent and systematic sanitary regulations.[78] But Spain's liberal experiment was short lived. The following year, French troops marshalled in the cordon along the border with Spain were used to restore the Bourbon monarchy to the Spanish throne, seemingly vindicating all that the opponents of quarantine had said.[79] However, critics of quarantine did not lose hope and continued to communicate internationally by correspondence and via a growing number of medical periodicals. One such was the British *Medico-Chirurgical Review*, founded in 1820 by a former naval surgeon, James Johnson. This journal had a distinctly reformist slant and favourably reviewed British and foreign works which sought the liberalization of quarantine. But like many other practitioners, its founder and editor saw that the best hope of change was to take a balanced view on the subject of contagion and seek reform rather than repeal of quarantine laws. For this reason, Johnson was contemptuous of Maclean, whose denial of contagion and vehement hostility to quarantine in all circumstances was damaging the case for reform.[80]

If this was evident in the 1820s it was even more apparent a decade later, as Europe faced a dreadful new threat in the form of epidemic cholera. Cholera had been making its way overland from India for some years, arriving in European Russia in 1830. Gilbert Blane warily observed that cholera followed trade routes much as plague had followed the Syrian caravans in former times and he advised his government to take similar precautions to those which it had taken against plague.[81] Indeed, in the face of the threat posed by cholera, most states fell back instinctively upon measures designed to combat the plague and established quarantines along their borders and at their ports. Even busy commercial cities such as Hamburg, in which there was some sympathy with quarantine reform, imposed severe restrictions on incoming vessels.[82] But a backlash was soon orchestrated by merchants and others who opposed quarantine on economic grounds. Critics argued that cholera had spread regardless of any barrier erected in its path and alternative explanations for the prevalence of the disease came to be favoured, ranging from noxious miasmas to pervasive immorality.[83] Many commercial cities thus began to dismantle or relax their quarantines and to rely on other means of preventing the disease, such as cleansing of filthy localities.[84]

Differences of opinion on the nature of cholera and how to prevent it existed in all countries afflicted or threatened by the disease, but over time distinctive trends began to emerge. In general, the severity of quarantine depended on the extent to which commercial and manufacturing interests held sway. In 1832 the bustling port of Hamburg dismantled the quarantine it had established the year before and took little action against cholera when it reappeared in 1848. But in Prussia, which was less dependent upon trade, the authorities continued to insist on the contagiousness of cholera and on the need for restrictions such as quarantine.[85] There were other considerations, too, not least the prospect of retaliation. Thus, despite its heavy reliance on maritime commerce, the British government continued to insist that quarantine was indispensable in order to prevent more damaging restrictions being imposed by other nations.[86] But if cholera had been a shock to the system, in 1832 it was not yet clear that it would become a perpetual threat. In view of this, most opponents of quarantine temporarily ignored cholera and focused again on the more familiar problem of plague in the Mediterranean, where myriad regulations continued to damage the business of European traders.

False dawn

During the 1820s, the Bourbon regime in France had been implacable in its defence of quarantine and had infamously employed it as a stranglehold around the new regime in Spain. After 1830, however, the country entered a more liberal phase of government under the 'Bourgeois Monarchy' of

Louis Philippe. The new Orléanist regime also enjoyed a relatively harmonious relationship with the Academy of Medicine, which had become hostile to quarantine since its misuse against Spain. At the same time, French merchants and diplomats in the eastern Mediterranean raised their voices against the increasing disruption caused by quarantine in the region and the high cost of detaining goods and persons in lazarettos.[87] After Muhammad Ali became pasha (Ottoman viceroy) of Egypt in 1805, he began to impose restrictions upon shipping from plague-ridden ports such as Smyrna and Constantinople, as well as isolating infected persons and their contacts within his dominions. Similar measures were taken by the pashas governing Mediterranean islands such as Cyprus and continued to be enforced by the rulers of Tunisia.[88]

Although they were unpopular with many European merchants, these initiatives were probably welcomed by companies that hoped to connect Europe with India by steamship. It was envisaged that passengers bound for India would disembark at Egyptian ports and later embark on other vessels from the Red Sea. This route was potentially very profitable, but regular outbreaks of plague in Egypt threatened to make the connection unreliable.[89] Quarantine afforded a certain amount of protection to the ports designated by the passenger lines but the Egyptian government had ulterior motives. It made use of sanitary restrictions to favour local shipping at the expense of European rivals and, in particular, to wrest trade from the Greeks, who had hitherto dominated commerce in the eastern Mediterranean. Such measures were consistent with the mercantilist policy through which Muhammad Ali aimed to restrict the importation of textiles (especially cheap British goods) in order to foster their manufacture locally.[90] He was well aware that Britain had killed off the Indian cotton industry by allowing its mass-produced goods to flood the Indian market while using tariffs to keep foreign goods out of its own. France, too, had made use of protective tariffs while publicly espousing the principles of free trade. Indeed, the Middle East was already feeling the effects of competition from mass-produced textiles and other products, which were being imported into the region in increasing volume. Much cheaper than local garments, factory-produced items manufactured in Britain were putting local weavers, spinners and dyers out of business.[91] Quarantine bolstered Egypt's meagre tariffs by making imported goods more expensive.

During the 1810s and '20s, the Egyptian quarantines proved irksome to European merchants but the situation did not become serious until 1831, when Muhammad Ali's army invaded the Ottoman province of Syria. Egyptian expansion posed a direct threat to British and French commercial interests, as well as to Britain's route to India. The other significant power in the region, Russia, saw assistance of the Ottoman Empire as a means of extracting territorial concessions and expanding its influence. Facing the military might of the European powers, Muhammad Ali was forced to abandon his formerly antagonistic stance and began to woo Franco-British support by promising commercial concessions. The same year saw the arrival of cholera in Egypt, which led Muhammad Ali to reassess precautions against epidemic disease. Now he turned to European governments for assistance, seeking their support for the formation of a permanent sanitary organization to replace the ad hoc arrangements hitherto used to combat disease.

Whether or not Muhammad Ali was pushed in this direction by the European powers remains unclear. The French physician Antoine Barthélémey Clot (later known as Clot-Bey) claimed that foreign powers had convinced Muhammad Ali of the utility of a permanent quarantine establishment. He had apparently supported such a move because he believed it to be in the interests of public health – not in a cynical attempt to raise revenue, like the quarantines recently re-established at Tripoli and Tunis. Although Clot-Bey was appointed to head Egypt's medical administration in 1836, he was a convinced anticontagionist and persistently, if unsuccessfully, attempted to persuade his master to abandon quarantine. He believed the pasha to be misguided; driven by the best of motives into the hands of unscrupulous foreign powers.[92] But Clot's version of events omits to mention that Muhammad Ali's sanitary policy was consistent with his other political objectives, such as gaining the assistance of Europeans in the reorganization of Egypt's army, navy and economy.[93] It seems highly likely that the pasha calculated that quarantine would benefit Egypt economically, while possibly protecting it against epidemic disease.

Whatever his motives, in August 1831 Muhammad Ali called upon foreign consuls in Egypt to help him draw up sanitary regulations that were likely to be effective against both cholera and plague. Two months later, the consuls formed a commission composed of representatives from Britain, Austria, France and Russia to fix the duration of quarantine in

Egyptian ports and to construct a lazaretto in Alexandria. The Consular Commission of Health began its working life in 1834 and was soon engaged in sanitary work against plague, which broke out in Egypt at the end of that year. This was the beginning of what was to become a very serious epidemic lasting until the end of 1835. As the disease increased in ferocity, the measures taken to combat it intensified, eliciting protests from Muslim clerics and other notables who complained about the isolation of families and the disruption of trade. But Muhammad Ali was convinced that such measures represented the only means by which plague could be controlled and, with the support of the European powers, he stuck to his guns.[94]

Further to the west, the new Tunisian ruler, Mustafa Bey (1835-7), entered into a similar agreement with foreign powers. Hitherto, quarantine had been the personal responsibility of the bey but in 1835 he agreed to a consular commission similar to the one formed in Egypt. The Commission initially had advisory powers only but European, and especially French, interests came to dominate. More so than in Egypt, the formation of the commission in Tunisia appears to have been the initiative of the European powers, which practically forced the bey to accept a commission after numerous complaints from their merchants about Tunisian quarantine ruining commerce. These complaints were particularly evident in the case of merchants based in Marseilles, who were in direct competition with those from Tunisia.[95]

The formation of the consular commissions was welcomed by the European powers, not only because they offered an additional means of defending Europe against Asiatic infections, but because delegates could ensure that sanitary measures were consistent with their commercial interests.[96] Indeed, the existence of these bodies aroused cautious optimism about the prospect of more extensive international cooperation and induced the French Ministry of Commerce to look again at the rules of quarantine then in force in the Mediterranean. The man charged with this task was M. de Ségur Dupeyron, Secretary to the Supreme Council of Health. Dupeyron examined a number of lazarettos personally and took note of their regulations on the length of quarantine and the terms under which it was imposed. Eschewing the speculation he thought characteristic of medical debates, he adopted an historical approach, seeing present arrangements in the light of epidemics over several centuries. Dupeyron concluded that there was a close link between commerce and plague, pointing to the fact that the disease seemed never to occur in those countries whose trade had been disrupted by war. All epidemics of plague in Europe also appeared to have spread outwards from the Levant, suggesting that the disease was contagious and that its source lay in Asia. The same was true of yellow fever, which appeared to radiate out from the Caribbean to temperate parts of the Americas and, sometimes, ports in Europe. Although sanitary precautions had been effective in certain cases, he felt they were unnecessarily oppressive because they were imposed in an unsystematic way. In view of this, he made a number of suggestions to establish what he regarded as a 'reasonable and uniform system'. This included quarantines of shorter duration; abolition of quarantines of observation (quarantines for vessels sailing from infected ports) against vessels coming from the West Indies and the USA with clean bills of health; and, most importantly, forbidding arbitrary increases in the duration of quarantine.[97]

When Dupeyron's report was published, the diplomatic climate was not conducive to an international agreement.[98] But by 1838 the French government had accepted the thrust of Dupeyron's report and proposed a conference of delegates from various European countries with ports on the Mediterranean. Its aim was to iron out the obvious differences between quarantine regimes in order to remove constraints on French navigation. The initiative was welcomed in Britain by free-traders such as the physician Joseph Ayre and the Benthamite MP Dr John Bowring, who had made no secret of their opposition to quarantine and the doctrine of contagion on which it was based.[99] There was no prospect of the British government abandoning quarantine but it agreed in principle to the French proposal,[100] despite warnings from some quarters that commercial interests ought not to be placed before those of the nation as a whole.[101] However, the government was anxious to secure a reduction in quarantine both for commercial reasons and because royal naval vessels and mail ships were often subjected to long delays in the Mediterranean.[102] Other countries also expressed an interest in the French proposal, the most significant being Austria-Hungary. The Empire had hitherto been staunch in its support for quarantine but it also had substantial commercial interests in the eastern Mediterranean. Indeed, the Austrians had protested for some years about 'impediments thrown in the way of navigation' in the Ionian Sea.[103]

These tentative steps towards an agreement on quarantine exemplified the system of international relations inaugurated by the Congress of Vienna in 1815 and which prevailed until the Crimean War of 1854–6.[104] It was fundamentally different to the situation before 1815, when colonial rivalry between the Atlantic nations intermingled with the continental struggles of the Great Powers. After its defeat in that year, France no longer saw itself as an imperial rival of Britain. Indeed, its colonization of Algeria between 1829 and 1848 gave France a greater incentive to work with Britain in order to moderate quarantine in Mediterranean ports.[105] In the forty years after the Vienna Congress, the Great Powers thus sought to work out their differences at the conference table rather than on the battlefield and, in such an atmosphere, there was less need or scope for the use of quarantine as a political weapon. There was no mention of quarantine in the Vienna settlement but the Congress concluded agreements on related matters such as freedom of navigation on the River Rhine. Like subsequent agreements on traffic on the Danube, this was intended partly to satisfy economic interests and partly to promote peaceful coexistence.[106]

The 'conference system' that evolved out of congress diplomacy remained dedicated to the peaceful solution of political problems. It was also more pragmatic and, in many respects, more successful, involving smaller gatherings of states which aimed to reach agreement on specific matters.[107] Although the system was predominantly driven by the commercial and colonial interests of Britain and France, agreement on such issues as quarantine must be seen in the light of this earnest desire to remove potential sources of conflict. The fact that the focus of sanitary discussions was the eastern Mediterranean made such an agreement all the more desirable, for the Levant was now a flashpoint in international affairs. Attempts to convene an international conference on quarantine coincided with the outbreak of another war between Turkey and Egypt, which again highlighted Russia's growing influence over the weakening Ottoman Empire.

Disagreement also occurred between Britain and France because of French support for Muhammad Ali, but the French were unwilling to risk war with the Austrians and British, who had sent an expeditionary force to the Levant. The tension was relieved when Egyptian forces retreated and by the Treaty of London (1840), in which the four principal European powers (Austria-Hungary, Britain, Prussia and Russia) jointly guaranteed the security of the Ottoman Empire. As the British Foreign Secretary, Lord Palmerston, put it, the aim of all the governments concerned was to 'agree upon a common course of policy, which may be calculated to accomplish purposes [the preservation of peace in the Levant] so essential for the general interests of Europe'.[108] The Straits Convention of the following year also made the prohibition of foreign naval traffic through the Bosphoros and Dardanelles a matter of international agreement rather than simply an Ottoman policy, as it had been before.[109] At the same time, relations between Britain and France became warmer following the dismissal of the Prime Minister Thiers in 1840 and the subsequent fall from power of Palmerston and the Whig government. The two countries sought to work together amicably and this gave additional momentum to discussions of quarantine.[110] According to Dr Gavin Milroy, one of Britain's most vocal and influential critics of quarantine, everyone who had studied the subject – statesmen, travellers, merchants and physicians – had come to the conclusion that an international agreement on quarantine was vital to their 'common welfare'.[111]

The Austrian chancellor Prince Metternich claimed that it was now possible to relax quarantine in the Mediterranean because Egyptian measures against plague made its spread westwards less likely. The prospect of similar regulations being introduced in the Ottoman Empire also gave grounds for optimism. In 1838 the sultan asked the Austrian government to send him several experienced officials to assist in establishing quarantine stations throughout the Ottoman provinces. Most parts of the Ottoman Empire had been severely affected by plague in the late eighteenth and early nineteenth centuries. In 1812 an estimated 300,000 people died during an outbreak in the greater Constantinople area and, as late as 1836, it claimed the lives of 30,000 people in the Ottoman capital. Although its virulence seemed to be decreasing, plague continued to visit Constantinople and the Balkan provinces almost annually until the middle of the century. Moreover, the Empire faced a new threat in the form of cholera, which first entered Ottoman territory in 1821.

Over the next three decades, seven epidemics of cholera swept through the Ottoman world, some arriving directly from India, others with pilgrims returning from the holy cities of Mecca and Medina.[112] This presented a great challenge to successive administrations which were attempting to modernize the Empire, for epidemics stunted population growth and disrupted international trade.¹¹³ In seeking European expertise, Mahmut II (1808–39) followed precedents set in other branches of state such as the Army. The attempt to construct a sanitary infrastructure across the Empire was also in line with the rapid growth of the Ottoman state during the nineteenth century, with regulations in all ports expected to conform to instructions issued in Constantinople.¹¹⁴ For the European powers, however, the creation of a 'Commission of Public Health' in Constantinople presented yet another opportunity to exert influence over the Sublime Porte and to secure concessions beneficial to European navigation.¹¹⁵ Although the influence of foreign representatives on the Constantinople Council of Health,¹¹⁶ as it became known, was sometimes rather less than the European powers had hoped, their role, like the Straits Settlement of the following year, was symbolic of the sultan's waning independence.¹¹⁷

The establishment of the Constantinople Council illustrated growing recognition of the need for international cooperation in the Middle East.¹¹⁸ But despite his initial support for a conference to discuss quarantine, Metternich and other foreign ministers were unable to agree where to hold the meeting. These wrangles were in no sense untypical, as both Metternich and Palmerston tended to favour conferences over which they had control.¹¹⁹ Talks resumed in 1843, again as a result of French initiatives and the British Foreign Secretary, Lord Aberdeen, one of the architects of the new entente cordiale, responded enthusiastically, declaring that 'great benefits would result from it to Mediterranean commerce and communications'. However, he thought that prior to offering an invitation to Russia and the Italian states it would be wise for Britain, France and Austria first to reach an agreement on key issues. He then hoped that Austria would exert its influence over the Italian states and induce them to cooperate. Aberdeen was keen that Russia be involved in the conference because it was a major regional power and any agreement was unlikely to be workable without it. He proposed the neutral port of Genoa as a venue.¹²⁰ Other departments of the British government were equally enthusiastic, noting that the mood seemed more conducive to progress. Mr J. MacGregor of the Office of the Privy Council for Trade declared that 'A very decided tendency has been manifested on the part of the principal Powers, to assimilate in some degree the periods of detention, and at all events to relax very considerably the severity of the restrictions on merchandize and vessels'. He noted that 'the general good understanding which now prevails between this country and foreign Powers ... encourage[s] the hope that the deliberations of such a conference ... would result in the adoption of that general system of Quarantine which is so desired'.[121]

The system of international diplomacy that developed after 1815 – with its overriding objective of preventing war in Europe – was a vital precondition for any agreement on sanitary regulation. The signs were certainly encouraging, with Britain and France showing their willingness to participate in a conference if it were convened in one of a number of neutral cities. However, the Austrians were slow to respond and, when they did, they did so with less enthusiasm than expected. Metternich considered a conference premature and insisted that the three main parties first reach an agreement over technical matters such as the minimum and maximum periods of quarantine necessary for humans, the terms for various types of merchandise, and the best methods of disinfecting objects thought susceptible of contagion. This was not unlike Aberdeen's proposal, but the Austrians maintained that they required a period of six months in which to consider the matter by themselves; Metternich also expressed his preference for any such conference to be held in Vienna.[122]

While France awaited a response from Vienna, the British government commissioned its own investigation into quarantine in the Mediterranean from Sir William Pym, the Superintendent of Quarantine at the Privy Council. In 1845 he made a detailed report on the numbers of persons and vessels quarantined at different stations, procedures for the handling of goods, charges levied, and so forth. Pym reached a similar conclusion to that of Dupeyron: that quarantine was necessary in some form but that it operated unsystematically. It was this arbitrariness, rather than quarantine per se, that posed the chief obstacle to trade in the Mediterranean.[123] On the basis of his investigation, Pym drafted a response to the issues raised by Metternich,[124] but the latter continued to prevaricate, telling British and French officials that he would consider the matter only when he had received information from the Austrian departments of the Interior and of Finance.[125]

Metternich seems genuinely to have wanted an international agreement, believing that it would be of great benefit to Austrian commerce. Indeed, records kept by quarantine stations in the eastern Mediterranean and the Levant confirm that Austrian ships were among those most commonly inconvenienced by quarantine.[126] Though still in its infancy, steam navigation had raised the volume of trade through the Danube and there was also increasing pressure from within Austria to relax quarantine regulations along the border of the Hapsburg Empire, on commercial and humanitarian grounds. The Austrian ambassador to Britain told Lord Aberdeen in 1845 that a commission had been established 'with the desire to diminish the expenses of the Cordon Sanitaire, which it is said has completely failed in preventing intercourse across the frontier, and which offers unnecessary interruption to traffic'.[127] Some prominent medical men, such as Professor Sigmund of Vienna, had even recommended that Austria should rely solely on sanitary improvements to keep epidemic disease at bay.[128] But Austria's extensive boundary with formerly plague-ridden Ottoman provinces meant that many were reluctant to abandon measures which had apparently protected them for many years.

Britain and France were motivated to seek an international agreement because of their respective commercial and imperial interests. Growing French involvement in Algeria and its trade with the eastern Mediterranean provided a powerful incentive to reform quarantine and in the 1840s France took measures unilaterally to reduce quarantine in its Mediterranean ports.[129] In Britain commercial interests were also becoming more influential and the repeal of the tariff on imported grain (the Corn Laws) in 1846 encouraged free-traders to seek the removal of other restrictions. Critics of quarantine estimated the annual cost to Britain at £2–3 million, and probably the same again from losses incurred by merchants in the Mediterranean.[130] In the late 1830s, Britain and other nations had also concluded a series of commercial and navigation treaties with the Ottoman Empire, the aim being to open up areas of trade formerly prohibited to foreign merchants and to reduce the tariff on imports into the Ottoman dominions.[131] The attempt to reach an agreement on quarantine that involved the Ottomans was therefore part of a process whereby Britain and other foreign powers were attempting to wrest concessions on trade and navigation.[132]

But commercial interests were not the only factors that induced Britain to seek international agreement over quarantine. Quarantine was still a great inconvenience to travellers through the Mediterranean, especially the growing number of Britons who made their way to and from India via the Levant. There were many complaints about the 'absurdities' and 'irregularities' of quarantine in Mediterranean stations, particularly

Alexandria.[133] Following the creation of the Consular Commission of Health, European merchants and diplomatic staff often protested that quarantine was enforced selectively and that the system was inefficient.[134] There was growing suspicion, too, that the Egyptian government was using quarantine deliberately to damage Britain's interests, perhaps with the connivance of other European powers. Relations between Britain and Egypt had deteriorated following the East India Company's acquisition of Aden, where, in 1836, it established a coaling station for steamships en route to India. Muhammad Ali resented the presence of a British garrison adjacent to his territories and was deeply suspicious of the Company's intentions.[135] In 1838 the British consul in Alexandria began to make formal complaints about the way that quarantine was performed in Egypt and demanded concessions for British vessels.[136] In the following year, the relationship between Britain and Egypt deteriorated further. An Egyptian army under the command of Muhammad Ali's son Ibrahim invaded the Ottoman province of Syria and a combined European force was sent to assist the Ottomans, thwarting Egypt's ambitions.

In these circumstances, it is hardly surprising that the Egyptian authorities made use of any opportunity to monitor the intentions of what they regarded as a hostile power. Steam-packet agents in Alexandria and Cairo frequently complained that sanitary fumigation was used as a pretext to intercept, delay and even to destroy diplomatic communiqués.[137] Dr John Bowring told the House of Commons in 1842 that 'Official dispatches were opened, perforated with awls, incised by chisels, dipped in vinegar ... and at length transmitted to their destination in a mutilated, and scarcely legible condition.' He claimed that 'There was no doubt that political objects were sought for in the maintenance of quarantine in the east; and it was equally certain that political interests were promoted by them, and that these, and not the health of nations, were the principal motives for the great severity with which the regulations were enforced abroad.' It was not only the Egyptians who used quarantine in this way, he insisted, but also – to his shame – British consular officials. Yet there was no country that abused quarantine as routinely and blatantly as Russia. Bowring described its quarantine officials as 'political functionaries' who 'arrested and released travellers at will. They took possession of all correspondence ... they checked or facilitated commerce according to the passing interests of the moment ... and in the name of public health', he declared, 'they had introduced a system of universal police and espionage'. In his view, it was imperative that the government reach an international agreement: a motion which was enthusiastically supported by members of the government, including the Prime Minister, that doyen of free-traders, Sir Robert Peel.[138]

The revolutions of 1848 distracted attention from efforts to bring about an international conference on sanitary regulation. However, the French reopened negotiations with renewed vigour and were successful in persuading eleven other Mediterranean states (including the Ottoman Empire) to agree to a conference in Paris, in 1851. Most countries sent two delegates, a diplomat and a physician: although it was considered necessary to reach agreement on technical matters, the diplomats were present to ensure that political and commercial issues were given due consideration. As the French Minister of Foreign Affairs insisted, it was necessary to find a modus operandi befitting an age of technical and industrial progress, and to strike a felicitous balance between the needs of commerce and of public health. Just as new modes of communication were erasing the tyranny of distance, he argued, it was now time to remove political and commercial impediments that stood in the way of international harmony.[139] The time was right diplomatically, too. Tension between Britain and France had evaporated following the removal of the Orléans monarchy in 1848,[140] while the triumph of reaction elsewhere brought stability and a desire to avoid conflict.[141]

The Paris conference is usually considered a failure because the delegates did not agree on key issues such as the transmissibility of cholera and because the resulting convention was signed by only three states – France, Sardinia and Portugal – and ratified by Sardinia alone.[142] Although the divisions were primarily between the Mediterranean countries, which were generally more reluctant to abandon quarantine, and Britain and France, which were eager for commercial and colonial reasons to liberalize it, the fault lines were numerous and often cut across one another. Despite Metternich's earlier optimism, Austrian delegates opposed any attempt to modify maritime quarantine and disinfection regimes in times of plague, and were particularly hostile to British-backed proposals to reclassify susceptible merchandise so as to downgrade the threat from cotton. Together with Russian delegates, they also opposed British proposals to abandon land-based cordons, which, however imperfect, were regarded as the only means of defending their empires against plague from the Levant. Yet Austrian (but not Russian) delegates backed the French and British position that cholera was not contagious in the same way as plague, and opposed the use of quarantines and sanitary cordons to prevent it.[143] Public opinion was important in influencing positions at the conference, often to the detriment of liberalization in the case of most Italian states.[144] Yet the delegates were unanimous in their support for the idea of an international agreement, as well as upon the desirability of certain measures such as the strengthening of sanitary surveillance in Egypt and the Ottoman Empire.[145] Despite earlier complaints about the abuse of quarantine regulations at Alexandria in particular, it now seemed that such measures had been effective, Egypt having been free from plague since 1844.[146]

Although the Paris conference reached no binding agreement, it was remarkable that it had been held at all. The previous century had seen widespread abuse of quarantine as an instrument of foreign policy and as late as the 1820s the Bourbon regime had used a sanitary cordon to crush the liberal experiment in Spain. Things had moved a long way since then but, more importantly, the spirit of Paris was to be raised on several occasions in the future and was finally resurrected in 1866, in the wake of another great wave of cholera. In the course of the next half-century, there would be ten further international sanitary conferences, the last of which resulted in widely ratified conventions. These developments were the fruit of a sporadically evolving internationalism of which we see the first shoots in the 1830s and '40s. But the Paris conference marks the end of an era as much as the beginning of a new one. It was held at a time of rising tension and the system of international cooperation that produced it was shortly to collapse following the outbreak of the Crimean War. Several decades passed before it was possible to reach a binding agreement on sanitary matters, largely due to competition between the imperial powers. On international bodies such as those in Alexandria and Constantinople, and at the sanitary conferences of the 1870s and '80s, Britain became increasingly isolated, as its rivals questioned its insistence upon the liberalization of quarantine. To understand why the world's leading commercial nation took this solitary course, we must look first at events in Britain, where sanitary arrangements were gradually harmonized with the prevailing doctrine of free trade.

CHAPTER 4

Quarantine and the empire of free trade

In November 1844, a steam-sloop, the *Eclair*, set out from the English naval base of Devonport for the Bights of Biafra and Benin where she joined a naval squadron engaged in the suppression of the slave trade.[1] On 28 September 1845, the vessel returned to Britain with less than a third of her original crew, most having succumbed to a fever contracted while ashore in Sierra Leone. In one sense, there was nothing remarkable about the fate of the *Eclair*, for mortality rates on vessels in the Royal Navy's West Africa squadron had previously exceeded 50 per cent.[2] But the fate of this particular ship gave rise to a controversy of international proportions and figured prominently in sanitary affairs for years to come.

The reason lies partly in the peculiar circumstances of the *Eclair*'s homeward voyage, during which the crew of the vessel were permitted to land on Boa Vista, one of the Portuguese Cape Verde Islands. Shortly after the vessel departed, Boa Vista was ravaged by a violent epidemic of what many believed to be yellow fever. The incident guaranteed the *Eclair* instant notoriety but in Britain the plight of its crew elicited sympathy rather than condemnation. Some attained the status of heroes, especially those who had volunteered to replace the deceased; the ship's commander also came from a distinguished and well-connected political family, which aroused the interest of newspapers and members of Parliament. But the outbreaks on the *Eclair* and Boa Vista were of more than local interest. This was the first time that a major epidemic had been attributed to the movement of a steam-powered vessel and the prospect of yellow fever spreading rapidly to Europe from Africa or the West Indies posed a challenge to those who wished to see quarantine relaxed or abolished.

The Boa Vista incident also occurred at a crucial juncture in the commercial history of the world. Britain's formal empire, which at that time was confined mainly to India, the Caribbean, Australasia and Canada, was now complemented by an 'informal' empire of free trade, policed by the Royal Navy.[3] These imperial networks formed the basis of what was to become the first global economy; a 'world system' based expressly on principles of free trade.[4] This system was forged in the wake of the abolition of slavery in the British Empire and its profitability depended not only upon greater freedom of trade but upon the trans-oceanic migration of labour from Asia to the impoverished former slave economies of the Americas. The movement of goods and peoples was eased by the growing use of steam vessels and railways, which increased the speed and frequency of communications between hitherto disparate territories. The Empire thus acquired greater coherence but its networks presented new opportunities for the passage of disease.[5] Epidemic diseases such as plague and yellow fever, which had been confined for decades to what many regarded as their natural habitats, were again on the move, shattering complacency and posing a threat to the new commercial order. Devising a sanitary regime compatible with this system was an enormous challenge and was to prove highly controversial in the years ahead.

The pest ship

After leaving Britain for West Africa, the *Eclair* spent five long months anchored off the southern coast of Sierra Leone near the island of Sherbro, with a view to blocking the passage of slaving ships. The location was considered one of the least healthy on the African coast and few vessels remaining near the island escaped a serious outbreak of fever. However, it was not until the *Eclair* sailed north to pick up supplies and fuel that it began to experience unusual mortality. Up to that point, nine of the crew, which originally numbered 146, had died, but they appeared to be victims of the common coast fever, which was regarded almost as an occupational hazard by those on the African service. The vessel was anchored for thirteen days and, for a single day, the crew was allowed ashore, during which time a number became helplessly drunk and incapable of returning for several days. On 23 July the *Eclair* finally weighed anchor and set course for the Gambia in the company of another steamer, the *Albert*, which had been involved in the ill-fated expedition up the River Niger in 1841. This well-publicized venture had attempted to wean local tribes away from slavery and open up tropical Africa to 'legitimate trade'.[6]

At this stage, only one new case of fever had occurred but, before the vessel reached the Gambia, on 10 August, seven men had died from a particularly virulent disease. The ship's master, Commander Walter Estcourt, wrote that 'So rapid has been the progress of this Fever, that men who appeared in good Health and Spirits have died in 3 or 4 days in agonies, with black vomit and all the other symptoms of the most malignant kind of that Fever which is Caught in the rivers of Africa.'[7] From the Gambia, where the vessel remained for five days, the *Eclair* sailed to the Cape Verde Islands, where Estcourt hoped that a change of air would aid the recovery of the sick. The *Eclair* reached one of the islands – Boa Vista – on 21 August and anchored offshore: by this time, one in six of the original crew was dead. Estcourt obtained permission from the Governor-General of the Cape Verde Islands to land the crew, with the exception of the African 'Kroomen' who remained on board to clean and fumigate the ship. Africans – and the Kroomen in particular – were then thought to possess immunity to the fevers of the region so were often charged with duties like fumigation, or sent ashore in malarious locations to gather fuel and supplies. The Kroomen – who were recruited from two tribes on the coast of Sierra Leone – were said to be fine physical specimens and were generally regarded as better suited to heavy work than Europeans. Furthermore, they were diligent and eager to learn, many showing a great capacity for seamanship.[8]

While the Kroomen remained on board the *Eclair*, the European sick were accommodated in a fort on a small, uninhabited island near the entrance to the harbour. The healthy European members of the crew were also allowed ashore and slept under canvas; a house was arranged for the use of the officers and Estcourt lodged with the British consul. But these measures had little effect on the health of the crew.[9] By 31 August, five more men had died and Estcourt recorded subsequently that although 'A few soon recovered so as to walk about . . . the Fever spread rapidly among the Crew and between the 1st and the 8th of this month [September] we have lost 3 Officers, 9 Seamen, 4 Mates, 3 Boys. . . . Making the total number of deaths since leaving about 35.'[10] Many of those who remained were already suffering from fever and were too weak to perform their usual tasks.

After taking the advice of three naval surgeons, Estcourt set course for Madeira which was famous for its salubrious climate, so that the crew would have the opportunity to recover before returning to England. By this time, the ship's assistant surgeon, Mr Harte had died of the fever and had been replaced on a voluntary basis by a naval surgeon, Mr Mclure, who was travelling as a passenger on another naval steamer, the *Growler*. The *Growler*'s own assistant surgeon, Mr Coffey, and several other members of its crew, also volunteered to serve on the *Eclair*. The vessel embarked on 13 September, Estcourt having been taken ill the day before; he and Mclure died three days before the *Eclair* reached Madeira.[11] By the time the vessel had reached Funchal, the fever had killed nearly two-thirds of the crew. But the Governor of the island was wise to the condition of the ship and refused the crew permission to land. The *Eclair* now faced a hazardous journey across the Bay of Biscay with a depleted crew, and with the storms of the autumn equinox fast approaching. Its prospects would have been grim but for the fact that seven merchant seamen taking leave on the island, and a young naval surgeon, Sidney Bernard, had offered their services to bring the vessel home.[12]

The *Eclair* reached Portsmouth on 28 September, its crew having suffered ninety more cases of fever (which included Bernard) and forty-five deaths. The vessel was not permitted to land its sick at any of the nearby hospitals, despite offers from the Naval Hospital at Haslar, but was instead ordered to undergo quarantine for twenty-one days – the period advised under the Quarantine Act of 1825. An infected ship arriving at Portsmouth would normally have been quarantined locally at the Motherbank but poor weather prevented contact from being established with the shore. After three days, the *Eclair* was directed to another quarantine station, at Stangate (also referred to as Standgate) Creek in Kent, where all the crew were forced to remain on board. During this time, more of the men fell ill and those already sick deteriorated rapidly; the only good news was that the thirty-six Africans had suffered just five cases of fever, none of them fatal.[13] The mounting death toll on the *Eclair*, which was now receiving a good deal of attention in the press, led the Admiralty to demand that the crew be taken off the vessel and, on 8 October, the sick were transferred to HMS *Worcester* and the convalescents to HMS *Benbow*. By this time, Bernard was already too ill to be moved and died the following day; fortunately, the other surgeon, Mr Coffey, was making a recovery.[14] In the next few days, most of the remaining sick began to recover, although one of the attending surgeons, Dr Rogers, who was working on the *Worcester*, had contracted the fever and the pilot, who had brought the ship up the creek, had died of what appeared to be the same disease.[15] By 13 October, there were no new cases on any of the vessels and Dr Rogers had begun to make a recovery.[16] On the 31st, the quarantine on the *Eclair* was lifted,[17] and those who remained alive were released from their duties and paid off.[18] The survivors amongst the twenty-nine volunteers who had assisted the vessel received promotion, pensions or other monetary rewards.[19]

One of the most important consequences of the *Eclair*'s tragic voyage was that it placed the spotlight firmly on Britain's quarantine laws which many regarded as archaic and inhumane. Immediately after the vessel was detained at Stangate, *The Times* thundered:

> The frightful condition of the seamen on board the steam-sloop Eclair must not be allowed to continue without some effort being made to save them from the death that appears to be impending over them. . . . If several individuals were known to be shut up in a house surrounded by flames, many would rush forward to attempt their release. . . . Such is the state of the care at the present moment with reference to the still surviving portion of the Eclair, who are left to become a prey to the ravages of disease, while within the reach of aid they are not allowed to grasp. . . . Is it to be tolerated that our absurd and antiquated quarantine laws are to prevent relief from being afforded, and that the crew of the Eclair shall be left to their fate for forty days . . . .[20]

The newspaper had captured the mood of the nation, which regarded the 'gallant' Bernard and other volunteers as Christian martyrs.[21] Their 'cool and disinterested heroism' contrasted starkly with the petty officiousness of the authorities who had enforced an 'absurd and antiquated' system.[22] With the doctrine of free trade in the ascendant, and agitation for the repeal of the Corn Laws reaching its peak, quarantine was widely regarded as an obstacle to commerce. The decision to place the *Eclair* in quarantine thus received extensive coverage, not least in the local press. Newspapers in Gloucestershire, the home county of the ship's commander, reported the 'fearful ravages' of what they described as a 'most malignant' fever. The Liberal *Gloucester Journal* expressed its concern that deaths continued to occur on board the vessel while in quarantine,[23] but the Tory *Gloucestershire Chronicle* put a more positive gloss on the actions of the authorities. While lamenting the tragic death of Commander Estcourt and other members of the crew, it presented the quarantine arrangements in the best possible light, claiming they had been liberally implemented so as to allow medical practitioners and supplies on to the vessel. The newspaper's coverage perhaps reflected the fact that its outlook was solidly protectionist, having been a supporter of the landed interest during the debates over the Corn Laws.[24]

Medical opinion was equally divided, although the controversy generated by the *Eclair* shows that the profession was increasingly critical of quarantine legislation, if not of quarantine per se. The *Lancet*, which was generally reformist in outlook, typified this view. 'It is heart-rending to reflect,' stated an editorial of 11 October, 'that, after pursuing their dreary track across the ocean for several weeks, towards the hope-inspiring shores of Great Britain, the miserable crew should be detained on board their vessel, become a focus of infection and death, for more than two days, before any relief could be administered to them.' The quarantine arrangements at British ports were inhumane and outdated, it argued, and the treatment of the *Eclair* compared unfavourably to the 'succour and assistance which was so generously granted by the Governor of a small Portuguese colony in Africa'.[25]

It was in the Navy, however, that criticism was strongest. The *Naval Intelligencer* censured the quarantine authorities for having allowed the vessel to remain at the Motherbank for three days after she had appealed for assistance. 'Naval medical men,' it reported, 'consider that a couple of balks ought to have been sent without delay to the distressed vessel, and the sick put into one and the able into the other.' 'Such measures,' it argued, 'would at once have checked the progress of the disease, whereas the neglect of such caution has already proved fatal to three or four since her arrival, and many more may, we fear, be sacrificed to this want of proper caution by the Government authorities.'[26] In naval circles there was a feeling that quarantine was unnecessary in this instance. Dr Richardson of Haslar naval hospital declared that 'Notwithstanding the extraordinary mortality that has swept off so large a proportion of the crew ... I entertain no fears of her being the means of introducing epidemic disease into this country; and were the sick placed in well-ventilated wards ... I anticipate no further risk to the attendants than would occur in wards set apart for cases of typhus fever.'[27] Richardson's letter to the Admiralty was quickly countered by Sir William Pym, the Superintendent-General of Quarantine, who reminded the Admiralty that

> the disease from which the crew ... have suffered is one against which Europe generally has established a quarantine; and being aware of the disastrous consequences to the mercantile interest of this country, in the event of the 'Eclair' being admitted to pratique – viz., the establishment of a rigid quarantine by most of the European Powers, but certainly the Italian States, upon all vessels arriving in their ports from the U.K., for this reason alone, and without entering into a discussion relative to the security of the public health, I object most decidedly against the release of the 'Eclair', as well as against the landing of the crew.[28]

The Director-General of the Naval Medical Service, Sir William Burnett, took great exception to the view expressed by Pym. 'I do not mean to deny the possibility of this or any other fever becoming infectious under such circumstances as attended that in the "Eclair",' he reported to the Lords of the Admiralty in November 1845, 'but there is not the slightest proof that it was so, while there are circumstances and proofs that inevitably lead to a contrary conclusion.' 'But be this as it may,' he continued, 'I have no hesitation in declaring my firm belief that the sick men of the "Eclair", when the ship arrived at the Motherbank, might have been landed at Haslar Hospital and placed in the well-ventilated wards of the establishment without the public health suffering in the smallest degree.'[29]

Burnett's report marked the start of a long and bitter dispute between himself and William Pym. Pym was an expert on yellow fever and had witnessed the disease at first hand on several occasions, including on the Mediterranean territory of Gibraltar during his service with the Army. In 1815 he had authored a treatise on the 'Bulam' fever, better known as 'yellow fever', in which he argued strongly that it was a separate disease, distinct from the common remittent fevers of tropical climates.[30] Pym believed that there were two types of fever present in West Africa. One of these was a remittent fever, which seamen often contracted when they went ashore to gather wood and water, and especially on boat service upriver. This fever was common wherever there was uncultivated and marshy ground, and enough heat to stimulate putrefaction. Putrefaction was thought to give rise to a noxious miasma ('malaria'), exposure to which caused fever, especially in men weakened by fatigue. It was said that this disease was found most commonly in tropical climates but also in low-lying parts of Europe, having been the cause of the great mortality in the British force which landed at Walcheren in 1809. This much was uncontroversial but Pym also believed that there was another disease, present in West Africa and the West Indies, which was very different, being in no way connected with malaria or insanitary conditions. This disease – yellow fever – was 'highly infectious' and its infectious power was increased by heat. It had peculiar symptoms such as black vomit, which did not exist in cases of remittent fever and it was distinguished by the fact that its victims, if they survived, never suffered a second attack.[31]

In mid-October 1845, Pym submitted a report to the Admiralty in which he stated that the fever on board the *Eclair* was certainly the 'yellow' or 'Bulam' fever. It was, he warned, 'of a highly infectious nature, much more to be dreaded than plague, and if imported into this country during the summer months would occasion a most frightful mortality, and would for a long time prove disastrous to commerce, in consequence of the rigid quarantine which would be established in all parts of Europe'. Although Pym thought that the fever was ordinarily a disease of warm climates, he believed it could be imported into Britain by means of an 'artificial warm climate having been kept up during the voyage by the fires of the steamers'.[32] In other words, the steam-sloops used on the African service could act as 'greenhouses' incubating exotic infections which might then take root elsewhere, given favourable weather. However, Burnett asserted that Pym had been 'most decidedly mistaken in assuming that there is any increase of heat in the deck of a steam-vessel where the men live, occasioned by the fires in the engine-room, which could maintain an artificial climate so as to foster and bring the fever to England from the West Indies and coast of Africa'.

The idea that naval steamers were themselves productive of disease was not one that the Admiralty or its medical officers were inclined to endorse. Burnett therefore insisted that the fever on the *Eclair* arose not from infection but from the employment of men upriver in an atmosphere prejudicial to their health. In his view, this cause was aggravated by the decision to allow the men ashore in Sierra Leone, where overindulgence in alcohol had weakened their constitutions.[33] Burnett claimed, with some justice, that his views on 'yellow fever' were shared by 'nineteen-twentieths of the medical officers of both services' and that the prevailing opinion was that the so-called yellow fever was but a 'mere modification of the bilious remittent so extensively known over all the tropical regions'.[34] But although Burnett was convinced that there was no threat to Britain from the so-called yellow fever, he acknowledged that it might be necessary, for commercial reasons, to reassure other nations that British vessels were safe. Rather than place seasonal limits on the voyages of naval vessels, as Pym proposed, he suggested that it would be better to establish a quarantine station on the Isles of Scilly, in Britain's south-western approaches, under the charge of a capable naval medical officer who would have a comfortable hospital ship in which to place the sick.[35] This was a pragmatic proposal, which clearly did not reflect his professional convictions. It was also a very different type of quarantine to that which then operated in Britain, because it allowed the crew to leave an infected vessel and be placed in a sanitary environment.

From one perspective, the dispute between Pym and Burnett can be seen as a reawakening of the old controversy over the nature and origins of yellow fever. The Naval Medical Service, like that of the Army and the civil medical profession, was split over the question of yellow fever, although the majority now inclined towards the position that it was not a separate disease.[36] However, the practical measures proposed by Pym and Burnett were shaped by political expediency as much as professional judgement. Pym's main concern as Superintendent-General of Quarantine was to ensure that Britain was protected against the importation of disease. He also had to ensure that the precautions were sufficient to deter other countries from imposing quarantine against Britain. This would have been very much in Pym's mind at the time of placing the *Eclair* in quarantine and in making his recommendations for restrictions on naval voyages. He had recently returned from a tour of Mediterranean lazarettos and his findings had been submitted to the Foreign Office and the Board of Trade a few weeks before the arrival of the *Eclair* at Portsmouth.[37]

As we saw earlier, Pym had also been involved in negotiations with foreign powers with a view to convening the international conference that would ultimately be held in Paris in 1851.[38] These discussions had been revived by Peel's administration in the early 1840s, which was concerned not only at the 'unnecessary severity' of restrictions imposed by Mediterranean states but at their 'uncertain and variable application'.[39] Pym had been asked by the Foreign Office to draft a detailed response to certain points raised by Prince Metternich, relating to such matters as the incubation period of plague.[40] This was completed just days before the arrival of the *Eclair* and, if Metternich were satisfied, there seemed to be a real prospect of an international meeting to discuss the issue of quarantine. In such circumstances, the arrival of the 'pest ship', as one newspaper called it,[41] was hardly to be welcomed, for it raised the possibility of yellow fever being introduced into the Mediterranean from British possessions in Africa and the Caribbean. Burnett, by contrast, had little professional interest in the subject of quarantine, except in so far as it represented an inconvenience to naval vessels. As he saw it, the measures taken against the *Eclair* were unnecessary and inhumane, having no firm medical basis. Moreover, at a time of increasing discontent in the Navy, low recruitment, and a high rate of desertion, quarantine was to be avoided if it meant keeping a crew on board an infected vessel.[42]

As Director-General of the Naval Medical Service, Burnett also had to be careful of its reputation and may have anticipated the storm that was about to break over the epidemic on Boa Vista. One of the main points at issue was whether naval medical officers had given a true account of the disease to Commander Estcourt and the Portuguese authorities, or whether they had deliberately concealed the existence of yellow fever. By denying that it was a distinct disease, and by stressing other reasons for the high mortality on board the *Eclair*, Burnett was attempting to defend the reputation of his service. But his defence of the medical officers rested on an unpopular observation: it meant apportioning some measure of blame to the ship's late commander, Walter Estcourt. If the fever was not yellow fever, then it had to be a form of the common coast fever, the unusual severity of which could only be explained by the insanitary state of the vessel.

Charity begins at home

Walter G. B. Estcourt was the fourth son of Thomas G. B. Estcourt, MP for the University of Oxford, and the head of a distinguished landed family.[43]

He joined the Navy at the age of twelve and enjoyed an unblemished career before the outbreak on the *Eclair*.[44] The subsequent controversy over Estcourt's command was sparked by Burnett's allegation that the sickness was largely due to 'irregularities' in the behaviour of the crew whilst ashore in Sierra Leone and that the subsequent increase in mortality at Boa Vista had been caused by 'the most intemperate use of spirits I have ever heard of'. Burnett's informant claimed that he had been offered 'a bucket full of spirits' by one of the ship's crew.[45] These allegations were made in a confidential letter to the Admiralty but, in May 1846, the matter was due to come before Parliament, which was about to conduct an inquiry into the outbreaks on the *Eclair* and Boa Vista. At this point, Estcourt's family became alarmed that the allegations against their son would be published in parliamentary papers and were concerned that he would become the victim of a smear campaign. The matter was taken up with the Admiralty, on behalf of the Estcourt family, by the Commander's brother-in-law, A.H. Addington, the Parliamentary Undersecretary for the Colonies.

The family was concerned, not only because Burnett's letter was to be presented to Parliament, but because the evidence of the *Eclair*'s First Lieutenant, and other documents exonerating the Commander, had been omitted. As Addington put it in a letter to the Admiralty, 'these omissions ... seem, prima facie, to bear hard upon the late Commander of the vessel'. He expressed his particular regret that the papers did not include a letter written by Captain Buckle of the *Growler*, who was Commander of the West Africa Squadron. The family was convinced that this would absolve Estcourt of the most serious charges made against him.[46] Captain Buckle, who was then taking his leave in Bath, and with whom the Estcourt family may have been in contact, had been asked by the Admiralty to comment upon discipline on board the *Eclair* and the conduct of its commander, in the light of the allegations made by Burnett. The Admiralty, however, may have been embarrassed to receive what amounted to a rebuttal of most of the allegations made by the Director-General. Buckle assured the Admiralty that Estcourt had done his best to keep the vessel in good order and that 'discipline was strictly maintained'.[47] He claimed that Estcourt could have done little to prevent the great mortality that occurred on the *Eclair*, ascribing it in large measure to the depressed state of the crew on hearing that their tour of duty was to be extended. He added that 'The circumstances also of not making a single capture until they had been five months on the station seemed to cause them to imagine that they were in an "unlucky ship", and thus added to the feeling of despondency.'[48]

This news was grist to the mill of those who denounced naval operations against the slave trade as useless and wasteful of British lives.[49] Critics alleged that it was common for naval vessels to wait for months without intercepting a single ship, although there were others who argued that the mere presence of the Navy deterred the trade. In any case, it was the problem of low morale that led Estcourt to take what Buckle described as the 'well meaning but mistaken' action of allowing the crew to go ashore in Sierra Leone.[50] The action was mistaken, he believed, because it exposed the crew to malarial vapours, and enabled the men to mix with the crew of the *Albert*.[51] Buckle described the crew of the *Albert* as 'very troublesome and ill disposed'; they had apparently taken advantage of being paid off to 'commit excesses' and induced the *Eclair's* men to follow their example.[52]

But if the fever had been contracted in Sierra Leone, Buckle was in no doubt that it found a hospitable environment below decks in the *Eclair*. He informed the Admiralty that 'the ventilation of her decks and the arrangement of her magazine, and the store rooms, for a hot climate were not so perfect as he [Estcourt] wished them to be'. This was not the Commander's fault, Buckle stressed, but a reflection of the fact that the ship had been hastily fitted out before leaving England. In these circumstances, there was little anyone could have done to improve ventilation.[53] Similar comments were made in an anonymous letter to *The Times*, which pointedly remarked that the engine rooms of steam sloops were stifling at the best of times, let alone on African service.[54] There was increasing criticism of the Navy's deployment to the unhealthy African station and many now believed that the welfare of sailors was a more important issue than that of slaves.[55] However, Estcourt's conduct was still questionable in view of the suspicion that he had deliberately concealed the true nature of the fever from the Portuguese authorities in Boa Vista. This appears to have been the main concern of the Foreign Office, which requested its own reports on the vessel's arrival at the island. The British consul in Boa Vista thus informed the Foreign Secretary, Lord Aberdeen, that Estcourt did not knowingly conceal the existence of yellow fever on the vessel but that he had been wrongly advised by his surgeons who had 'misunderstood the malady which prevailed on board their ship, declaring from first to last that it was nothing but the common Coast Fever'. This opinion received confirmation from some physicians on the island, who 'insisted that the symptom alluded to [black vomit] was not confined to, or indicative of Yellow Fever; and that the illness of their patients had not been of that Character.' A Portuguese physician, Dr Almeida, also confirmed that the disease was not yellow fever and that it was not particularly contagious.

But while the consul stopped short of suggesting that these physicians had deliberately concealed the existence of yellow fever to prevent a quarantine being imposed, he was highly critical of the 'false estimate which was formed by all the medical men of the dangerous and malignant Character of the disease'.[56] The *Eclair*'s surgeons attributed the fever to the fact that its crew had lived for many months on a ration of salt meat and few vegetables. This 'low' diet had left them debilitated, depressed and unusually susceptible to the malarious vapours of the coast. Estcourt had readily endorsed this opinion and had requested the ship's original surgeon, Mr Kenny – who had more reservations – to give it 'every publicity with the least possible delay; conceiving as I do, that such rumours [of yellow fever] imply a doubt in the good faith of British Officers ...'.[57] Once the *Eclair* had put its crew ashore, it was in no one's interest to confirm the existence of yellow fever: quite apart from the honour of the Navy and the integrity of its surgeons, the Governor-General of the Cape Verde Islands would have been severely censured for allowing the crew to land if it transpired that the disease was yellow fever. This may explain why, when more formal inquiries were made into the origin of the outbreak, the Governor-General continued to insist that the disease was not contagious.

Ruinous expense

At the end of 1845 and the beginning of 1846, the connection between the *Eclair* and Boa Vista outbreaks was still tenuous. Of more immediate concern to the British government was the damaging quarantines imposed against British vessels by some of the Mediterranean states. The first of these were imposed even before the Boa Vista outbreak became common knowledge, although news of the latter did much to excite alarm in the Mediterranean and many states resolved to maintain their restrictions. The first state to impose quarantine on British vessels was Naples, which, on 17 October 1845, notified the Foreign Office that British vessels sailing from the south coast between Portland and Dover would be refused admittance to any of its ports. Vessels arriving from other parts of England were to be subjected to a quarantine of twenty-one days and those arriving from the Atlantic to a period of observation lasting fourteen days. Ships sailing from Scotland and Ireland were exempt from these restrictions and those sailing from Gibraltar were to be assessed as individual cases. The decision was made after the Neapolitan ambassador in London read about the *Eclair* in English newspapers. Captain Gallway, the British ambassador in Naples, complained to the Earl of Aberdeen that 'the absurdity of this measure would render it ridiculous if it was not fraught with such seriously vexatious results to British commerce, . . . British vessels to repair to the foul bill lazaretto at Nisita, and there at ruinous expense, to unload their cargoes and send them round to Naples in boats, when they shall have undergone sufficient purification'.[58]

Pym and his masters at the Privy Council now knew that the measures taken against the *Eclair* had been insufficient to allay fears that English dockyards had been infected with yellow fever. The fact that the Neapolitan ambassador in London had heard of two cases of yellow fever on board another ship recently returned from West Africa – the *Growler* – induced the Privy Council to take more general measures.[59] On 21 October, the Council directed the Commissioners of Customs to instruct their officers 'to be most particular in future in the examination of the masters or officers commanding vessels (more particularly steamers) arriving from the coast of Africa or the West Indies, by putting the Quarantine questions to them, and taking down their replies, so as to have their papers to refer to if necessary'. If it became apparent that vessels carried persons who were sick of a disease resembling the yellow fever, they were immediately to be placed in quarantine.[60]

The restrictions placed on British shipping prompted a flurry of diplomatic activity between London and Naples. The British tried to impress upon the Neapolitan government that there had been no new cases of fever among the crew of the *Eclair* and their attendants since 9 October, and that only six people had been taken ill since the vessel arrived in the United Kingdom. The Neapolitans were also informed that the cases among the crew of the *Growler* had shown none of the symptoms of yellow fever and that the disease was 'most decidedly not of an infectious nature'.[61]

These assurances seem to have been sufficient to persuade the Neapolitans to reduce their period of quarantine and to rescind the total ban on vessels from the south coast.[62] This was some consolation to the British government, but it was overshadowed by news that most of the other Italian states had followed Naples in placing restrictions on British shipping. It was thought that only a clear statement from the Privy Council that Her Majesty's ships and dockyards were in a sanitary state would 'dissipate the unfounded apprehensions entertained by foreign Governments, of the nature and progress of the disease in question'.[63] But what the British authorities perceived as an overreaction was considered a wise precaution elsewhere.[64]

Quarantine remained the sanitary policy of choice for most Mediterranean countries because it entailed a relatively small financial outlay, which was usually more than matched by the charges levied.[65] Towards the end of November, the Foreign Office had managed to obtain statements from the Privy Council that certified the sanitary condition of the dockyards and HM ships; these had been countersigned by the Neapolitan ambassador to London. Copies were sent to Naples and all the other Italian states,[66] and the assurances proved sufficient for the Neapolitans and the Sardinians to raise all restrictions against British vessels; the other states soon followed suit.[67]

During December, most of the remaining quarantines against British vessels were lifted, ending a brief but damaging disruption to commerce with the Mediterranean. The outbreak on the *Eclair* had highlighted the threat that British steamships posed to the rest of Europe and similar restrictions could not be ruled out in future, especially during the summer. Indeed, the incident remained very much in the public eye over the coming months following allegations that the *Eclair* had imported yellow fever into Boa Vista. Within a few weeks of the initial outbreak – which occurred amongst Portuguese and native troops stationed near the crew of the *Eclair* – the disease assumed epidemic proportions: two-thirds of Boa Vistans contracted the disease, which claimed the lives of 266 of the 3,075 native islanders and thirty-two of the sixty-nine European residents. Most of the latter, including the Governor-General and his family, fled the island as soon as they could. According to the British consul, the disease had 'prevailed mostly amongst the lower orders, and consequently there was much misery, even for the want of common food'.[68]

In Portugal news of the outbreak created great alarm, recalling memories of the dreadful cholera epidemics which had afflicted the country in 1832–3. Acknowledging uncertainty about the nature of fever on Boa Vista, there were calls for an urgent investigation into its mode of propagation and the possibility of its spreading to the mainland.[69] However, it was already known that the symptoms of the disease were black vomit, pains in the head, back and thighs, suppression of urine and broken blood vessels: the characteristic features of yellow fever.[70] The British consul also believed that the disease was contagious because those who nursed the sick had invariably fallen ill. He further implied that naval doctors had deliberately misled the Portuguese authorities as to its contagiousness:

> if the authorities of the place could have imagined that the fever on board the 'Eclair' had exhibited a dangerous type before her arrival, they never would have granted the vessel pratique.... The medical men who belonged to that vessel are now numbered with the dead, yet it must always be a cause of regret that they were not more circumstantial in their statements, particularly to the medical officers who were deputed by the authorities to confer with them.[71]

The disaster was compounded by the fact that all the Cape Verde Islands were placed under quarantine, the disease having appeared on the neighbouring island of Sao Nicolau. Quarantine hindered efforts to aid the islanders, especially attempts to bring in food and medicine, which were in short supply.[72]

As the epidemic appeared to be the fault of the Navy, it was suggested that the British government should pay compensation.[73] These demands were made most loudly by the poor of the island, who according to Portuguese accounts had suffered terrible hardship.[74] However, to lay the blame for the epidemic at Britain's door would have been profoundly embarrassing for Portugal, a long-standing ally and trading partner. The official Portuguese investigation into the epidemic was therefore politically charged, especially as some medical practitioners on the island had endorsed the popular view that the disease was yellow fever and had been introduced by the British.[75] The fever experts sent by Lisbon came to a very different conclusion and reported that the disease was not contagious.[76] The island's chief physician also changed his mind about the nature of the fever. Initially he had thought there was a connection between the outbreak and that on the *Eclair*, and ordered that all communication with the island should cease. But after the arrival of the doctors from Lisbon, he 'concluded that the disease that devastated the island of Boa Vista was endemic and not epidemic, and that there are grounds for rejecting the view that it was the same as that which attacked the crew of the Eclair'.[77] This volte-face was accompanied by a declaration from the public health authority of the island that the malady was not contagious.[78] These conclusions effectively absolved the Navy and the Governor-General of blame and kept relations between Britain and Portugal on an even keel.

Despite this, the British were still sufficiently concerned by the implications of the outbreak to commission their own inquiry. By the summer of 1846, there were demands in Parliament for an investigation to ascertain the nature of the fever and whether it was contagious.[79] The *Eclair* also featured regularly in parliamentary debates on the subject of quarantine, having become a *cause célèbre* of free-traders. The leading exponent of the case against quarantine in the Commons was Dr John Bowring, who had been a vocal critic of quarantine in the Mediterranean. Bowring was closely associated with MPs such as Richard Cobden and John Bright who represented Manchester manufacturing interests, and had been one of the most influential opponents of the Corn Laws.[80] He opposed quarantine on both commercial and humanitarian grounds, seizing on the plight of the *Eclair*'s crew as another example of the 'great sacrifice of human life' in the name of quarantine. He believed that the proper course of action would have been to transfer the crew to a healthier atmosphere, whereas they had been left to fester in the pestiferous air of the ship. Moreover, there was no threat from the crew as the disease was almost certainly non-contagious, he argued.[81] Bowring was supported in Parliament by the Radical MP Joseph Hume – another veteran free-trader – who claimed that over 120,000 guineas were spent every year in maintaining the system of quarantine in Britain.[82] The government expressed its sympathy with Bowring and his supporters but pointed out that 'The preservation . . . of our own commerce in the Mediterranean required that we should deal with the matter judiciously and deliberately'.[83] In other words quarantine was sometimes a regrettable necessity, to reassure foreign powers that there was no sanitary threat from British ports.

Amidst these heated debates, the Privy Council commissioned a report on the subject from Dr McWilliam who had come to prominence during the Niger expedition. The expedition had to be abandoned because of enormous casualties from fever and McWilliam acquired the status of a hero after piloting his ship downstream when its master and engineer had perished.[84] This exploit made him one of the best known naval figures of the 1840s but it was McWilliam's subsequent report on the fever that commended him to government. In it, McWilliam insisted that the disease which had ravaged the expedition was not yellow fever – as many claimed – and he denied that it existed as a separate disease. McWilliam therefore seemed a good person to investigate an epidemic of a disease which the government dearly wished not to be contagious. If the existence of yellow fever on Boa Vista could be plausibly denied the government could avoid embarrassment and claim that vessels from its colonies in the Caribbean and Africa posed no risk to health in Europe.

McWilliam's inquiries lasted some months and involved questioning hundreds of the island's inhabitants. Most of the questions were designed to ascertain whether the fever spread because of contact with infected persons.[85] But the conclusion reached by McWilliam was not at all what the government expected. He averred that the island had previously been healthy and that the disease which had afflicted it was a contagious malady imported by the crew of the *Eclair*.[86] This seemed to confirm Pym's account of the fever, justifying the imposition of quarantine on the *Eclair* when it returned to England. But McWilliam's opinion differed from Pym's in one crucial respect: his report stated that the disease was not yellow fever. 'If a disease such as that described by William Pym be really endemic on the coast,' he argued, 'surely we ought to hear more of it, considering the large squadron which is kept there.'[87]

McWilliam believed that the fever on the *Eclair*, and later on Boa Vista, was the common remittent fever of the African coast but that it had acquired a contagious character on account of the unhealthy state of the vessel and its crew, and the insanitary conditions in which the sick were housed on the island. He suggested that the great mortality amongst the islanders was due to their poor state of nutrition and lack of medical assistance.[88] McWilliam was also critical of the course of action taken by Pym once the vessel reached England. He believed that there was no need for quarantine because it was too cold for the disease to remain in a contagious state. Pym, he argued, had contradicted his own published works, in which he had stated that yellow fever was destroyed by cold. 'One would have ... hardly expected that the Superintendent-General would have much dread of the danger of infection, from the crew of the "Eclair" being landed in the month of October,' he declared, especially if they had been admitted to the 'cool, and well-ventilated wards of the finest hospital in the world' (Haslar).[89]

McWilliam's pronouncement in favour of contagion was no doubt unexpected given his report on the Niger expedition (often cited by Bowring and other opponents of contagion), but he remained true to the naval line in his insistence that yellow fever was not a separate disease and in his objections to the quarantine imposed on the *Eclair*. Nevertheless, Pym seems to have welcomed McWilliam's report and in a letter to the Privy Council declared it to be 'a most valuable document, and that the variety of uncontrovertible [sic] evidence brought forward by him relative to the disease in question, has finally decided and set at rest a most important and long contested question relative to the nature and history of yellow fever, more particularly as to its infectious power'.[90] Pym's letter sidestepped the controversy over whether or not yellow fever was a disease *sui generis* – which McWilliam denied – and highlighted those parts of McWilliam's report that confirmed his own theory that the fevers on the *Eclair* and Boa Vista were contagious. He saw McWilliam's admission of contagion as a vindication of his own decision to quarantine the *Eclair*; a decision for which he had been widely criticized. 'Looking to the result of the hospitality at Boa Vista,' he remarked, 'was it not properly refused in England?'[91]

Sir William Burnett was less favourably impressed by McWilliam's report, which was presented to the House of Commons in March 1847. While admitting that McWilliam had taken great pains to acquire information on the outbreak, he informed the Admiralty that he could not agree that the fever was contagious or that it was imported into Boa Vista by the *Eclair*.[92] He persuaded the Admiralty to commission a second inquiry by another senior naval doctor, Gilbert King, the Inspector-General of Naval Hospitals. King's investigations were much less thorough than those of McWilliam and there is no indication that he conducted interviews of the kind that informed the latter's report. However, in his assessment, which reached the Admiralty in October 1847, King argued strongly that the Boa Vista fever originated on the island and was due to unusually heavy rain which had flooded parts of the island, making them miasmatic.[93] This opinion concurred with that of the Governor-General of the Cape Verdes, that 'the disease had its origin in the great falls of rain which took place at a very advanced period of the season'.[94] On the basis of his investigations on Boa Vista and his experience of an epidemic in Bermuda, King went on to conclude that 'yellow fever' was not a contagious disease.[95]

King's report was welcomed by Burnett and the Admiralty as it ensured that Parliament would receive a 'balanced' view of the matter. Indeed, the selective use of medical opinion and expertise would characterize official disagreements over quarantine in the years ahead. But Pym's reaction to the report was scathing. 'After having most carefully perused this Report,' he informed the Privy Council, 'I must candidly confess that I cannot ascertain that he [King] has added the slightest information on the subject . . . on the contrary, I am of opinion . . . that he has still more mystified the subject, and that his conclusions are more likely to confuse and perplex than give information to young surgeons as to the nature and history of yellow fever.' He pointed out that King had referred only to a few slight cases of fever which he had seen in 1846, and which conveniently showed no symptoms of black vomit. Nor did he make any reference to the outbreak on the *Eclair*. 'It would appear,' he concluded, 'from Dr. King's queries and statements, that he had only one instruction, viz., to prove that the fever on shore was not connected to the ship, and that it had its origin from local causes, malaria and marsh miasmata.'[96]

The aftermath

The free-trade, anti-quarantine lobby, represented by the likes of Bowring and Hume, continued to attribute the outbreaks of fever on the *Eclair* and other African steamers to the confined and poorly ventilated conditions to be found below decks.[97] But the contents of the reports on the *Eclair* and Boa Vista fevers did not become widely known until a few years after their publication, when they were subjected to detailed scrutiny by the newly formed General Board of Health. Established by the Public Health Act of 1848, the General Board of Health under the presidency of the civil servant Edwin Chadwick was hostile to quarantine and openly espoused free trade. In its first year, it had prepared a report on the subject in order to pre-empt opposition from the authorities at the Privy Council.[98] This report declared emphatically that there was no evidence for contagion and, even if there were, quarantine had been proven useless.[99] But these conclusions did not receive unanimous support from the medical profession. While many practitioners agreed with the Board on the issue of quarantine, most were reluctant to endorse its strident anticontagionism, being prepared to admit that diseases such as plague and cholera might become contagious under certain conditions.[100] Perhaps because of the criticism it received from the medical profession, the Board's second report on quarantine – which dealt specifically with yellow fever – was less dogmatic. Drafted by Chadwick, Lord Shaftesbury and Dr Thomas Southwood Smith, the report rejected theoretical speculation and stuck to the facts, as they saw them. But the Board's approach to the evidence was selective, to say the least. Most of the 'facts' assembled in the report were such as to deny the contagiousness of yellow fever and the usefulness of quarantine. This can be seen from the Board's treatment of the *Eclair* and Boa Vista epidemics, which occupied a sizeable portion of the report. The great significance of these outbreaks for steam navigation in the Atlantic explains why the Board chose to focus on yellow fever rather than on plague – the usual subject of debates over quarantine.[101]

To begin with, the Board did its best to discredit McWilliam's opinion that the disease had become contagious and had been imported into Boa Vista. It did so by casting doubt on the reliability of the witnesses interviewed by McWilliam, most of whom came from the lower orders of society and were deemed to have a pecuniary interest in claiming that the disease was imported.[102] Even if the witnesses were correct, it argued, there was no evidence that unambiguously proved the theory of contagion. The occurrence of fever among soldiers who had mixed with the crew of the *Eclair* could be attributed just as easily to their filthy and ill-ventilated quarters, it claimed. The Board also ignored evidence that the disease had been carried to another of the Cape Verde Islands, declaring emphatically that this had not occurred.[103]

As far as the Board was concerned, the unusual severity of fever on the *Eclair* could be explained by circumstances peculiar to that vessel and its crew. It cited the opinion of Dr King that the ship's hold was in a 'pestiferous state',[104] and a report by one Captain Simpson who had visited the

Eclair while it was anchored off Sherbo Island and Sierra Leone. Simpson claimed that the crew were in poor health owing to their exposure to heavy rains and a 'state of excitement' that he attributed to alcohol. He also noted that the ship had taken on green wood as fuel, which was 'very unhealthy to burn'.[105] The epidemic on the island was regarded as coincidental and was attributed, as in Dr King's report, to the unhealthy state of the atmosphere following heavy rain. The Board also claimed that there had been sporadic cases of fever before the arrival of the *Eclair* and that there had been another epidemic of fever the following year, which suggested that local conditions, rather than importation, were to blame.[106]

The General Board of Health therefore concluded that there was no evidence to support the theory of contagion in the case of yellow fever. As for quarantine, it opined that such measures could offer no protection, as numerous epidemics had occurred on the Atlantic seaboard and in the Mediterranean despite the fact that quarantine was generally enforced. The Board also claimed that its conclusions were in line with the opinions expressed by the medical profession as a whole.[107] This was something of an exaggeration but medical opinion was certainly moving towards the view expressed by the Board. Doctors in Britain were swayed by the sheer number of reports from medical officers in the colonies that attested to the non-communicability of yellow fever.[108] But professional opinion was far from unanimous. While many declared themselves 'non-contagionists' with respect to yellow fever, some were not prepared to deny that it might be communicable in certain circumstances.[109] A vocal minority insisted that contagion was the principal explanation for most outbreaks of the disease on British vessels. The *Lancet*, for instance, carried an article by the Southampton-based doctors John Wilbin and Alexander Harvey, who attributed a recent outbreak of fever on a Royal Mail steam packet to contagion.[110] Significantly, though, the article was criticized by the vessel's own surgeon and by other medical officers working on the mail ships.[111] These critics included the influential Gavin Milroy, a former medical officer with the government's mail packet service, and co-editor of the reformist *Medico-Chirurgical Review*. His views endeared him to the General Board of Health, which appointed him Superintendent Medical Inspector during the cholera epidemics of 1849–50 and 1853–5.[112] The Board of Health was also able to cite the proposal made at the recent International Sanitary Conference in Paris, that the period of quarantine imposed against ships from countries subject to yellow fever should be substantially reduced.[113]

In 1851 the prospect of an international agreement on quarantine still seemed bright. The leader of the anti-quarantine lobby in the House of Lords, the Earl of St Germans, saw the Paris conference as a great opportunity to reach an accord with other nations and he urged the government to take more vigorous action to secure it. He chided the Conservative Foreign Secretary, the Earl of Malmesbury for seeming less enthusiastic about the promotion of free trade than his predecessors, Viscount Palmerston and the Earl of Aberdeen. He also alluded to the fact that the Manchester Chamber of Commerce had been unhappy with the recent decision to impose quarantine on some vessels from the Baltic, which had carried persons suffering from cholera. This, he claimed, was indicative of the government's weak commitment to the freedom of commerce. Malmesbury replied that only the infected persons had been quarantined and he refused to rule out quarantine in the absence of agreement on the subject of contagion amongst the medical profession. He also had a more circumspect view of the conference and stressed the difficulty of securing an agreement among the various nations concerned.[114] He was proved correct, for no binding agreement was reached at Paris and quarantine regulations in most countries remained much as before.[115] Nor was Malmesbury's position significantly different from that of previous foreign secretaries. Lords Palmerston and Aberdeen had performed the same delicate balancing act between the interests of commerce and Britain's relations with foreign powers. They never lost sight of the fact that quarantine regulations were necessary to prevent more damaging restrictions against British vessels.

But this is not to say that the *Eclair* and Boa Vista controversies had no effect on sanitary policy. The public outcry caused by the decision to place the vessel in quarantine, and the subsequent death of several members of its crew, made it more difficult for British governments to take such measures in future. Quarantine continued to be imposed on vessels deemed infected with yellow fever,[116] but the 1825 Quarantine Act was now interpreted more liberally. The deaths that occurred on the *Eclair* when it was placed under quarantine in Britain were cited again and again in parliamentary debates,[117] and British arrangements were modified significantly in view of this. The first ship to be quarantined at Southampton in the wake of the scandal was RMS *La Plata*, which, in 1853, was subjected to a quarantine of just two days – far shorter than the *Eclair*. An order in council was issued to the effect that whenever a vessel was placed under quarantine for yellow fever in the United Kingdom, all passengers who had already recovered from the disease would be permitted to land.[118] Milroy praised this as an enlightened measure that marked a significant reduction in the 'vexatious restrictions upon the passengers and crew of vessels for the alleged protection of the public health, but which are known to be utterly futile'.[119] Such restrictions were especially annoying to the sick who were heading for colder climes in order to recover their health.[120] However, a fatal case of fever among one of *La Plata*'s crew, who died after being released from quarantine, led to 'intense excitement' among the population of Southampton, who feared the disease would spread throughout the city. In view of this unrest, the duration of quarantine was increased, a period of ten days being imposed in the case of RMS *Parana* the following year.[121] Even so, the length of quarantines remained significantly lower than the discretionary maximum of twenty-one days permitted under the statute of 1825.

It was in Britain's West Indian colonies that a liberal stance in respect of quarantine was most evident. Despite the *Eclair* and Boa Vista scandals, the prevailing view in the Caribbean was that yellow fever was not a contagious disease. The Board of Health's report on the Boa Vista epidemic made it possible for the colonial authorities to deny any connection between such outbreaks and the landing of infected crews, as is evident from a dispatch relating to a suspected outbreak of yellow fever on the *Dauntless*, which arrived at Barbados in November 1852. This missive, from the Governor of Barbados to Sir John Packington MP, noted that 'The public attention in Barbadoes ha[d] recently been drawn to the case of Her Majesty's ship "Eclair," and the spread of the African fever at Buonavista ... in reference to the case of the "Dauntless", and declared that in Barbados, too, there was no evidence to link cases on the island to importation.[122] In other words, reports that yellow fever had not been imported into Boa Vista were used to vindicate the Governor's decision not to quarantine the *Dauntless*. The Governor also pointed out that the ship's captain had quickly tackled the disease using hygienic and sanitary measures. The captain, in turn, gave thanks to the Governor for 'the generous manner in which the refuge and resources of this colony were at once thrown open to us, when under an attack of pestilence so deadly as might well have made a less enlightened government and community ... hesitate before admitting such applicants within their shores'.[123]

Most Britons working in the colonies shared Milroy's view that quarantine was a 'vexatious restriction' which hampered trade and disrupted their passage home. Indeed, his opinions were echoed by many prominent medical practitioners in the West Indies, including Mr Watson, the Deputy-Inspector of Fleets, and Drs Dempster and Bowerbank of the Jamaican Central Board of Health, all of whom insisted that yellow fever was not contagious except in exceptional circumstances. In line with these opinions, the authorities on most British West Indian islands seem, at this point, not to have imposed quarantine against yellow fever. Vessels arriving with suspected cases on board were able to send their sick to the nearest army or naval hospital, where they were treated alongside other patients. Healthy passengers and crew were permitted to land rather than forced to remain on the vessel.[124] Normally, proximity to sources of infection predisposed states – like those in the Mediterranean – to take strict measures against infection,[125] but the ever-present danger of yellow fever in the Caribbean may actually have reduced fear of the disease. More importantly, to admit that yellow fever was highly contagious would have been to accept the need for quarantine during several months of every year, entailing enormous disruption of trade. Thus, it was only in the case of diseases exotic to the Caribbean – such as cholera – that quarantine was strictly enforced.

In the West Indies, the policy of not quarantining vessels infected with yellow fever continued for years after the Boa Vista affair, despite the protests of some members of the medical profession and several violent epidemics of yellow fever in British territories.[126] As we shall see in the next chapter, the West Indies were ravaged by three waves of yellow fever in the coming decade: the first in 1852–3, the second in 1856, and the third, and most deadly, in 1863.[127] And yet, in the face of increasing criticism, the governors of most islands continued to insist that quarantine was useless against yellow fever because the incubation period was uncertain. Indeed, a committee appointed by the Royal College of Physicians later endorsed these opinions when invited to consider the subject of quarantine by the Colonial Office.[128] As the medical profession was divided on the questions of yellow fever and quarantine, the arguments of the colonial governors held sway and most continued to regard quarantine as an unnecessary restriction on trade. Another reason for their opposition may have been the labour shortages which crippled the sugar colonies following the abolition of slavery. Their economies were still in a fragile state in the 1850s and quarantine added time and expense to the costly expedient of bringing indentured labourers halfway round the world from India and China.[129] Vessels from India often stopped at one location in the Caribbean – Trinidad, for example – before proceeding to another, such as British Guiana. If one of these locations was rumoured to be infected, quarantine for yellow fever would have been added to the burden of the measures that occasionally needed to be taken against cholera.

Before 1840, most British politicians and most of the British public did not wish to see the nation's quarantine laws repealed or even significantly relaxed. Despite the gloss of liberalism, those who pressed vigorously for the abolition of quarantine tended to be regarded as exponents of vested interests who were prepared to sacrifice the public's health for selfish reasons. Such attitudes lingered on, especially in some of Britain's maritime cities, but from the mid-1840s there was a significant change, with opposition to quarantine becoming more respectable and more widespread. A more liberal sanitary regime came to be identified with the national interest and was justified as progressive and humane. This change cannot be attributed to any single factor or event but it was a transition in which the *Eclair* and Boa Vista epidemics were pivotal. These related incidents served to unite a humanitarian interest in the welfare of sailors with a revivified campaign against quarantine by free-traders; the one reinforcing the other. They contributed to the liberalization of quarantine in Britain and its West Indian territories and produced a climate of opinion in which alternatives to old-style quarantine were seriously entertained.

This modification of quarantine arrangements became possible because the balance of public and professional opinion had tilted further towards reform and because of the constraints entailed by quarantine. It was not simply freedom of trade that was at issue, or the violation of abstract liberal principles, but the 'vexatious restrictions' that quarantine placed on the armed forces and the movement of colonial administrators and labour. Their free circulation within the Empire was vital to the economic order of which Great Britain was the hub. The military, naval and colonial lobbies within the medical profession were thus virtually unanimous in condemning quarantine and often became spokesmen for the organizations they served. Complaints about the detrimental impact of quarantine on imperial communications had often been made in the past but new technologies had raised expectations and had made possible a new vision of the relationship between Britain and its overseas territories. Rapid communication by steamship and electric telegraph fostered a view of the British Empire as a coherent whole. This may explain why the British government and colonial administrations were increasingly prepared to reform or abolish quarantine in the case of yellow fever.

In the debates about how far to go in this direction, the *Eclair* and Boa Vista incidents loomed large for many years. The former had become emblematic of the inhumanity of quarantine, while the inquiry into the latter was arranged in such a way as to justify its abolition. Another notable feature of these debates is the way in which medical expertise and evidence were employed by the protagonists. Both the British and the Portuguese seem to have chosen medical experts to conduct investigations in the belief that they would reach opinions which would be diplomatically convenient. When the conclusions arrived at proved inconvenient, an alternative expert was quickly dispatched. Subsequent reports, like that of the General Board of Health on quarantine, also made selective use of previous inquiries to shore up their positions, and government officials picked out those details which seemed to justify their preferred course of action. It was not the first time that medical opinion had been used in this way but experts now played a greater role politically than at any time in the past. It was a sign of things to come. As the foundations of an international sanitary system began to be laid in the years ahead, disputes over quarantine and similar measures were increasingly conducted by specialists of various kinds. And yet the resolution of such disputes still owed more to diplomacy than to the force of scientific logic.

CHAPTER 5

Yellow fever resurgent

In the 1840s and '50s many countries on both sides of the Atlantic relaxed their quarantines against yellow fever. There had been no outbreaks of the disease in Europe since it appeared briefly in Gibraltar in 1828 and its absence thereafter engendered a feeling of security. It was as if the beast had returned to its lair in the tropics. France went so far as to unilaterally abandon quarantine against yellow fever in 1847, while in Britain and its Atlantic empire such precautions were either relaxed or abolished. Some North American ports continued to be affected by yellow fever but not as regularly or as severely as before and many American states also permitted their sanitary barriers to lapse. But in the mid-1860s this era of lax or non-existent regulation came to an end. Countries began to re-examine their sanitary arrangements and considered new measures to replace or supplement existing ones.

This change of heart followed a massive surge in the prevalence and incidence of yellow fever. From the 1850s, the disease spread rapidly through the western Atlantic and threatened other parts of the world linked by trade to its 'reservoirs' in South America and the Caribbean. The advent of steam power in the 1830s reduced journey times across the Atlantic from around thirty to fifteen days, and by the 1880s this had been halved again due to improvements in propulsion such as the screw propeller.[1] Yellow fever could now be carried across the Atlantic before its characteristic symptoms had time to appear and posed a growing menace to commerce throughout the region.

Tropical emanations

Concern about the predations of yellow fever grew steadily on the western side of the Atlantic following several well-publicized epidemics in the 1850s. The first of these occurred in 1849–50, when 90,000 people were afflicted by the disease in Rio de Janeiro, and over 4,000 died.[2] Although the proportion of deaths to cases was lower than in many previous outbreaks, Brazil had suffered few epidemics in the recent past and the devastation caused by yellow fever caused shock waves to reverberate around the region. By 1852 the disease had appeared in the Caribbean and the following year a severe outbreak occurred in New Orleans.[3] The epidemic in New Orleans killed 7,849 people and prompted a major rethink of sanitary arrangements in this and other American ports.[4] Over the next few years the disease continued to spread throughout the Americas and crossed the Atlantic Ocean to Europe. In 1858, a correspondent to one of Britain's leading medical journals, the *Lancet*, declared: 'Twenty years ago it was generally believed that yellow fever, in its virulent epidemic form, was confined to inter-tropical regions; but of late years it has crept along the coast of America to places where it was formerly altogether unknown, and has now attacked in a malignant form a city of Europe.'[5]

He was referring to the severe outbreak of yellow fever which began in Lisbon in September 1857 and raged until the end of December. By the time the epidemic had ceased, there had been over 12,000 cases and 5,000 deaths, while around a fifth of the city's 250,000 inhabitants had fled in terror. The Lisbon epidemic and the growing frequency of outbreaks in the Americas were a sharp reminder of the havoc wreaked by the disease in the 1790s and early 1800s. Some Europeans reassured themselves that Lisbon, like Rio, was a filthy place, with steep, congested alleys and a dirty, degraded population, quite unlike more civilized lands. Others blamed the epidemic on an unusual amount of rainfall, which had the left the atmosphere inundated with noxious particles. Yet there was no escaping the connection between the outbreak and navigation, particularly between Lisbon and ports known to be infected in Latin America. Some suspected the crews of vessels sailing from Rio might have carried the infection on shore but merchandise such as hides and skins were also implicated.[6] Either way, the alarm created by the epidemic was sufficient to induce many neighbouring countries to take precautions.

In France, the country's long-disused lazarettos were reopened and prepared for the reception of large numbers of sick and suspected cases in the event of the disease spreading north. Dr François Mélier, the Inspector-General of the French sanitary services, made a tour of all the ports and framed rules for the establishment of quarantine. Even after the epidemic in Portugal had subsided, the French took no chances and prevented a ship from Senegal landing its passengers at Brest on account of its having entered the River Tagus during its voyage. Although no yellow fever cases were reported, the ship contained 186 persons classed as invalids and this would have been sufficient to cause concern. The French regulations harked back to the international sanitary convention drawn up in Paris in 1851, which, while it lacked the status of a legal document, was sometimes used to confer legitimacy on the actions of individual states.[7] In the case of yellow fever, it decreed that ten days were to elapse without the disease occurring on board a ship between its touching an infected port and the removal of quarantine.

Although the period of quarantine imposed in Brest was relatively short, it alarmed many ardent champions of free trade. The French measures were implemented in January 1858 even though Lisbon had been officially declared free from the disease on 24 December. This raised the spectre of a return to the panic-driven and opportunistic quarantines of old. The question of culpability was also very much to the fore. As we saw in the previous chapter, in 1845–6, the inhabitants of the Portuguese island of Boa Vista had demanded compensation from Britain for the contagion allegedly introduced by a British naval vessel. Now, the Portuguese authorities were pointing to another British ship, the mail packet RMS *Tamar*, which had touched at Lisbon on her voyage back from Rio de Janeiro. In this instance, however, the vessel and its captain were soon exonerated by an independent expert, Dr Guyon, Inspector-General of the Sanitary Department of the French Army, who had been sent to Lisbon in October to take part in a congress considering the causes of the outbreak. He remained there until the end of December and then sailed to Southampton, where he interviewed Dr Wilby, the port's medical superintendent. He discovered that the *Tamar* had returned to Southampton after calling at Lisbon and, according to Wilby, she 'had not a single case of fever, either on her outward voyage to or from Rio [de] Janeiro'.[8] Guyon concurred that the vessel could not have imported the disease into Lisbon but in a subsequent meeting in London, with Dr William Pym, Superintendent of Quarantine with the Privy Council, he expressed concern about the large number of vessels which had arrived in Southampton from Brazil reporting cases of sickness and death on their return voyage. During his stay in Southampton, one ship, the *Orinoco*, had been permitted to enter the harbour despite having reported twenty-eight deaths and carrying a considerable number of sick and convalescent passengers.[9]

Precautions in Southampton were not especially lax, but the observation made by Guyon suggested that sooner or later the matter of yellow fever would have to receive attention. There was a growing and embarrassing disparity between British arrangements and those recently implemented by the French authorities. The French were the first to depart from a laissez-faire approach to commercial navigation and had even begun to construct permanent lazarettos for yellow fever cases, one being located seven miles from Brest.[10] As if to reinforce the message, in the coming months further cases were reported on British vessels in the Caribbean. The disease appeared at St Thomas and affected many ships in the island's harbour. RMS *Tyne*, a mail packet steamer, also experienced a severe attack of yellow fever while docked at Rio de Janeiro.[11] In view of the frequency of intercourse between Britain, the Caribbean and South America, British ships were particularly vulnerable not only to yellow fever but to the protective measures imposed by other states. There was considerable annoyance when the authorities in Lisbon followed the example of the French and began to impose quarantine on vessels from ports in which yellow fever had been reported. The first target of the new restrictions appears to have been the *Tyne*, when it arrived from Rio. Passengers were taken from the vessel to the lazaretto in Lisbon but subsequently even this was considered too dangerous and the city's Board of Health declared that no vessels from Brazil which had reported a case of yellow fever on board would be allowed to land unless they had performed rigorous quarantine at some other lazaretto.

While these precautions may seem reasonable in view of the destruction caused by yellow fever in 1857, they were strenuously opposed by the British. The Royal Mail Company's agent and the British Minister in Lisbon raised the matter with the authorities there, with a view to easing the restrictions. Measures taken in Lisbon were portrayed in Britain's press as unnecessary and crude diversions from the 'real' problems revealed by

1 This watercolour reinforces popular images of the Black Death, which had become synonymous with all that was sinister about the Middle Ages. However, the term 'Black Death' did not come into general usage until the nineteenth century and its association with ignorance and barbarism was a construction of the later eighteenth century. For many people in Asia, North Africa and Eastern Europe, visitations of plague remained an awful reality.

2 St Carlo Borromeo (1538–84) was Archbishop of Milan during the plague epidemic but also a major figure in the Counter Reformation and the suppression of heresies such as witchcraft. Here, he is depicted risking his life in an attempt to minister to the poor and the sick. The broader significance of this image is that it evokes a new spirit which regarded the prevention of disease as a civic duty.

3 One of many posters issued by the Italian states in an attempt to prevent the spread of plague by restricting trade. It indicates the perceived importance of commerce as a vehicle of infection.

ORDERS
CONCEIVED
AND
PUBLISHED
By

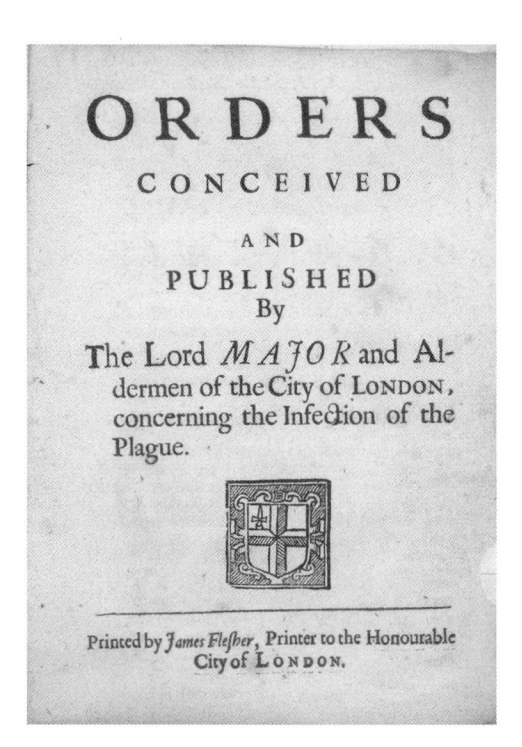

The Lord *MAJOR* and Aldermen of the City of LONDON, concerning the Infection of the Plague.

Printed by *James Flesher*, Printer to the Honourable City of LONDON.

4 The publication of plague orders in London in 1665 was typical of the approach then taken to the prevention of plague in England and most other parts of Europe. Measures tended to be ad hoc and took the form of municipal ordinances or royal commands. However, the experience of plague in the 1660s led some continental countries to enact national quarantine legislation and most countries, including Great Britain, followed suit in the 1700s.

5 The slave markets were among the principal hubs of infectious disease in the Americas but attempts to block infection by quarantine and similar measures were ephemeral. After yellow fever crossed the Atlantic from Africa in the 1640s, some North American ports began systematically to impose restrictions on trade and navigation in an attempt to keep the disease at bay.

6 This illustration indicates several of the features which permitted yellow fever to become established in the sugar colonies: the water pots, often impregnated with sugar residue, in which mosquitoes were able to breed and the lack of vegetation which might have supported birds and other animals which could have preyed on these disease-bearing insects.

7 Sickness and trade invariably went together and this plate shows how both were institutionalized in the Dutch East Indies. Batavia, which was the main port of the Dutch East India Company, was notoriously unhealthy and was regularly afflicted by epidemics of fever and dysentery.

8 During the eighteenth century, some Italian states cooperated with each other by issuing passes certifying freedom from infection, which reduced the disruption caused by sanitary measures. However, in most parts of Europe, quarantine was increasingly deployed in an aggressive manner to damage the trade of rival nations.

9 This illustration shows the importance of the Thames to the commercial fortunes of Britain, but the river was also often a conduit for plague. In the etching, a river god representing the Thames sits on a chariot, centre, accompanied by nereids in the water. Also standing in the water at the front of the chariot is Captain Cook, behind him are Sir Francis Drake and Sir Walter Raleigh, and to the right is a man identified in the British Museum catalogue (and elsewhere) as Dr Charles Burney, the scholar and composer of music. Cook, Raleigh and Burney are clothed. Left, figures representing Europe, Africa, America and Asia. Above, Hermes or Mercury with the caduceus and Roman tuba (or similar).

10 The lazaretto at Genoa was a fairly imposing building. It was enlarged and fortified on several occasions, partly in order to deter attacks from Barbary pirates.

11 During the 1700s many merchants began to protest about lazarettos and their impact on trade. They argued that such establishments were intended as much for generating revenue as preventing the spread of disease. The authorities in Naples were notorious for their readiness to impose quarantine and remained so through to the late nineteenth century.

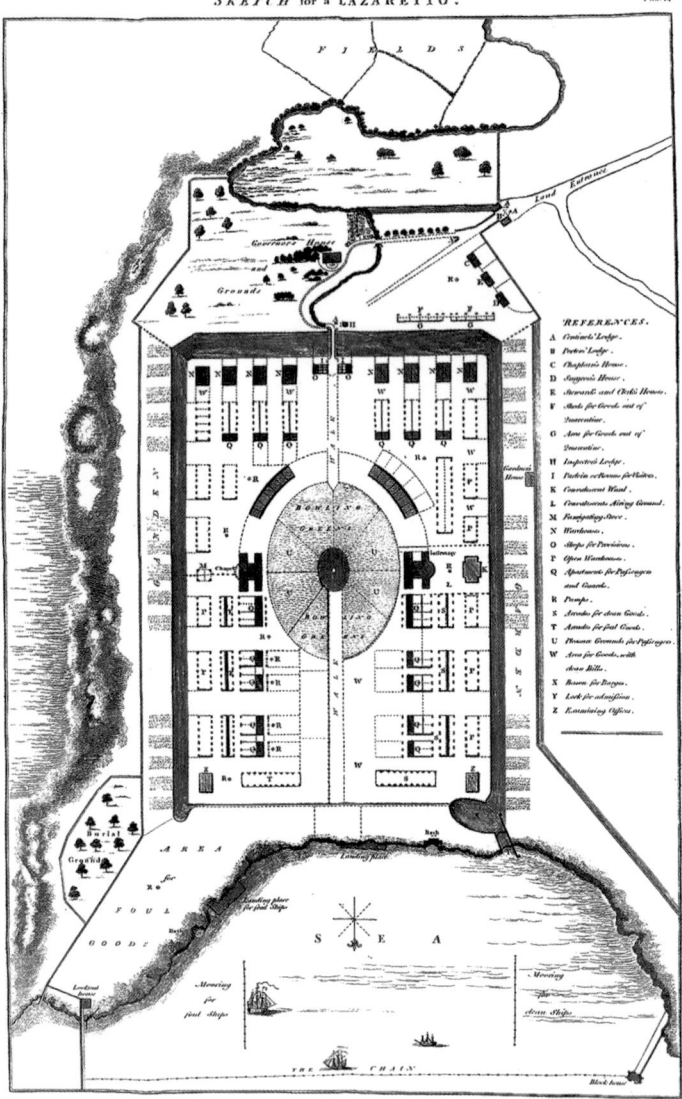

12 John Howard is better known as a prison reformer but he also directed his attention to abuses in other institutions, including lazarettos in the Mediterranean. In this sketch, he reveals his plan for an ideal institution in which to keep passengers and goods free from harm. Howard called for the establishment of such a lazaretto in Britain in order to prevent the detention of British ships in the Mediterranean whenever plague ravaged the Levant.

13 Sir James McGrigor (1771–1858) is one of Britain's best known military physicians. He joined the Army in 1793, at the beginning of hostilities with Revolutionary France, and was appointed Surgeon-General of the Army in Spain and Portugal in 1811. In 1815, he became head of the newly created Army Medical Department – a post he held until 1851. McGrigor was a great reformer but also an ardent supporter of quarantine and the theory of contagion that underpinned it.

14 This scene from the yellow fever epidemic in Barcelona in 1821 evokes images reminiscent of the plague in earlier centuries. It was the most serious episode of yellow fever in Europe for some years and caused widespread alarm. However, the danger posed by the epidemic to other countries was probably exaggerated, partly in order to justify the French military cordon around Spain and the subsequent invasion which restored the Bourbon dynasty.

15 A.B. Clot (1793–1868) was a brilliant graduate of the Montpellier Faculty of Medicine. He was appointed physician to the Egyptian Army by the Ottoman viceroy of Egypt, Mohammad Ali, who later conferred on him the title of 'bey' in recognition for his distinguished service. He subsequently became head of Egypt's medical administration. During his tenure, Clot urged his master (unsuccessfully) to resist foreign interference in Egypt's sanitary affairs.

16 Sir William Burnett (1779–1861) was a leading figure in British naval medicine and served as Physician-General (later Director-General of the Naval Medical Department) from 1832 to 1855. He campaigned for better conditions for young surgeons and to keep the Navy as free as possible from the constraints imposed on its operations by quarantine. He was opposed to the view that yellow fever was contagious in ordinary circumstances.

17 James Ormiston McWilliam (1808–62) was one of the most distinguished naval surgeons of his generation and well known to the general public because of his heroic exploits in the Niger Expedition of 1841, when he piloted his ship, the *Albert*, down river after the captain and engineer were lost to fever.

MEDICAL HISTORY

OF THE

EXPEDITION TO THE NIGER

DURING THE YEARS 1841-2

COMPRISING

AN ACCOUNT OF THE FEVER

WHICH LED TO ITS ABRUPT TERMINATION

BY

JAMES ORMISTON M'WILLIAM, M.D.

SURGEON OF H.M.S. ALBERT AND SENIOR MEDICAL OFFICER OF THE EXPEDITION

WITH PLATES

LONDON
JOHN CHURCHILL PRINCES STREET SOHO
MDCCCXLIII.

18 McWilliam's medical history of the Niger expedition was widely praised and made his reputation as an expert on tropical diseases. On the strength of this treatise, McWilliam was appointed to investigate the outbreak of fever on the Portuguese island of Boa Vista but his conclusions proved inconvenient to the Admiralty and another expert was subsequently dispatched.

THE ROYAL WEST INDIA MAIL STEAMER "TYNE" ON SHORE AT ST. ALBAN'S HEAD.—(SEE PRECEDING PAGE.)

19 Mail ships frequently carried cases of yellow fever back to Britain from the Americas and tropical Africa, resulting in frequent disease scares in European ports. The *Tyne*, pictured here, also achieved notoriety when it ran aground on the Dorset coast on its return from Brazil in December 1856.

20 This painting expresses the artist's horror at witnessing many deaths from yellow fever in Buenos Aires. It shows a woman lying dead of fever while her infant son seeks succour from his mother. Behind her, two doctors enter the room, arriving too late to save the child's parents. This painting depicts a real scene from the epidemic and the two doctors in the picture later died from the fever. It is estimated that well over 13,000 people perished in the city from fever during 1871.

21 The increasing speed of vessels making trans-Atlantic crossings meant that infected persons might not display symptoms of yellow fever before they disembarked. A few cases of yellow fever did subsequently occur among passengers who had travelled from the Americas to Europe but they were few in number. Yellow fever epidemics in Europe during the 1850s and 1860s generally occurred when cargoes from the Americas were unloaded in European ports. Disease-bearing mosquitoes sometimes lurked in the merchandise and the holds.

22 The construction of the Panama Canal was blighted with epidemics of yellow fever until the disease was eradicated by a concerted campaign against mosquitoes undertaken by the US military under the direction of Surgeon-General William Gorgas.

the epidemic. In the *Lancet*'s opinion, it was a case of shutting the stable door after the horse had bolted, the epidemic having been used as a pretext 'for the imposition of additional absurdities in the way of quarantine'. The implication was that the measures were punitive in nature and designed to divert attention from the city's failure to remove the 'miasmatic nuisances' which the *Lancet* believed to be the true cause of the epidemic.[12] Indeed, for the next few years, most medical practitioners remained unconvinced of the need for more stringent regulations against shipping from the West Indies and other places regularly infected with yellow fever. This position was heartily endorsed by the government, which was anxious to prevent restrictions on trade and naval operations.

Yellow fever was obviously a growing problem but the solution, in the opinion of most officials and medical writers, was better sanitation in ports and on ships rather than the imposition of quarantine. Small steam-powered ships seemed particularly vulnerable to diseases such as yellow fever because their ill-ventilated engine rooms created enormous heat, capable of propagating and sustaining tropical infections. Although little was then known about the health of merchant seamen and the conditions in which they worked, it had been apparent for some time that merchant vessels were badly affected by yellow fever. In his description of the epidemic that occurred in Demerara in 1851–2, the physician Dr Blair noted that the disease had appeared first on merchant ships, and that those carrying coal and patent fuel suffered especially badly. In early 1853, yellow fever also appeared on merchant vessels in Kingston, Jamaica, prompting a committee of the island's Central Board of Health to draw up measures to deal with outbreaks on ships. These concentrated on improving sanitary conditions on board vessels and the removal of crews to well-ventilated quarters on shore. Although all unnecessary contact with infected ships was to be avoided, there was no mention of quarantine.[13]

In 1861, however, the British authorities were beginning to review their formerly liberal attitude to yellow fever. Just a few years after the Lisbon epidemic, yellow fever broke out again on European soil, this time in St-Nazaire. In the past, ships had often reached France with yellow fever cases on board but the disease remained confined to passengers and crew. On 14 September 1861, for example, a vessel laden with Cuban sugar arrived in Nantes with several such cases on board, eighteen people having died on the voyage.[14] But the outbreak in the harbour of St-Nazaire later the same month was different in that several deaths occurred among porters and stevedores as well as the crew. A doctor who had arrived to assist local practitioners in treating the sick also perished: all told, there were twenty-six deaths out of forty reported cases.[15] To some observers this course of events provided evidence that the disease had been spread by human intercourse, suggesting that quarantine was a wise course of action. However, Dr Mélier, who conducted a rigorous investigation of the outbreak, concluded that this was not the case. In his view, the disease was not straightforwardly contagious but had nevertheless been imported. He concluded that the infection known to be raging at Havana must have been trapped in the holds of sugar vessels and preserved on the journey across the Atlantic. This conclusion was not unlike the speculation which surrounded the Boa Vista scandal but Mélier had systematically examined all the evidence for contact-contagion and found it wanting. Yet it was equally unlikely that the disease simply arose from local circumstances, as anticontagionists typically claimed. By the 1860s, the clinical and post-mortem features of yellow fever were pretty much agreed and the pathological signs and symptoms of the victims at St-Nazaire matched those observed elsewhere. The importation of an infectious atmosphere seemed to be the only plausible explanation and suggested the wisdom of maintaining quarantine measures. Indeed, the imposition of strict quarantine on other vessels in the sugar fleet was credited with having prevented further outbreaks of the disease in 1861.[16]

Although the finer points of the discussions taking place among medical practitioners were probably unknown to most of the public, the occurrence of two outbreaks of yellow fever on European soil in the space of a few years created great alarm. In August 1862 there was great consternation in London following the arrival of a ship that had run the blockade of southern ports imposed by the North during the American Civil War. Rumours spread that the vessel had arrived at Gravesend with yellow fever on board but had been allowed to proceed up the Thames to the Victoria docks where it unloaded its crew and cargo of cotton.[17] In Plaistow and the area around the docks, there was 'considerable excitement' as fear of an epidemic of yellow fever spread. In the event, the anxiety proved groundless but it shows that there was real anxiety about the spread of yellow fever from tropical to temperate lands. 'Enormous losses' among merchant seamen at ports such as Havana and Barbados ensured that these fears intensified in the months ahead.[18] There were also reports of severe outbreaks of yellow fever along the African coast, which was linked by trade to ports such as Liverpool.[19]

The British government now seemed to be leaning in favour of quarantine and this aroused opposition from those committed to a more liberal policy. As one medical practitioner based in the West Indies noted: 'It seems that the question as to whether yellow fever is infectious or contagious or not, is not yet settled in England.' He added, 'Out here it is held, universally, I believe, not to be so. At Havanna [sic], Vera Cruz, Barbadoes, Demerara, Trinidad, Jamaica, and St Thomas – places where the authorities are very strict, so much so that they quarantine for the measles, a disease particularly mild in the West Indies – they grant *pratique* to ships having yellow fever on board.' In his view, it was the 'height of cruelty' to shut people in pest-houses and he hoped that the British government would listen to practitioners who had extensive experience of the disease.[20] But the government was more inclined to listen to public opinion, which was swinging decidedly towards quarantine, at least in the principal ports. Thus, when yellow fever became prevalent in Rio at the end of December 1862, the government decided to place all vessels sailing from there under quarantine.[21]

The issue of quarantine was highlighted again in 1865 in the second yellow fever outbreak in Europe since the Lisbon epidemic. It happened in the rather unlikely setting of Swansea on the southern coast of Wales in early October, following the arrival of a ship – the *Hecla* – which had sailed from Cuba. When it docked, only one of the crew was sick and he died the day after reaching the port. Within a fortnight, however, yellow fever appeared among some of the town's inhabitants who had been in direct contact with the ship while unloading its cargo of copper ore.[22] Despite previous arrivals with cases of yellow fever on board, this was the first time that the disease had spread from ship to shore and it caused consternation throughout the country. Proponents of the contagion theory pointed to the fact that the vessel had not been subjected to quarantine, while opponents blamed the epidemic on an 'unusual approximation of the temperature of Swansea during the summer to that of the . . . regions of the yellow fever zone'.[23] Ultimately twenty-seven townspeople were infected with the fever, fifteen of whom died. As in St-Nazaire, the outbreak was subjected to close medical scrutiny, both by the local Medical Officer of Health, George Buchanan, and Sir John Simon, the

Medical Officer of the Privy Council. Like Mélier, they concluded that yellow fever was not a straightforwardly contagious disease but that the outbreak had been caused by an imported infection, which thrived in the bowels of cargo ships. Yet, whereas Mélier had adopted a conservative solution in the form of quarantine of merchandise and persons, Simon preferred to concentrate on the cargoes which might have been contaminated by infectious particles. In his view, this was a problem which could be addressed in much the same way as other sanitary nuisances, rendering the quarantine of persons unnecessary.[24]

Despite Simon's disinclination to use quarantine, pressure was mounting on the Privy Council to do so. The lay public were already frightened at the possibility that yellow fever might be imported and the inhabitants of port cities trading regularly with the Caribbean remained nervous. Quarantine was also receiving support from some very unlikely quarters. Medical periodicals such as the *Lancet*, which had not so long ago denigrated quarantine, were now urging the government to take decisive action. Following reports of yellow fever on the Caribbean island of St Thomas, in 1866, and the return of ships from there with yellow fever, the journal declared it was not improbable that the disease would once again erupt in Britain, most likely in the port of Southampton, which was in frequent contact with the 'yellow fever zone' through the mail packet service amongst other concerns.[25] Shortly afterwards it was announced that a mail ship, the RMS *Atrato*, had arrived at the port from St Thomas with thirty-five cases of yellow fever on board, having suffered fourteen deaths on the passage. The Privy Council hesitated but the port's medical superintendent took the initiative and removed healthy passengers and crew to the quarantine station at the Motherbank.[26] Just as the vessel was being released from quarantine, another ship arrived from St Thomas also with cases of yellow fever on board.

Stung by criticism of its hesitation in the case of the *Atrato*, the Privy Council decided upon stringent quarantine immediately.[27] It now veered to the opposite extreme and was criticized for reverting to 'inhumane' practices. British procedures certainly fell far short of those established in France since the St-Nazaire outbreak, where both the healthy and the sick were removed from infected vessels. If quarantine was to play an important part in controlling yellow fever, as many evidently desired, then it had to be carried out 'with the least hardship to the individuals affected by it'.[28]

The Privy Council seems to have been attuned to these feelings and soon aligned its measures with those of France.[29] It now left no one in doubt of its intention to form an impenetrable barrier to the 'wave' of yellow fever radiating from the Caribbean to the English coast. Other countries, including France, had already forbidden their vessels to touch at the island of St Thomas and the government thought it wise to force all vessels sailing from there to anchor some distance from shore while they awaited medical inspection.[30]

Towards a new sanitary system

In order to cope with the resurgence of yellow fever, most Atlantic countries had reverted to the use of quarantine but Britain had done so with obvious reluctance, its hand forced by the weight of public opinion and the expectations of neighbouring countries. But since all states were adversely affected by curtailments of trade, the desire grew for a new system of regulation which placed less reliance on quarantine. Some progress in this direction was enabled by the gradual redefinition of yellow fever from a 'contagious' to a 'filth' disease. Although it did not entirely lose its identity as a communicable disease, or for that matter as an infection originating in the tropics, it came increasingly to be associated with poor governance and civic neglect. This transformation cannot be traced to any single event but it is evident in the response internationally to the outbreaks of yellow fever in the early 1870s in South America. In 1871, reports began to reach North America and Europe of 'the great ravages' made by yellow fever in Buenos Aires, a city which lay to the south of the normal yellow fever zone.[31] It was not the first time that the disease had broken out in the Argentine capital – serious epidemics had occurred in 1821 and 1857 – but the growth of steam navigation in the interim meant that countries immediately to the north and south of the 'fever belt' were at greater risk than before. Until the 1870s, most Argentines tended to view Brazil as the most likely source of infection, blaming shipping from that country for the epidemic of 1857, for instance.[32] But in 1870 the threat came unexpectedly from the east, from the landlocked country of Paraguay which had been devastated by its defeat by Argentina, Brazil and Uruguay in the War of the Triple Alliance (1864–70). The armies of all the protagonists had been ravaged by cholera and at the end of the war it was joined by yellow fever.

Merchants and civic authorities in Buenos Aires feared that the disease would travel downriver and infect their once salubrious port, leaving businesses ruined and the city's population decimated. But defending the city from yellow fever presented the authorities with a dilemma. Quarantine against vessels sailing from Paraguay was the most obvious course of action but if their port imposed quarantine unilaterally it was likely to lose trade to the Uruguayan capital of Montevideo across the River Plate. With this in mind, in December 1870 the President of the Junta de Sanidad for Buenos Aires, Col. J.M. Bustillo, wrote to his counterpart in Montevideo proposing a uniform set of arrangements which would disadvantage neither port. The reply from Montevideo was positive and it was proposed that members of both sanitary juntas should meet to work out the arrangements in detail. A quarantine of twelve days was agreed for all vessels arriving directly from Paraguay.[33] Efforts to coordinate these measures began in late January but broke down some weeks later when the first cases of what appeared to be yellow fever were reported in Buenos Aires. Within a month, the Uruguayan authorities were maintaining what one Montevidean newspaper described as 'strict vigilance' against all vessels sailing from Buenos Aires.[34] In the coming weeks the number of cases reported in Buenos Aires increased rapidly and with them the number of fatalities. By 26 March, 100 deaths were being reported every day and, in the same week, the Uruguayan government announced the closure of its ports.[35] The newspaper *La Democracia* justified the action on the grounds that nothing less than drastic measures would satisfy the terrified citizens of Montevideo.[36]

In Buenos Aires itself, the authorities took equally robust action, including the quarantining of afflicted families in their houses – a decision condemned by some as inhumane. Belatedly, too, attempts were made to improve the sanitary state of the city, which had deteriorated badly in recent years.[37] But these efforts did nothing to abate the pestilence. In mid-May the *Buenos Aires Standard* reported that nearly 4,000 people had died from yellow fever in the previous week, with an estimated 7,000–10,000 sick among the 30,000 or so inhabitants remaining in the city. By this time, most of those capable of doing so had fled. As a result of this and the embargo placed upon Argentine ships by Uruguay, business had been 'entirely suspended'. The paralysis of commercial life was such that the government was forced to declare a public holiday for the rest of the month.[38]

Most observers agreed that yellow fever had been imported into Buenos Aires and the disease was often referred to in Argentina and Uruguay as a 'plague' or 'contagion'.[39] Further afield, countries like Italy which had a close connection with Argentina also took precautions in the form of quarantine.[40] Buenos Aires had a large population of Italian origin and the Italian authorities feared that many would attempt to return to their native land. But there was also a widespread belief that the city had squandered its natural advantages. Once famed for its invigorating atmosphere, Buenos Aires had descended the sanitary ladder owing to a 'reckless and obstinate disregard of the commonest rules of hygiene'. The city stank and its soil and water were said to be so polluted as to be beyond salvation. Nothing less than the wholesale clearance and relocation of the population seemed capable of restoring its reputation as the healthiest of cities.[41]

The causes of yellow fever had been hotly debated for more than a century but hitherto the debate had revolved principally around whether yellow fever was contagious or predominantly a climatic disease. There was still a strong conviction that it was a tropical malady but poor drainage and sanitation were increasingly implicated in its propagation. Hot weather partly explained the normal distribution of the disease, but many claimed that Caribbean and Latin American ports were being more regularly infected because of poor governance, which had allowed sanitary conditions to deteriorate. In Brazil and the Caribbean, it was alleged, deep layers of filth had accumulated over many years with the arrival of human cargoes of slaves from Africa.[42] The lingering taint of slavery, not to mention the economic disruption caused by yellow fever, made its removal a matter of urgency. Thus, in 1873, when yellow fever broke out in Rio de Janeiro, the city's Imperial Academy of Medicine called upon government to adopt permanent measures for sanitary improvement. It demanded better drainage and the supply of clean water to every house; the government was also asked to extend the operations of the City Improvement Company to include certain densely populated suburbs.[43] The influential medical school based in Salvador, Bahia, in the tropical north-east of Brazil, had reached a similar conclusion. The 'tropicalistas' as they came to be known, were reclassifying many diseases formerly regarded as tropical; they attempted to show that yellow fever was not the inevitable product of hot climates but could be prevented by sanitary reform.[44] In 1873 their journal, the *Gazeta Medica da Bahia*, recommended a combination of measures to prevent disease spreading from ships in Brazilian harbours and to improve the hygiene of ports.[45]

Keeping Brazilian harbours free from yellow fever was widely seen as vital to the country's economy. In the 1870s, trade with America and Europe was booming. The country's main port, Rio de Janeiro, was exporting increasing quantities of cotton, precious metals, meat, coffee and many other commodities. The United States was the chief importer of Brazilian coffee, followed by Britain and France, whereas Britain was pre-eminent as the destination for most other exports.[46] Recognition of the growing value of this export trade to the economy led to an effort to improve sanitary conditions in cities such as Rio, which was now regarded internationally as the central hub for the dissemination of yellow fever.[47] Sanitary reform in the city had been paralysed by controversies over proposed improvements to its water supply and drainage,[48] but in the 1870s there was greater determination to press ahead, as well as to clear away some of the shanty towns in which the worst sanitary conditions existed.[49] The stimulus to these reforms was the eruption of yellow fever in the middle of the decade. An epidemic in 1873 left 3,659 dead and in September 1874 there were further serious outbreaks in Rio, Fortaleza, Pernambuco and North Paraiba. Vessels from these ports were subjected to strict quarantine on their arrival in many countries which traded regularly with Brazil.[50] The following year, the Netherlands and other European countries took similar action when another serious epidemic was reported at Rio,[51] and when the disease returned to the city in 1876, many foreign vessels fled the harbour, some without their cargo.[52] Although Rio's sanitary authorities were determined to tackle yellow fever, the rest of the world was not easily convinced. Thus, when the *Gazeta Medica da Bahia* reported that the government had resolved to open infirmaries in the afflicted localities, there was astonishment that foci of infection in the city were being deliberately multiplied. Moreover, there appeared to be little evidence of progress in the drainage and ventilation of the most crowded parts of the city.[53]

Over the next few years, yellow fever was reported regularly in South America and the Caribbean, as well as in some of the southern ports of the United States.[54] Alarm was also created in Britain when merchant ships sailing from Rio and the West Indian islands arrived with cases of yellow fever on board. The authorities responded by placing the sick crew members in isolation, in one case in a workhouse infirmary from which the inmates had been temporarily evacuated.[55] Passenger steamers moving to and fro across the Atlantic were another cause for concern. Most passengers who became ill were detected before they reached Britain and placed in isolation when they arrived but a few managed to pass unnoticed. One man who had fallen ill on a voyage from the West Indies came ashore, later to die in London's fashionable district of Belgravia. The fever expert Dr Charles Murchison examined him and confirmed that he had died from yellow fever. Yet his death did not cause alarm because no further cases were reported and the unusually cold April weather dampened fears that the disease had spread.[56] Apart from these isolated cases, no more outbreaks of yellow fever were reported in European ports during the 1870s. Even in the American South, where yellow fever erupted periodically, epidemics were relatively small and confined to port cities in Georgia and Florida. Although these outbreaks affected hundreds of people, they were nothing by comparison with the calamity that would soon befall the southern states.

In May 1878 yellow fever was reported in New Orleans, having apparently spread from Havana; from New Orleans, the disease travelled quickly up the Mississippi Valley devastating the states of Louisiana, Mississippi and Tennessee. By the time the epidemic had subsided in October 1878, over 20,000 people had died. This was the first serious outbreak of yellow fever in the USA since the 1850s and in terms of the sheer number of deaths, the worst in the nation's history. The epidemic was part of a wave which began in the late 1860s, radiating out of Rio in the first instance, with ports such as Havana and New Orleans later acting as regional hubs.[57] The mounting frequency and intensity of epidemics in many port cities during the 1860s and '70s may have been due partly to climatic factors, as the mosquito vector of yellow fever is very sensitive to changes in temperature. However, the rapid dissemination of the disease was facilitated by the massive growth in trade which the region was experiencing following the development of economies in the USA, Cuba (then a Spanish colony) and the Latin American republics. Steam navigation also made it harder to scrutinize passengers and crew for cases of infectious disease, as the shorter journey times allowed many to land before they had developed symptoms. The incubation period of yellow fever is between three and six days, only slightly less time than it then took the fastest vessels to sail the Atlantic. Steamships sailing from many Caribbean ports to the Gulf Coast of the USA, and even to many of its Atlantic ports, could also arrive in less than a week.

In many of these vulnerable cities, sanitary conditions and urban infrastructure had failed to keep pace with rapid economic development. Most were poorly drained and full of uncovered pools in which the mosquito vector could easily breed. This was especially true of the poorest quarters, which were often situated near badly drained land adjacent to docks and industrial facilities. Sanitary reformers were therefore correct in linking yellow fever with poor governance, but their understanding of the problem was quite different than it is today. A commission of inquiry which reported on the yellow fever epidemic in Demerara in 1866, for instance, concluded that it was probably a faecal disease, like cholera or typhoid, capable of occurring even at high altitudes and in cool climates.[58] In his report on an epidemic afflicting Savannah in 1876, Alfred Woodhull also blamed the outbreak on disregard of sanitation. At first, he had been inclined to the belief that the disease had been imported into Savannah on Spanish vessels from Cuba, which had been carrying ballast used to build railway embankments. But he later concluded that there was no evidence of importation and, even if there had been, no system of quarantine could have prevented it. The most likely source of the epidemic, in his view, was an open sewer which went under the deceptive name of the Bilbo Canal. His report, which was submitted to the United States Surgeon-General, revived discussion of the causes of yellow fever and led many to speculate that the disease might not necessarily be of exotic origin. It seemed increasingly likely that yellow fever could flourish only in insanitary conditions.[59]

By the time yellow fever struck the American South in 1878, the first steps had barely been taken to address what was already recognized as a major public health problem. Inspired by the formation of the Metropolitan Board of Health in New York City and the propaganda of the American Public Health Association, many states had begun to create boards of health. Before the disease arrived in New Orleans, the states of Virginia, Georgia, Mississippi and Tennessee all possessed sanitary authorities but they had achieved little due to lack of funds. As well as a shortage money, the boards had no power beyond the capacity to advise the state on what action to take.[60] The epidemic of 1878 exposed these inadequacies and demonstrated the need for state boards to inform each other about the progress of disease. Just as steam navigation had carried yellow fever by river and sea, the growth of the railroad network ensured that it would become widely diffused inland. This probably accounts for the unusual severity of the epidemic of 1878 and the unprecedented alarm and unrest it created. People gathered covertly, attacking trains and railroads and turning away strangers fleeing from the disease; to prevent these attacks, state governments were forced to erect quarantines at their borders, disrupting rail travel and bringing business to a standstill.[61]

The epidemic of 1878 left southern public health officials briefly united in their desire for reform. Public concern also placed Congress under pressure to introduce a national quarantine system, but this was successfully opposed by certain commercial interests which advanced a states' rights defence.[62] Although the bill for a federal quarantine system was rejected, a weaker piece of legislation created a National Board of Health in March 1879. This short-lived body was supposed to coordinate and assist state boards of health but it also attempted to build a system of medical inspection and disinfection for suspect cargoes. The National Board thus helped several states to deal with further outbreaks of yellow fever but it sometimes encountered stubborn resistance. In Louisiana, the President of the State Board of Health feared that outside interference would undermine his authority and imperil commerce.[63]

As memories of 1878 faded, the fragile consensus on public health reform began to disintegrate and most states returned to their insular and antagonistic ways. And yet, there was one lasting change that proved highly beneficial to sanitary reform at local level: the growing involvement of the business community in public health. Hitherto, business interests had stood aside from such matters and had even blocked sanitary initiatives but the epidemic of 1878 persuaded many that it was in their interest to take an active role in reform. The inland trade had been brought to a halt by the disruption of the railroads and many countries, alarmed by the severity of the 1878 epidemic, had imposed restrictions on the importation of American goods. Britain, for example, used the provisions of its 1825 Quarantine Act to place vessels from southern ports under quarantine and to destroy or disinfect suspect items.[64] This response appears to have been intended largely to reassure the population of port cities, although it was condemned by some in the medical profession as an unnecessary and retrograde step.[65] The justifiability of these actions may have been debatable from a medical point of view but they made a profound impression upon businessmen in America. They began to recognize that the poor sanitary condition of many southern cities had given them a bad name overseas, as well as in the northern states of their own country. New Orleans, for example, had lost an estimated $10 million in trade because of the epidemic of 1878.[66] If poor sanitary conditions were responsible for yellow fever, as many in the medical profession evidently believed, then such losses might be prevented by depriving the disease of the conditions it needed to propagate.

Surprisingly, perhaps, some businessmen also began to favour the idea of a national quarantine system. As a rule, commercial enterprises tended to prefer quarantines that were under the control of local administrations, which could be more easily manipulated than national governments. But in many southern ports quarantine had operated in such a relaxed manner as to provoke retaliatory action. Port cities in the southern states had also imposed quarantines for purely economic reasons, hoping to protect their own businesses and to gain from an increase in revenues through quarantine charges. It was to put an end to these 'commercial quarantines' that large corporations demanded a rational system, ideally organized at national level without favour to any particular place or interest. The railroad companies were at the forefront of this campaign, having been most severely affected by the chaos of 1878, but they received support from many cotton merchants, import and export houses and navigation companies.[67] Nevertheless, there were still businessmen and officials who feared that a national quarantine system would act in the interests of northern states. The Louisiana Board of Health complained that the National Board was dominated by the 'powerful railroad lobby of Eastern capitalists' who used quarantine to close southern ports so that trade would be channelled through Atlantic cities connected by rail to the interior.[68] Critics of quarantine claimed that other advanced nations like Britain had lost confidence in quarantine and were relying increasingly on sanitary measures instead. In their view, the cleansing of cities rendered quarantine redundant.[69] However, the fears aroused by the creation of the National Board of Health proved groundless. Starved of funds, it was too weak to intervene effectively in the affairs of individual states and was dissolved in 1883.[70]

In the absence of an effective national system of quarantine, it is instructive to see how American ports attempted to balance commercial interests with their legal responsibility to protect public health. Although many physicians in southern ports remained implacable in their hostility to quarantine, most medical authorities believed that such restrictions – in conjunction with sanitary measures such as disinfection of suspect cargoes – offered the best means of preventing yellow fever.[71] The majority of American ports acted accordingly, ensuring that harbours were kept clean but also that quarantine and disinfection of suspect vessels were utilized against ships sailing from ports likely to be infected. Let us consider for a moment the case of Baltimore, one of America's largest Atlantic ports. Sitting in an advantageous position on the Chesapeake Bay, Baltimore is a gateway to both the South and the Midwest and a natural point for the redistribution of many goods. By the mid nineteenth century it had also become a major centre for canning and packaging and refining, including the processing of seabird droppings (guano) from Peru, one of several commodities then linked to yellow fever.[72] Baltimore had formerly been noted for its liberal attitude to quarantine; a stance which had led to its being regarded as a pariah or a paragon, depending on one's point of view. But at a time when quarantine and medical inspection were high on the international agenda, it was no longer possible for any port to act as a maverick and continue to thrive economically.

Baltimore was also an important node in a network of transatlantic trade. Steamships regularly arrived there from Havana, Martinique, Cartagena, Rio de Janeiro, VeraCruz, Puerto Rico and other South American and Caribbean ports which were often infected with yellow fever. At the same time, Baltimore enjoyed frequent commerce with European cities including Liverpool, Swansea, Antwerp, Amsterdam, Bremen, Marseilles, Gibraltar and Lisbon. Some of these ports had suffered from yellow fever in the past and all were nervous in the wake of the terrible epidemic of 1878. In addition, there was a great deal of coastal navigation, with schooners and brigs plying their trade between Baltimore and other eastern ports. The city's harbour authorities had to strike a delicate balance between free trade, public health and the reassurance of trading partners.

In the 1880s this appears to have been achieved with little disruption to commerce, using medical inspection of vessels from suspected ports only. There are no indications that Baltimore fell foul of other ports for its lax sanitary arrangements and yet remarkably few vessels were subjected to quarantine or disinfection. Figures for 1883 and 1884, which seem to be typical years in terms of volume of shipping and the prevalence of infectious diseases, show that a total of only forty-eight steam vessels were subjected to quarantine and/or fumigation. Cases of smallpox, measles and diphtheria, which occurred all year round, were removed from ships to isolation facilities but the vessels carrying them were not placed in quarantine. Disruption to trade from these measures was largely confined to those times of the year when yellow fever was either present in ports further to the south or considered capable of spreading in North America. These months were by far the busiest commercially and the potential of yellow fever to disrupt the vitality of the port remained. And yet, in 1883, just 6 per cent of steam vessels inspected by the port's officials were subjected to sanitary measures and only 12 per cent of those from ports likely to be infected with yellow fever.[73]

In the peak months for quarantine in 1883, which coincided with the maximum risk from yellow fever, the longest duration of quarantine was five days, imposed on a vessel from Cienfuegos in Cuba. Most other ships placed in quarantine were from the same port, the rest coming from other Caribbean ports such Guantanamo (Cuba) and Veracruz (Mexico). In these cases, the duration of quarantine ranged from one to two days and some vessels from infected ports escaped quarantine altogether because they had performed lengthy quarantines in ports en route to Baltimore. In 1884, when a slightly lower volume of shipping passed through the port, the percentage of all ships that were subjected to some form of sanitary measure was a little higher than the previous year, at 10.5 per cent, and nearly 21 per cent for ships reaching Baltimore from ports regularly infected with yellow fever. During the peak yellow fever months, most of the vessels that were fumigated or quarantined came from Havana, Veracruz and Cartagena (Colombia). Nevertheless, only one vessel was detained for longer than forty-eight hours.

Taking both years together, it is clear that disturbance to trade in Baltimore was minimal but the system operating there and in other American ports was only as good as the intelligence they received about the prevalence of disease. By the late 1870s, some writers were calling for the establishment of an international commission that would make sanitary surveillance more reliable.[74] This was the main reason why the President of the United States, the Republican Rutherford B. Hayes, was so determined to convene an international sanitary conference – the first to be held on that side of the Atlantic. As the State Department put it:

The extensive prevalence of yellow fever in certain parts of this country during the past two years, and the almost continual existence of the danger of the introduction of such contagious and infectious diseases as yellow fever and cholera, by vessels coming to this country from infected ports abroad, gave rise to ... legislative measures [in the USA] but the difficulty in their application has been chiefly owing to the fact that in certain foreign ports where infectious or contagious diseases have existed ... the local authorities have shewn some hesitation as to cooperating with the Consular and medical officers of the United States in carrying out regulations deemed essential by this government as a sanitary safeguard.[75]

The US government had little hope of creating an international code but it wished to go further in this direction by framing a common system for the notification of communicable diseases.[76] In January 1881 the conference desired by the US government was convened in Washington, DC, in the Reception Hall of the State Department. By this time Hayes had left office and another Republican, James Garfield, had begun his short presidency. Unlike some of the earlier conferences, this one had more claim to be a global affair, with delegates from Central and South America and Asia sitting alongside those from Europe, who had formerly dominated. Most delegates paid lip service to the desirability of international notification but they found it hard to agree on how it should be implemented. A committee established by the conference to consider the matter put forward the radical proposition that an international sanitary committee should be established on a permanent basis to oversee a cadre of physicians with expertise in matters of sanitation and epidemiology. These were to be appointed to ports and other major cities around the world where they would act as independent sanitary inspectors. They would not only inspect ships before they left ports but also ascertain the sanitary state of each locality. In addition, all signatory nations would be expected to issue a weekly bulletin reporting the prevalence of disease in their countries. This was not to the taste of all those attending, some of whom preferred the scheme advocated by the Mexican delegate, Dr Ignacio Alvarado, who proposed that each country appoint a resident physician to carry out inspections alongside a physician chosen by the host nation. In the event of a disagreement, an 'umpire' would be selected from the physicians appointed by other foreign powers. These physicians would comprise an international sanitary board in all major ports, presided over by the highest civil authority. If the Board could not agree on the appointment of an umpire, then it would decide disputed questions by a majority vote. One obvious attraction of this scheme was that local administrations retained rather more power than if a truly independent, permanent international body were established.[77]

The failure of the conference to agree on a binding system of disease notification is hardly surprising in view of the different interests of the countries concerned. Even in the USA, port authorities continued to act independently and, in some cases, irresponsibly when it came to both disease reporting and quarantine procedures. The relative scarcity of yellow fever during the 1880s meant that disputes between state boards of health were rare, but in 1888 the disease broke out again in Florida. When yellow fever was later reported in Mississippi and Alabama, there were fears that another region-wide epidemic might be imminent. But, by this time, a new national body had taken charge of quarantine, picking up some of the responsibilities of the old National Board. The Marine Hospital Service (MHS) came to the aid of Florida and gave the state authorities valuable advice about how to deal with the epidemic. However, the MHS lacked the power to coordinate sanitary measures and some towns and cities imposed quarantine against each other on the basis of rumour rather than accurate information. This panic caused enormous disruption to commerce and led, in the following year, to a conference in Montgomery, Alabama, which drew delegates from most of the southern states. Its purpose was to establish regional regulations that would be adhered to by all signatory states. An agreement was reached and remained in force for the next decade, helping to resolve periodic disputes between sanitary boards about the terms of quarantine imposed against each other's states.[78]

The fact that the agreement appeared to work meant that there was little desire for a new national quarantine system. In any case, the MHS was progressively taking over more of the work than would have been expected of such a body. In 1890 it was given powers to supervise new regulations which governed the transmission of disease between states. Wilful carriage of diseases such as yellow fever and cholera across state borders was made a criminal offence.[79] Three years later, the MHS was granted full authority over quarantine in the USA. Although states still had their own quarantine rules, the Surgeon-General of the MHS now had the power to intervene and formulate additional regulations if necessary.[80] For many states, this was as far as they wished to go in the direction of centralization and a bill from Senator Caffrey which proposed to grant additional powers to the MHS in 1898 was criticized on the basis of states' rights and of the argument that local officials knew best. One witness to the House of Representatives committee which considered the matter objected that 'there are differences, such as climatic conditions and differences in shipping, and things of that character, which require entirely different treatment; and I think it is fully appreciated that that is better known to the native physicians . . . than to a Federal official'.[81] He added that the bill was opposed by the American Public Health Association and the New York Academy of Medicine for the same reasons. These bodies supported an alternative bill drafted by a Republican senator, John C. Spooner,[82] which proposed an advisory board consisting of delegates from each state. The general feeling of the southern states was that they already had an effective system of quarantine and did not require national regulation. Indeed, delegates from across the South had recently met at Mobile, Alabama and had produced a convention that was binding on all the South Atlantic and Gulf states.[83] The Health Officer for Florida was even more forthright and denounced the Caffrey bill as 'an invasion of the reserved rights of the people of each State of the Union'. He warned of an 'army of [federal] sanitarians that, like the locusts in the field, eat up our substance and usurp our liberties'.[84]

Speaking in favour of the Caffrey bill, other doctors claimed that their profession was broadly behind the proposals and that there was public support for a 'national head of sanitation'. It is not clear how they reached this conclusion but they would have been aware of growing anxiety over immigration and disease. The supporters of quarantine measures could always rely on 'nativist' sentiments to bolster their public appeal.[85] Walter Wyman, Surgeon-General of the MHS added that the alternative scheme proposed by Spooner would be likely to produce a great deal of resentment, as any advisory council would be dominated by interior states, which outnumbered coastal ones. The latter might find rules drafted by the former oppressive. In his view, the Treasury Department ought to have jurisdiction over quarantine as it had oversight of all other matters relating to commerce.[86] Wyman also did his best to whip up support for the

Caffrey bill in the lectures he was invited to deliver around the country. In an address to the Commercial Club of Cincinnati in October 1898, he spoke of the temptation for local officials to yield to commercial interests and argued that such vacillation was normally counterproductive, the consequent invasion of disease and measures imposed by other ports bringing about a 'crippling of commerce'. The only way to guard against this, he argued, was to have quarantine under federal control, where it would be less amenable to vested interests.[87]

The reaction to the Caffrey bill displayed two very different views about how best to regulate commerce in the interests of public health. Although the debate had certain features which were unique to the political ecology of the United States, the same basic divergence was evident in discussions of sanitary policy around the world. Essentially, the difference was one of perspective. National governments and their officials were mindful of commercial considerations but they had to look at the bigger picture. The interests of mercantile and shipping groups and those of manufacturing industries and agriculture were not always identical. As had evidently been the case in the American South, they also divided locally, using quarantine to gain an unfair advantage over rivals. Such disruption might temporarily advantage one group of businessmen but the retaliation it engendered was detrimental to the economic vitality of the nation as a whole. Added to this, national governments had to consider commerce within the larger framework of international relations. The views of central government officials thus tended to be better informed, more farsighted and diplomatically nuanced than those of local administrations. While businessmen increasingly saw an advantage in standardizing arrangements to prevent unnecessary disruption, unless they were international concerns their perspective tended to be more limited. In the USA, long-standing suspicion of northern motives and of federal authority served to entrench opposition to Treasury control and to limit the desire for sanitary federation. But the Surgeon-General took a broader, long-term view of the question of quarantine and Wyman reminded the meeting in Cincinnati that there had been only nine years in the present century when the United States had been free from yellow fever and that on many occasions it had caused almost complete suspension of traffic. To avoid this situation in future, he thought it was necessary not simply for the states of the Union to be brought under a federal body but for the

American government to take a more interventionist role in the sanitary affairs of neighbouring states.

By way of example, Wyman referred to Cuba, which had become a de facto colony following America's victory in the 1898 war against Spain. Indeed, the desire to control yellow fever may have been one of the reasons the United States chose to intervene in Cuba's conflict with its former ruler.[88] Formal complaints had been made to the Spanish government about the sanitary condition of Havana since 1896 but to no effect. The outbreak of war in 1898 provided a pretext for more direct intervention. Wyman confidently asserted that, 'with American predominance in the Island of Cuba there will be little or no difficulty, even before the ports and cities are placed in a better sanitary condition, of this Government exercising a surveillance over vessels leaving for the coast of the United States proper'. America's growing interest in the region, he argued, would awaken 'an international sentiment' that would lead to the removal of the presumed sources of yellow fever throughout the Caribbean. The implication was that if they could not be shamed into action, foreign governments could be forced into it – a point hammered home by Wyman at a meeting of the Pan-American Medical Association in Mexico.[89] Although he does not seem to have contemplated further invasions in the name of public health, diplomatic pressure and the threat of force were clearly on the cards.

And yet the first American attempts at sanitary intervention in Cuba were less successful than Wyman had hoped. It was only after the US Yellow Fever Commission set about proving the mosquito-transmission theory of the Cuban doctor Carlos Finlay that an effective programme of measures was devised. By concentrating efforts on mosquito destruction, the Commission showed by 1901 that the disease could be eradicated, establishing an entirely new paradigm for the control of yellow fever. The Cubans rightly felt that they were never given the credit they deserved, either for Finlay's discovery or for their collaboration with the Americans in removing the disease. And, although the American occupation of Cuba came to an end with the disappearance of yellow fever, the sanitary state of the island remained a bone of contention. The re-emergence of yellow fever in epidemic form in 1906 prompted another American invasion and, from that point on, it was evident that Cuban independence was dependent on keeping the disease at bay.[90] Other Latin American countries took note.

The sanitary corollary

As might be expected, the spectre of yellow fever loomed over the proposal to unite the Atlantic and Pacific Oceans by a canal through the Isthmus of Panama. Work on this project began in 1879, soon after the formation of the French Panama Canal Company under Ferdinand de Lesseps. The original plan was to dig a canal connecting the two oceans at sea level but, by 1887, it became apparent that this would be impossible and the decision was made to build where necessary over gradients using locks. Construction continued along these lines until 1889 when the Company went bankrupt and was placed into receivership. A new company was formed in 1894 and continued working until 1902 when it agreed to sell to the United States. American interest in the project had been growing for some time as the USA extended its influence in Latin America. But following the Spanish-American War it became an article of faith that the canal should be built and placed under American control. Apart from its strategic uses, it was envisaged that the canal would do much to stimulate American industry, particularly in the Pacific states. This was the genesis of the Isthmian Canal Commission (ICC), which was created by Congress in 1899. The ICC soon began negotiations with Colombia, which then controlled the isthmus, with a view to gaining permanent control of a strip of land along either side of the canal. Matters were complicated by Panama's declaration of independence from Colombia in November 1903 but, within a month, the new Panamanian republic granted the USA the use, occupation and control of a strip of land ten miles wide along the canal, which was to pass from Limon Bay in the west to Colon in the east.[91]

Work on the canal then began in earnest in 1906, when construction passed into the hands of the US Corps of Engineers. This led to the Canal Commission being reorganized with Col. G.W. Goethals of the Corps of Engineers at its head. In 1908 President Roosevelt abolished the Commission and gave Goethals full executive authority. The next three years were devoted to preparing the engineering and construction works, during which time Col. William Gorgas (a veteran of the Cuban campaign) and a team from the US Sanitary Service set about the cleansing of the Canal Zone with a view to destroying mosquitoes likely to convey diseases such as malaria and yellow fever. In what subsequently came to be celebrated as one of the greatest sanitary achievements in history, yellow fever was eradicated from the Canal Zone.[92] At the same time, the railways necessary to supply the works were modernized, a workforce was assembled, and accommodation with food and clean water supplies was arranged. In another four years, construction work on the canal was complete and it opened to the first steamer in August 1914.

But concern over the sanitary implications of joining the two oceans was raised long before the canal was opened. As early as 1893, the US Quarantine Act stated that medical officers of the Public Health (PH) and Marine Hospital Service should be dispatched to the office of the American consul at Panama City on the Atlantic side of the country and to the consul in Colon on the Pacific. Over the coming years, the US government came to believe that the Panamanian authorities were not exercising due diligence when it came to preventing the infection of their ports by vessels from South America and began to lobby for control over quarantine in the country's ports. This was granted and the quarantine duties of the Panamanian authorities were turned over to officers of the US PH and MHS attached to the two consulates. When the ICC took over, it assumed responsibility for quarantine and requested that the medical officers be turned over to the Commission.[93] In the years before the canal was completed and before the zone around it had been cleared of yellow fever, the ICC was very sensitive to accusations that the disease was endemic. It blamed other countries for having lax sanitary precautions which permitted yellow fever to leave their ports and find its way to Panama.

This view of South American ports seems to have been widely held, and in 1903 the medical officer in charge of Angel Island quarantine station near San Francisco wrote to his superiors in the MHS drawing attention to poor sanitary conditions on the west coast of Central and South America. But more important still, he argued, was the connection with Panama, as steamers from the Pacific Steam Navigation and South American Steamship companies arrived weekly at Panama from ports in Chile, Peru, Ecuador and Colombia. Passengers from vessels owned by these companies often transshipped at Panama to those of the Pacific Mail Steamship Company which plied regularly between ports there and the western coast of the USA. There were thus many ways by which yellow fever might spread to the northern Pacific coast.[94] Although the Pacific Steam Navigation Company and the Pacific Mail Steamship Company contributed to the cost of maintaining some quarantine stations used by their ships, these precautions do not appear to have inspired much confidence.[95] Yellow fever was also a pressing concern for US naval officers in the Canal Zone. Rear Admiral Henry Glass singled out ports in Ecuador as the most likely source of yellow fever in Panama and suggested that the ICC send medical officers to every major port in Ecuador with a view to inspecting all out-going passengers for suspected cases of yellow fever. He hoped that such an arrangement would at least provide advance notice of any vessels likely to be carrying cases of the disease.[96] For this, a good telegraphic service was essential and the cables now linking South America to the Canal Zone and the canal to the USA would permit a system of early warning to work effectively.[97] The US envoy in Panama, William Buchanan, agreed and urged the Secretary of State to 'spare no effort to prevent any appearance of contagious diseases here'. He echoed calls for skilled medical inspectors to be dispatched to Peruvian, Venezuelan, Colombian, Ecuadorian and West Indian ports.[98] The Secretary of State concurred and immediately informed Buchanan that the new arrangements had been sanctioned. Another US public health official was to be sent to Colon to assist in the inspection of incoming vessels and others leaving for South American ports. Wisely, the man chosen for Colon was Dr Claude Pierce, who was known to be immune to yellow fever.[99] The government of the new Republic of Panama endorsed these proposals and looked forward to the sanitary assistance of the Americans.[100]

There was now a concerted attempt to bring quarantine arrangements in Panama more firmly under the control of the Treasury Department. The situation in Panama was increasingly anomalous as the MHS extended its control over quarantine in most foreign outposts of the USA, including the Philippines, the Hawaiian Islands and Puerto Rico. As a memorandum from the service observed, 'It only remains for this function to be extended to Panama to make the chain of quarantine authority uniform between the two oceans.'[101] The memorandum also invoked the precedent set in the USA itself, where state after state had turned over its quarantine functions to the Treasury Department, leaving only two or three operating independently. Treasury officials declared the measure to be in the interests of both the Canal Zone and the USA. One possible advantage of a consolidated system under the Chief Quarantine Officer at Panama was that officers of the MHS in South American ports would be in more regular contact with officials in the Canal Zone.[102] However, other government departments, most notably the War Department, remained to be convinced and could see no particular advantage in a single authority. It seemed inconceivable that quarantine in Panama would be allowed to weaken, for shipping from the Canal Zone would face severe restrictions in American ports.[103]

Despite these objections, the Treasury's case was supported and the administration of quarantine in Panama was placed under its direct control. In 1913 an executive order was issued to bring this about and its provisions came into effect when the canal opened the following year. Among other things, the order dispensed with the need for ships touching at Panama or passing through the canal to obtain an additional bill of health on top of that issued from the port of departure. This measure would speed up the transit of goods and passengers to the USA, while apparently maintaining effective safeguards for public health. All ships passing through the canal and those docking at Panamanian ports would be subject to quarantine if any cases of yellow fever or other infectious diseases were found. Quarantine officials there were empowered to take whatever measures they believed necessary to deal with ships and their cargoes.[104] By 1918, they were inspecting over 700 vessels each quarter but in the first three months of that year just eighteen of the vessels inspected were detained, with twenty-eight being permitted to transit the canal in quarantine.[105]

The frequency and duration of quarantine were of particular concern to enterprises such as the United Fruit Company of Boston, which had plantations in many Caribbean and South American countries. Delays caused by quarantine or rough handling during inspection and fumigation could easily damage fruits or cause them to rot.[106] The fortunes of such companies depended on minimal disruption to commerce and this in turn depended on the consolidated system of disease notification which linked the Canal Zone to South American ports. The volume of traffic – which amounted to several thousand large vessels every year – was such that it would have been difficult to manage quarantine for a high proportion of vessels. Normally, the Chief Quarantine Officer of the canal could rely on reasonably accurate information about the state of a ship's health reaching him in good time. Telegraphic messages enabled officials to single out suspect vessels and thereby save time as well as reducing the chances of evasion. But in years when yellow fever was prevalent, the expected benefits from this system failed to materialize. The canal's quarantine authorities were swamped with infected and suspect vessels, causing considerable delays and commercial losses.[107] The USA therefore redoubled its efforts to improve notification of disease from foreign ports.

The desire for more accurate intelligence about disease was also the main reason why the USA was keen to convene the first specifically Pan American sanitary conference in 1902. This conference and those that came after it were part of a wider Pan American movement driven by the USA, which aimed to embrace all the independent nations of the western hemisphere. The movement had several objectives, not the least of which was preventing war between nations, but it also strove for greater standardization with regard to transport, commerce, banking and public health. The conference of 1902 established the world's first permanent international sanitary organization, the Pan American Bureau of Health, the main aim of which was to improve information about disease and sanitary regulations in the American republics. In 1905, a second conference drew up a sanitary convention which was signed by the USA, Honduras, Guatemala, Ecuador, Costa Rica, Peru and Salvador; it was a slightly amended version of a convention which had recently been signed at Paris by the European powers. The main purpose of the Washington Convention was to reduce quarantine and unnecessary interference with commerce. This objective was made easier by the growing consensus that yellow fever could be spread only by mosquitoes and not by person-to-person contact. Although vessels from suspect ports still needed to be fumigated to kill infected mosquitoes, the lengthy delays formerly caused by quarantine could be eased if accurate information was available.

The third International Conference of American States, held at Rio de Janeiro in 1906, ratified the Washington Convention and codified existing measures to prevent cholera, plague and yellow fever. Each signatory nation now undertook immediately to notify others of the appearance of authenticated cases of any of these diseases.[108] Subsequent conferences brought further standardization regarding such measures as the fumigation of vessels and cargoes harbouring yellow-fever mosquitoes.[109] However, some states were initially reluctant to cede their authority to a body which was dominated by the USA, and Brazil did not even send a delegate to the first two conferences. Sanitary measures in countries such as Argentina, Cuba and Costa Rica also remained at odds with the Washington Convention.[110]

Whilst most Latin American states wanted the benefits of a more uniform sanitary system, some had misgivings about the growth of American power in the region and to them the terms of the Washington Convention and its successors smacked of sanitary imperialism. Growing US interest in the sanitary affairs of other American republics was one component of the more assertive foreign policy pursued by President Theodore Roosevelt. In 1904 Roosevelt radically revised the Monroe Doctrine which had formed the backbone of US policy towards Latin America since 1823. The Doctrine originally sought to prevent European interference in Latin America, which was perceived as a direct threat to US interests and tantamount to a declaration of war. Effectively, the Doctrine sought to divide the Atlantic into American and European spheres of influence, with the USA undertaking to refrain from involvement in European politics. But the Roosevelt Corollary, as it became known, amended the Doctrine in order to justify US intervention in any Latin American country in cases of 'flagrant and chronic wrongdoing'. This principle seems to have embraced sanitary matters, too, and in the coming years it became common for American personnel to be dispatched to Latin American ports in order to ensure adequate sanitary standards on ships bound for the USA. Coupled with better intelligence, such actions permitted the USA and its quarantine authorities in the Panama Canal to operate a more liberal regime than might otherwise have been possible. American diplomatic pressure was therefore exerted in two different directions. On the one hand, it was necessary to ensure that adequate sanitary precautions existed in countries that traded regularly with the USA. On the other, countries had to be deterred from imposing sanitary regulations which threatened the fragile consensus that had emerged from the international conferences of the early 1900s.

The problem faced by the USA was in some ways analogous to that which confronted the European powers in Middle Eastern states such as Egypt and Ottoman Turkey. The international boards of health established in Alexandria and Constantinople provided a sanitary safeguard for the West while ensuring that commerce and imperial communications were not unduly damaged. The situation in the western Atlantic differed because of the dominance of the USA. This meant that the Pan American sanitary conferences were dictated largely by the interests of a single state, rather than providing a theatre for imperial rivalry, like the sanitary conferences convened in Europe. Throughout the region, American dominance had become synonymous with the interests of large commercial enterprises and Cuba's Director of Health, Dr Juan Guiteras, lamented that, 'In my rather long experience with serious epidemics I have never once failed to find important business interests enlisted against us.'[111] He had long been a critic of American sanitary policy in the region and disliked the fact that the USA dictated policies to countries such as Cuba. Guiteras had also served on the committee of the Pan American Sanitary Bureau, which was under the presidency of the USA.[112]

But the pressure for a more liberal sanitary regime did not emanate from the USA alone. Many Latin American countries were undergoing a process of modernization in which commercial elites – sometimes with the help of the military – were bent on the consolidation of land holdings and were removing impediments which hindered their growth as export economies.[113] Their aims in sanitary policy therefore harmonized with those of the US government. In 1918, for instance, quarantine and fumigation procedures on both sides of the Mexican border were relaxed considerably following complaints from Mexican as well as American chambers of commerce. Although the US Public Health Service was not prepared to countenance the removal of all such precautions, those deemed 'illogical and unnecessary' were abolished.[114]

The same considerations bore heavily on the sanitary establishment of the Panama Canal. It was vital that measures in force in the Canal Zone were neither weaker nor stricter than those of its commercial rival at Suez, as exporters in some countries had the option of sending goods to the United States via either canal. If shipping lines faced lengthy delays caused by quarantine, they would be factored into their costs. Conversely, if the quarantine arrangements in either canal were seen to be lax, additional quarantines would probably be imposed at ports of destination. As the Chief Quarantine Officer of the Panama Canal made clear, arrangements for quarantine and fumigation had to be finely calibrated to balance these conflicting interests:

> It is important that everything pertaining to shipping going through the Panama Canal be placed on the highest state of efficiency, as the Panama Canal must compete with the Suez Canal for trade, and any deficiencies in quarantine – not only at the Canal but at any ports where vessels using the Canal may touch, will count against this route.[115]

It was also necessary to allay the anxiety of governments in Asia that yellow fever might spread through the canal to points further east.[116] In 1911, for example, the British Government of India discussed the possibility that yellow fever might be introduced into its territory by trading vessels sailing from endemic areas via the Panama Canal. Alarm had been raised by the 'father' of tropical medicine Sir Patrick Manson in a paper read before the Epidemiological Society of London, and the medical entomologist Sydney Price James of the Indian Medical Service was afterwards dispatched on a fact-finding mission, visiting the Canal Zone and Caribbean ports, as well as Asian ports such as Singapore and Hong Kong. He eventually concluded that India was not directly at risk from yellow fever but that it might become so if Hong Kong or other East Asian ports became infected. In view of this, he argued, it was necessary to have better information about the spread of the disease in the Americas, which would require a British Indian Medical Service officer to be stationed in the endemic zone of Central America and possibly in Hong Kong and Singapore.[117] The Americans also decided to intervene more directly in Latin American countries with a view to diminishing the risk of yellow fever spreading to Asia. In July 1914, at the behest of William Gorgas, then Surgeon-General of the US Army, the Rockefeller Foundation set out to eliminate yellow fever throughout the Americas, its first target being Guayaquil, the principal port of Ecuador and the only place along the Pacific coast where yellow fever was endemic.[118]

However, it would be some time before such efforts were able to make an impact on the region as a whole. In the next few years, epidemics of yellow fever were reported in Guatemala, Peru, Brazil, Honduras, El Salvador, Nicaragua and Mexico. Although the Rockefeller Foundation extended its operations to these countries,[119] US commerce remained vulnerable to foreign quarantines and embargoes aimed at shipping through the canal. The US government responded by increasing the number of its medical representatives in Central and South American ports, with a view to furnishing more 'accurate and continuous reports of diseases and sanitary conditions so that quarantine could be reduced to a minimum'. It was also hoped that these miniature 'boards of health' would provide advice and assistance to local authorities and shipping interests, so as to get all agencies working together to combat yellow fever.[120] Once these new arrangements came into force certain liberties could be granted to vessels from non-infected ports, and foreign governments began to apply for exemptions. In 1921, for instance, the Foreign Minister of Venezuela enquired whether it would be possible for the sanitary restrictions currently imposed at Colon to be lifted for vessels from his country owing to the fact that yellow fever was not prevalent there. He added that there had never been a single case of any contagious disease discovered at Colon on ships sailing from his country.[121] Fumigation measures were also aimed at suspect vessels only, which cut down massively on delays to shipping. Businesses such as the United Fruit Company began to inform the Chief Quarantine Officer of the canal about their own procedures for fumigation,[122] whilst ships were not fumigated for yellow fever unless it was believed that mosquitoes were on board.[123]

The relaxation of sanitary measures for shipping passing through the Panama Canal showed the confidence which the American authorities now had in their systems of sanitary intelligence. The USA had championed a move from quarantine to surveillance since the early 1880s and had exerted pressure on Latin American countries bilaterally and internationally to achieve these aims, marking an extension of the Monroe Doctrine into the realm of public health. But whatever their reservations initially, most Latin American governments came to see that it was to their advantage to play along, extracting concessions from the USA in return for providing speedy notification of disease and allowing US public health officials to work in their ports. This relationship was not unlike the 'marriage of convenience' which existed between the Rockefeller Foundation and some of the Latin American republics, in which early resistance to cooperation in public health was largely if not always completely overcome.[124] But in many other parts of the world sanitary diplomacy was influenced to a much greater extent by the vagaries of imperial politics. In the western Atlantic the USA was able to establish sole hegemony, but the British did not have things quite so much their own way east of Suez. It is to this troubled region that we now turn.

CHAPTER 6

A stranglehold on the East

In 1865 the world watched with horror as cholera broke out among pilgrims gathered at Mecca. The epidemic killed around 30,000 people – nearly a third of those attending – and formed the nucleus of a pandemic which would soon engulf the world. In Russia, with its large but widely dispersed population, around 90,000 perished; in North America, where the disease affected mostly port cities, the figure was nearer 50,000; but in the Austro-Hungarian Empire, embroiled in war with Prussia, fatalities exceeded 165,000. Although the death toll from cholera appears high by today's standards, in most countries it was generally lower than that for common endemic diseases such as tuberculosis. Indeed, fear of cholera was generally far in excess of the risks which it actually posed. It had such a powerful grip on the imagination – and on the Western imagination especially – because of the horrific and ignominious death which awaited most of its victims. Cholera 'seized' these unfortunates with little warning and after the first 'premonitory' symptoms they found themselves convulsed by abdominal spasms and emptied by profuse diarrhoea. Nervous collapse and death usually followed.[1]

Cholera was shrouded in an aura of sinister mystery. It was widely believed that it had originated in the foetid swamps of deltaic Bengal but beyond that its causes and mode of spread were unknown or, at any rate, hotly debated.[2] Cholera's oriental associations conjured images of darkness and degeneracy. It seemed to thrive in overcrowded and squalid localities, whether in India or, more worryingly, in Europe itself. The cleansing of these places had become a matter of urgency. But the uncertainty surrounding cholera meant that few people were sure that sanitation alone could keep the disease in check. Some medical specialists attributed cholera to noxious vapours from decaying matter, while others thought it arose from a heightened state of electricity in the atmosphere or from the ingestion of minute organisms as yet unknown to science.[3] If the medical profession was unable to agree, then what confidence could the public have in their sanitary safeguards? The question of what caused cholera also had important implications for commerce. The assumption that the disease was contagious necessarily entailed prevention in the form of sanitary cordons and quarantine. As well as impinging on trade, such measures were notoriously difficult to implement and not unlikely to cause civil unrest. But if cholera was not contagious – or only partially so – then such precautions were either redundant or were required only in exceptional circumstances. More attention was thus likely to be given to environmental or hygienic improvements. Neither option was unproblematic for governments, for both were likely to incur substantial costs of a financial and political kind.

In previous waves of cholera, advocates of the contagion theory had seen its spread across vast oceans as evidence of its transmissibility. Most governments concurred and when cholera threatened they ordered restrictions on vessels from infected ports. But in the 1830s, '40s and '50s, there was comparatively little concern with ship-borne infection from India, even though it was generally regarded as the fount of cholera. The same was true of the Mecca pilgrimage. Although Mecca was struck by cholera in 1833, 1836 and 1847,[4] the routes traversed by the disease were circuitous and its passage often slow. In the early 1830s, for example, cholera seemed to reach Europe haltingly through Russia, having first traversed South and Central Asia. Similarly, the International Sanitary Conference of 1851 made no special recommendations to control Indian commercial navigation or the Haj, despite acknowledging that it was a source of cholera outbreaks throughout the Middle East. The 1865 epidemic elicited a very different response because the progress of cholera was much faster than before and because its routes were more clearly defined. The disease leapt from country to country with what one report described as 'unexampled rapidity', its progress accelerated by steamships and railways.[5]

Over the previous three decades, the expansion of steam navigation had made it possible for ships to reach their destination more quickly and more directly than under sail. Between 1830 and 1870, journey times were practically halved on many routes. Steam power also enabled the construction of larger and stronger vessels made of iron, which were capable of carrying more people and thus of spreading more germs. Railways, too, provided new avenues for infection, the most important of which, from an epidemiological point of view, was the Alexandria–Cairo–Suez railway, which connected ports in the Mediterranean to those in the Red Sea. Many of the Indian pilgrims said to have brought cholera to Mecca in 1865 had opted to take the railway in preference to traditional caravans across the desert. The impending completion of the Trans-Caucasian railway also threatened to carry cholera from Persia – where it now occurred frequently – into Eastern Europe.[6]

The arrival of cholera in Mecca thus caused greater concern in 1865 because of the likelihood that it would spread rapidly and undetected from Arabia to points further west. It was this fear which prompted Napoleon III of France to revive the international sanitary conferences which had lain dormant since the early 1850s. Most European countries responded to the call, as did the Ottoman Empire, Persia and Egypt. There was a sense that the world had been transformed by new technologies of communication and that it could only be managed through greater international collaboration.[7] The third international sanitary conference was one facet of this new internationalism. While it might have been an internationalism of convenience, acknowledgement of mutual interdependence resulted in some serviceable agreements, among which were the International Telegraphic Union (1865), the Universal Postal Union (1874), the International Union of Weights and Measures (1875) and the fixing of the prime meridian in Greenwich as the basis for international time zones (1884).[8] In these circumstances, it was not unduly optimistic to try for a similar agreement on sanitary matters.

The proposed venue for the third international sanitary conference was the Ottoman capital of Constantinople, which most governments regarded as appropriate considering that the spotlight was now firmly upon the Orient.[9] By the 1860s, Western countries were starting to reap the benefits of several decades of sanitary reform. Adult mortality rates were beginning to fall and this was widely attributed to environmental improvements and other forms of sanitary protection. However, European observers could see little evidence of sanitary progress in the East and this led many to believe that polities such as the Ottoman Empire were not only staging posts for cholera but breeding grounds for the disease itself. The only way of protecting Europe against the threat of cholera seemed to be to encourage these empires to bolster their sanitary regulations, thereby reducing the necessity of stringent quarantines or embargoes further to the west. For many Western delegates, therefore, the intention was to create in the Middle East a buffer zone to prevent the passage of disease from the Orient. But in an age of competing empires this was easier said than done. A great deal was at stake, particularly for Great Britain, whose maritime commercial interests and connections with India would be disproportionately affected by any attempt to regulate movement within or through the Middle East. In view of this, the technical decisions made at Constantinople and subsequent conferences carried a lot of political weight.

Unfortunately for Britain, most delegates at Constantinople wished to tighten regulations in Egyptian and Turkish ports in order to control pilgrimages and commercial traffic.[10] The risk posed by cholera lurking in merchandise such as cottons or wools, however, was thought to be negligible because only a few commodities were regarded as susceptible to contamination by faeces, now widely regarded as the primary medium for the disease. This new understanding of cholera – which grew gradually out of the epidemiological work of the London physician John Snow in the 1840s and '50s – had profound implications for commerce, and for navigation more generally. Snow had traced the spread of cholera through water which he believed had been contaminated with cholera germs in human faeces. Even though direct human-to-human transmission seemed unlikely, the chance of an infected individual contaminating food or water supplies in a ship or port was very real.[11] However, some aspects of Snow's theory – especially the identity and nature of the causal organism – remained controversial, not to mention its practical implications. It was necessary for the Constantinople delegates to consider these problems in detail and they formed four separate committees, each of which made a report to the plenary session.

Three of these reports had a bearing on matters relating to commerce: the report on 'general preventative measures'; the report on quarantine; and the report on measures to be taken in 'the East', meaning the area from Egypt to India. Most countries attending the conference had delegates on each of the committees but some countries were not represented on all of them. Although there was some disagreement, the general inclination of the delegates was to tighten sanitary measures and lengthen quarantine, albeit in a manner thought to be consistent with the latest scientific knowledge. This meant blocking the ways in which the disease could be transmitted and keeping food and water supplies free from contamination in localities such as ports. The importance of general sanitary conditions was emphasized in the special report on the prevention of cholera, which stressed that the disease had to be tackled at its source. In particular, this meant giving more attention to hygiene and sanitation in ports and on ships. The Government of India was already considering amendments to its Native Passenger Ships Act of 1858 which seemed to meet most of the requirements stipulated by the delegates. However, the conference urged that the Act's provisions be extended to ships flying all flags sailing out of Indian ports, and not just those under a British flag. The conference also stressed that merchant vessels had to be regulated by the governments of each country and that they should award prizes for inventions designed to improve the health of ships.[12]

The majority of conference delegates agreed on the need for quarantine, too; or if they did not, they did not say so. It was acknowledged that quarantine had not always worked in the past – especially the land cordons used to prevent cholera in Russia and Central Europe – but if enforced diligently and systematically, it seemed that it could be effective. It was said that a number of ports in Greece and New York City had managed to intercept cases of cholera on several occasions and had thus often been free from the disease while neighbouring areas were affected. The conference also heard that some eminent merchants now gave their blessing to quarantine based on 'scientific' principles. Delegates resolved that 'Restrictive measures, known before hand and suitably applied, are much less prejudicial to commerce and international relations than the disturbance which affects industry and commercial transactions in consequence of an invasion of cholera.'[13]

Prior to the Constantinople conference, most countries had imposed a maximum quarantine of five days on ships from ports infected with cholera but afterwards this period was to be doubled in some cases. It was now generally agreed that the incubation time of cholera could last up to seven days, so the ten-day period seemed to guard against extraordinary cases. The conference proposed a two-tier system of quarantine for all vessels except pilgrim ships: a quarantine of observation lasting up to ten days, depending on the length of the voyage; and, if necessary, a rigorous quarantine of ten days' duration if cases appeared during observation or in the case of ships with foul bills of health. Under quarantines of observation, ships carrying suspect bills of health would be kept in isolation but none of their crew, passengers or cargo would be disembarked. Under rigorous quarantine, all would be taken from the ship to a lazaretto and the vessel and its contents would be disinfected. Goods would be cleansed by airing, washing with water, or fumigation with chemicals such as chlorides of lime and soda.

The period for land quarantine was set at eight days except during pilgrimages or movements of troops, when it could be longer.[14] The conference also recommended the establishment of two quarantine stations on the Arabian coast of the Red Sea for the reception of pilgrims bound for the Hejaz (western Arabia). There, pilgrims were to be subjected to a quarantine of observation lasting twenty-four hours or to a rigorous quarantine of fifteen days if cholera appeared among them. In the event of cholera being detected among pilgrims, the conference recommended complete suspension of maritime traffic between Egypt and the Hejaz. The British Government of India, meanwhile, was urged to take steps to prevent the spread of disease within its territory, particularly in the most important pilgrim port of Bombay.[15] What the conference proposed, in effect, was a stranglehold on the East.

The suspicion of foulness

After the Constantinople conference, the Consular Commission, or Board of Health, at Alexandria was forced to play a more active role than before; the same was true of the Sanitary Council in Constantinople itself. Both these bodies focused their attention firmly on India as the presumed source of infection. Initially, the regulations which most affected India were those framed by the Ottoman Empire and the international members of the Sanitary Council at Constantinople. This council controlled most of the ports along the Red Sea, including those used by pilgrims on the Haj. To protect its territory from further invasions of cholera, the Ottoman authorities declared in 1867 that a ten-day quarantine was to be imposed against Indian ships on which cholera had appeared or which had proceeded from an infected port, with a reduction of three days for vessels that had been at sea for thirteen days or longer. An exception was made for ships which Ottoman sanitary officers regarded as being in poor hygienic condition. A fixed number of passengers per ton was established along the lines of the Government of India's Native Passenger Ships Act, which had attempted to regulate the sanitary state of pilgrim and emigrant vessels following a series of notorious cases in which passengers had died in squalid conditions.

These measures and the proposed quarantine stations for pilgrims in the Arabian Sea created some difficulties for the British Government of India. It seemed likely that few vessels would leave Indian ports with a clean bill of health because cholera was 'more or less endemic throughout India'. British officials also resented the fact that India had been singled out in the Ottoman regulations and called for them to be applied to all ships carrying pilgrims to Mecca, particularly those from Dutch colonies such as Java.[16] However, the British Indian authorities were not unduly alarmed. They did not believe that the establishment of a quarantine station near Jeddah would interfere greatly with navigation as most of Jeddah's trade was with India and merchant vessels frequenting the port would still be permitted to start their homeward voyage without having to conform to the quarantine restrictions applicable to vessels leaving Jeddah for Egyptian ports.[17] Thus, it seemed that the disruption likely to be caused by the new Ottoman regulations would be minimal. Although stricter enforcement of rules on board pilgrim ships would raise costs for shipowners and pilgrims, from a purely commercial point of view it would not entail great hardship. Indeed, merchandise carried into Ottoman ports was generally to be exempt from fumigation or burning, except for articles such as unwashed cottons and woollens, and undressed skins and rags. Bills of health which determined whether a vessel sailed from an infected, suspect or clean port were also to be issued by the competent authorities in India rather than by Ottoman officials.

The temptation, of course, was to give Indian shipping the benefit of the doubt, but British officers who were too liberal in granting clean bills of health threatened to jeopardize the reputation of the Indian administration and invited retaliation. The first such incident occurred in the Arabian port of Aden, which had been under the jurisdiction of the

Government of Bombay since 1837 and was to remain so until it became a colony in its own right in 1937. In 1871 the Ottoman Board of Health alleged that Aden's health officer, Dr John Turner, had granted clean bills of health to vessels arriving there from infected ports and that measures had to be taken to prevent this from happening, on pain of stringent restrictions against shipping from India. Turner protested his innocence but even the Government of Bombay was forced to admit that his explanation was unconvincing.[18] To prevent such a situation arising again, the Bombay government suggested to the Government of India that a quarantine station be built at Aden. All ships entering the port from infected places would thus be scrutinized before they set sail for Ottoman ports.[19] The authorities in Aden concurred. As the Assistant Political Resident put it, 'considering the immense importance of keeping this Port of Aden free from even the suspicion of foulness, I am of opinion that a Bill of Health should be taken by all vessels leaving it *in whatever direction*'.[20]

The British authorities in Calcutta did not care for this suggestion. Although they had not protested against the new regulations drafted after the Constantinople conference, they were becoming concerned about what appeared to be a shift in the stance of the Ottoman Board of Health and its Egyptian equivalent. In 1870 these bodies had issued more stringent regulations against ships from ports potentially infected with cholera, fearing that the Suez Canal, which had been opened in late 1869, would permit the rapid spread of the disease from pilgrimage sites in Arabia. These fears were well founded. In 1871 and 1872 cholera spread north from the Hejaz into Syria and southern Russia and across the Red Sea into the Sudan. Egypt was threatened from the north and the south, and imposed ten-day quarantines on all ships arriving at Suez from areas deemed infected. Much to the annoyance of the Government of India, these included Aden and, in the coming years, the Egyptian regulations became stricter still, with a quarantine of around thirty days being imposed on all ships said to be in an unhygienic state or from ports considered likely to be infected.[21] In the case of Indian ports such as Bombay, this raised the prospect of perpetual disruption to shipping.

Although the British consul in Egypt regarded the actions of the Egyptian Board of Health as justified, the Government of India and other parties were irritated by what they regarded as unwarranted interference with commerce. The major shipping companies – the Peninsular and

Oriental Company and the British India Steam Navigation Company – complained bitterly of these new restrictions on their business.[22] Having previously cooperated with the Ottoman and Egyptian administrations, the Government of India began to protest against quarantine of any kind, proclaiming it to be useless and medically unjustified. But it did not seem to have the courage of its convictions. In 1870 it had passed the Indian Quarantine Act which enabled provincial governments to draft quarantine regulations to protect ports in their territory, subject to the Government of India's approval. It is not altogether clear why the government did so. It may be that the measure was intended to reassure the Ottoman and Egyptian authorities that India was doing its best to prevent the spread of disease through the coastal trade. However, the most likely explanation is that some provincial governments had requested it. Port authorities had been demanding quarantine for a number of years, fearing that cholera would be brought into their cities by pilgrims returning from Mecca and that yellow fever would be carried by migrant labourers returning from the West Indies and Mauritius. Calcutta had already prevented some pilgrim ships from docking when the authorities suspected cases of infectious disease on board.[23] These measures were ad hoc in nature and of dubious legality, so the Quarantine Act of 1870 was intended to rectify the situation by placing quarantine and medical inspection on a firmer footing. However, the Act was hastily prepared and the Government of India soon came to regret it. It had been passed while its Sanitary Commissioner, J.M. Cuningham, was on leave in Britain and Cuningham opposed any measure that gave credence to the theory of contagion in relation to cholera or plague. Such diseases, he believed, were solely the result of insanitary conditions.

The reasons for Cuningham's vehement opposition to theories of contagion (and quarantine) have attracted considerable attention, not least because Cuningham appears to have changed his mind within a short space of time. In 1867, while still an officiating sanitary commissioner with the Government of India, Cuningham endorsed the view that cholera was spread by human contact. A case in point was the recent epidemic at the pilgrimage site of Haridwar on the Ganges, which appeared to spread along the routes taken by returning pilgrims. Although there was still no consensus on this issue, Cuningham was by no means alone in believing that wind direction or local sanitary conditions had little bearing on the outbreaks.[24] He therefore recommended that military cantonments and municipalities guard their localities against infection by erecting sanitary cordons. However, he stopped short of advocating quarantine as a general course of action and thought it would be impracticable on a larger scale, as well as productive of serious unrest. Like many other British officials, he was mindful of the lessons of the Indian Mutiny and Rebellion of 1857, which had been provoked partly by interference in Indian customs and religious practices. But within a few years Cuningham had become an ardent opponent of quarantine in any form, as well as of the theory that cholera was spread by human contact.

The historian Sheldon Watts has claimed that Cuningham was pressurized into taking this view. According to Watts, Cuningham's 1867 report met with a 'frosty reception' in London and when he returned to Britain on leave in late 1867 he was 'won around' to the government's view that quarantine was an 'evil'. However, Watts has adduced no evidence to prove either that Cuningham's report was received negatively or that the British government exerted any pressure on him. He also asserts that the British government had recently altered its stance on cholera because it was concerned about the effects of quarantine on shipping passing through the Suez Canal, which was due to open in 1869. No evidence is offered to support this contention, either.[25] Indeed, there appears to have been no serious concern over measures at Suez until *after* it opened; nor is there any definite indication of why Cuningham changed his mind. It is possible that he calculated that a different view might serve him well if he wished to be installed permanently as Sanitary Commissioner in India, but in the absence of evidence we can do no more than speculate. Equally, we cannot discount the possibility that he changed his mind because of the accumulation of statistical evidence which seemed to indicate that cholera was not a straightforwardly contagious disease. One of the most influential reports in this respect was that of a fellow Indian Medical Service officer and Edinburgh graduate James Lumsdaine Bryden, who believed that cholera spread in India with monsoon winds. Although later ridiculed in Britain, Bryden's report was then considered authoritative by many in India.[26]

From 1868 Cuningham sent instructions to provincial sanitary commissioners to collect more such data, dealing harshly with some (though by no means all) officials who continued to insist on cholera's contagiousness.[27] Why Cuningham did so is unclear, especially as he went further than his political masters would have wished. Although the Government of India was beginning to doubt the wisdom of land quarantines, it was not until the 1870s that it finally urged that they be abandoned.[28] Even then, the government's attitude to land quarantine remained inconsistent and pragmatic; sometimes rejecting Cuningham's advice, they would favour that of army officers who wished to see large-scale sanitary cordons established to protect crucial military areas such as the Punjab.[29] Cuningham thus tended to be *in advance* of government in advocating a retreat from quarantine and was far less flexible than some members of the viceroy's council in this. As we shall see, his dogmatic resistance to theories of contagion troubled the government in London, too.

It therefore seems likely that Cuningham's change of heart was a matter of conviction rather than a calculated career move. This is not to say that he was ignorant of the political implications of his views, far from it, but there is no evidence that he was placed under pressure to change them or that he changed them in expectation of personal advantage. The Government of India was still a long way from having a fixed position on maritime quarantine and the fact that it had passed the Quarantine Act is a testament to this. Cuningham, however, was less pragmatic than the government he served. To admit the utility of quarantine in India, he argued, would strengthen the case of those who wished to impose it against Indian shipping. The 1870 Act had already induced the governments of Burma and Madras to introduce quarantine measures in their own ports and Bengal was in the process of formalizing its existing provisions under the new legislation. However, the Government of India was in a position to stop Bombay from following their example and Cuningham was determined that it should.

The city of Bombay was the most important of India's ports. During the American Civil War (1861–5) it had grown rapidly as Britain began to obtain most of the raw cotton it needed for the clothing mills of Lancashire from western India.[30] Cotton was one of those articles, along with wool, that had traditionally been regarded as likely to harbour diseases, including cholera and plague. Textile manufacturers in Britain and cotton farmers in India thus faced heavy losses from the disruption of navigation through the Red Sea and Suez Canal. Once Cuningham had returned from leave, he advised the government to make no more concessions regarding quarantine. Apart from its potential to disrupt trade, further restrictions on the pilgrimage to Mecca were likely to annoy pilgrims and result in losses for

Indian shipping companies. With the backing of its Sanitary Commissioner, the Government of India began to insist that there was no need for quarantine in Bombay or Aden. It considered the Native Passenger Ships Act, which compelled pilgrim vessels to touch at Aden, and the arrangements already in force in that port to be adequate. The Bombay government continued to disagree. It professed concern about the spread of infectious diseases such as scarlet fever from Europe, arguing that growing traffic through the Suez Canal increased the risk of this occurring. It also demanded the appointment of a permanent port surgeon at Aden who would have the authority to declare other ports in the Red Sea and along the coasts of Arabia and the Gulf infected and to place quarantine upon vessels leaving them.[31]

It is hard to say whether the Government of Bombay was really concerned about the prospect of infection from Europe but, given the enormous sanitary problem existing in Bombay, it seems unlikely that it was bothered about a few cases of scarlet fever. The most likely reason is that the government wanted to boost confidence in the sanitary credentials of its port and the vessels leaving it. Bombay was not the only port targeted by the boards of health in Constantinople or Alexandria but it was the source or destination of much of the shipping passing through the Suez Canal and a hub for business with ports in the Arabian Sea and Persian Gulf. It was also the principal port of embarkation for pilgrims undertaking the Haj. Sanitary measures formulated by those two boards thus affected Bombay disproportionately. Some of these measures also seemed to be driven by panic rather than by reliable evidence. The Indian Ocean region was now linked by a telegraphic network which enabled reports of disease outbreaks to be communicated in minutes to the authorities at Alexandria or Constantinople. Consuls would be informed by their delegates on the Board of Health about these outbreaks and notified other consular or imperial outposts accordingly. This new communications network – laid by companies such as the British–Indian Submarine Telegraph Company – enabled sanitary measures to be formulated more precisely so that they would not disrupt shipping unnecessarily. But the information conveyed was not always accurate and periods of quarantine fluctuated as risks were revised up or down.[32] The effect on shipping – especially lines such as the British India Steam Navigation and Peninsular and Oriental companies – could be disastrous; so too with merchant navigation operating out of ports in the Arabian Sea. It was clear that the authorities in Alexandria had little confidence in the sanitary credentials of ships leaving Bombay, or, indeed, its dependency, Aden. Despite what seems to have been general agreement on the need for bills of health to be issued to all shipping leaving the port, nothing was done.[33]

Mercantile interests in Bombay thus felt the full force of the vagaries of this system, as did all the commercial and passenger liners that embarked or ended their journey there. The authorities in Bombay were desperate for some means of controlling the system and they believed that having quarantine facilities of their own would improve confidence abroad and reduce periods of quarantine for vessels leaving the port. In this sense, the views of the Government of Bombay were closer to the government in London than the authorities in Calcutta. In May 1874, just two months before another international sanitary conference was due to be held in Vienna, the British Foreign Secretary enquired about the nature of sanitary laws in the Indian dominions (including Aden) following a request from the Turkish ambassador. As the conference approached, Indian sanitary arrangements were coming under scrutiny and London was concerned lest lax precautions in the Arabian Sea result in harsh measures against British shipping in the Red Sea and the Mediterranean.[34] Far from being implacably opposed to quarantine in Indian ports, the government in London wanted the Government of India to impose some sanitary restrictions in order to satisfy foreign powers. It knew that if the Indian government failed to take international concerns seriously, all British shipping in the Mediterranean and Red Sea was likely to be subject to harsh retaliation.

The Egyptian sanitary authorities and most of the foreign delegates to the Board of Health at Alexandria were growing impatient with what they regarded as prevarication on the part of the Indian government. The British delegate to the Board informed the British consul in Alexandria that

> a feeling of uneasiness prevails at the Egyptian Board of Health as to the present state of the public health in India consequent on the effects of the famine, and I beg to state that, under these circumstances, it is most essential that every vessel arriving at Suez from India should produce the original bill of health issued at the first port of her departure, in order to

afford the Board the whole of her sanitary history from the commencement of the voyage.[35]

He explained that it had been common practice for vessels leaving India to deposit their original bills of health at Aden and take fresh ones from that port. This meant vessels from ports infected with cholera could arrive at Egypt without effective sanitary surveillance. The Turkish consul at Bombay also requested that he be given the right to check all bills of health issued at the port; a request which the Secretary to the Government of India thought 'utterly unprecedented'. It was 'virtually an imputation that our health officers are not to be trusted', he complained, and advised the Government that the request be resisted.[36]

The Government of India believed that the issue of famine had been raised with the sole objective of forcing it to adopt more stringent regulations at Aden and other ports under its control. It was well known that famines, such as that which had recently ravaged the Deccan,[37] were usually accompanied by outbreaks of epidemic disease, including cholera. But the viceroy protested to the Secretary of State that 'There is no cause whatever for fear of infectious disease from the famine. The general health in India is excellent. Proposed quarantine at Suez for all vessels would be seriously inconvenient.'[38] The Government of India was convinced that famine had simply been used as an excuse by the French and other powers – acting through the Egyptian Board – to impose damaging restrictions on British navigation. Its Sanitary Commissioner urged the government to resist the Egyptian order 'to the utmost'. 'There is not the smallest excuse for it,' he insisted, claiming that 'the Egyptian Government had planned a scheme for detaining passengers in Egypt so that hotel-keepers would benefit thereby'. Here he was referring to the Board of Health's proposal to detain all passengers arriving in Egypt from India for three days, which would cause them to miss the steamer at Alexandria.[39]

This was very unwelcome news to the Government of Bombay but it was more inclined to comply with the demands of the Egyptian authorities than the Government of India. It did not want stringent regulations in the major ports under its control – particularly Bombay, Karachi and Aden – but wanted to do just enough to prevent more damaging restrictions in the Red Sea and Suez. Bombay's British and Indian merchants feared continued disruption unless a more permanent solution was found and the Government of Bombay believed that the best way of achieving this was to accede to some of the demands of the Egyptian Board of Health. Indeed, in Aden, the port's Health Officer and the harbour police were already placing vessels in provisional quarantine until they were inspected.[40] Whilst there was no basis for this practice in law, it had the tacit sanction of the authorities in Bombay. Following news of an outbreak of cholera in the Persian port of Bushire, for example, all vessels arriving in Aden from that port were directed to a specially demarcated area of the harbour until they were inspected by the Health Officer. If free of infection, vessels were to be permitted to dock; if not, they were to be refused, or would have to undergo a period of quarantine at sea. At this time, there was no permanent quarantine station at Aden, although one of the islands in the harbour was occasionally used when cases of smallpox and measles occurred on troop ships.[41]

By the end of 1872 quarantine rules had been drawn up for Aden with the Government of Bombay's approval. The rules aimed to satisfy the boards of health at Alexandria and Constantinople and at the same time to keep restrictions on shipping to a minimum. They were drawn up in consultation with the Alexandria Board of Health by a committee which included the Harbour Master, the newly appointed Port Surgeon, the French consul and, tellingly, an agent from the Peninsular and Oriental line.[42] But although the overriding concern was to satisfy the authorities in Alexandria, the power to impose quarantines of up to ten days also enabled Aden (and by extension Bombay) to take retaliatory measures. In 1873, when cholera was prevalent along the African coast of the Red Sea, the Egyptian Board of Health imposed quarantine on all vessels passing through the Red Sea, regardless of their port of origin. Aden responded by placing quarantine on all ships arriving from Egypt, which soon led, in turn, to Egypt removing restrictions against Aden.[43] The measures at Aden therefore seemed to be having the desired effect, and by 1874 the Bombay government wanted to place them on a more permanent footing, introducing a bill into the provincial legislature to establish a quarantine station under the provisions of the Act of 1870.[44]

The return of the plague

The Bombay initiatives caused great concern in British India's capital, Calcutta. The Sanitary Commissioner suspected that the Government of

Bombay was trying to introduce quarantine through the back door and asked for further details of the regulations applying to ports in its jurisdiction.[45] Although a persistent cause of anxiety on Cuningham's part, the issue of quarantine in Bombay was pressing in 1874 because an international sanitary conference had been convened at Vienna to reconsider measures recommended at Constantinople. Russia had called for the conference after experiencing grave inconvenience in its Black Sea trade, which was now regularly interrupted by having to perform quarantine at Ottoman stations in the Bosphorus.[46] With Russia's support, it seemed possible that sanitary restrictions might be eased, so the last thing Indian shipping interests needed was for India to appear inconsistent in its stance towards quarantine. The anticipated reforms did not materialize, however, and the conference ended without agreement. In any case, these discussions were soon eclipsed by an outbreak of plague across the Arabian Sea. Reports of plague in the Ottoman dominions and Persia raised the prospect that Aden and perhaps even Bombay might be infected as a result of their frequent trade with the region. Plague had been absent from Egypt and the Levant since 1844, the disease having disappeared from Eastern Europe three years before. In the 1850s and '60s, however, it returned. The presumed origin of these outbreaks was the Chinese province of Yunnan, but in 1853 an outbreak was reported as far distant as Mecca and five years later it had reached Tripoli. These were relatively small epidemics, and the connections between them were obscure. Of more concern were the outbreaks which occurred in the port of Basra, in 1867, and elsewhere in the Ottoman province of Mesopotamia during the 1870s.[47] This time, the disease did not remain localized and by 1876 it had spread to south-western Persia, creating tremendous alarm in Russia and throughout the Middle East. Although sanitary measures in these countries were no business of the Government of India, foreign governments saw India and its dependencies as a stepping stone from which plague might easily jump to Europe.

As one of the ports in regular contact with Mesopotamia and Persia, Aden was said to present a special danger and the Egyptian Board of Health imposed restrictions on any vessel which anchored there. The frequent traffic between Bombay, the Persian Gulf and Mesopotamia also raised the prospect of India's greatest port becoming infected, with disastrous consequences for trade and civil order. In order to reassure foreign powers, the Bombay government again called for a local quarantine Act.

British officials in Baghdad and Tehran also urged the Government of India to allow Bombay to quarantine vessels from infected ports in Mesopotamia.[48] Like the authorities in Bombay, they believed that such a measure would allay the anxieties of Egypt and lead it to relax restrictions on ships from the Persian Gulf. But the Government of India still refused to listen. Citing the opinions of Cuningham, it insisted that plague epidemics in Mesopotamia and Persia were due to poor sanitation and that there was no evidence that the disease was contagious. Cuningham had also noted that plague had occasionally broken out in the Indian hill districts of Kumaon but that it had remained there, despite the fact that quarantine had never been imposed.[49]

The laissez-faire attitude of the Indian administration prompted the Egyptian government to take action against British shipping in the Red Sea. In May 1875 Messrs Gray, Dawes & Co., agents for the British India Steam Navigation Company, wrote one of several letters to the British Foreign Secretary protesting about what they regarded as abuses of quarantine by the Egyptian and Ottoman authorities. They expressed their gratitude to the Foreign Secretary for his part in the removal of some of the more onerous restrictions at Suez, but the Company had re-established its line to Karachi and the Persian Gulf only to be faced with quarantine at Red Sea ports such as Jeddah, which were under the control of the Ottoman Empire. There was no foundation for such measures, it claimed, other than the 'rumoured existence of disease in Mesopotamia'. Indeed, they had been calculated with the sole object of damaging the Company's interests:

> Our Red Sea agents affirm that the blow is aimed at us because our steamers running to the Persian Gulf and calling at Red Sea ports interfere with the traffic of the Khedive's vessels; and Native merchants of position quite unknown to us have addressed to us letters upon the subject of the quarantine, pointing out the losses they have been subjected to, and urging us to try and get the position honestly dealt with in order that the communications may be maintained.[50]

Later that month, the British delegate to the Egyptian Board of Health intimated that the problem of quarantine at Suez might be averted if the Company were to employ surgeons on its ships. If they reported the vessel healthy, ships would be permitted to pass through the canal in quarantine provided they did not communicate with any Egyptian port. But still the Company believed that its interests were being compromised on false pretences. It protested that the proposed measures were likely to raise costs by an additional £1,200 per annum and that it was unnecessary for ships to pass through the canal in quarantine if they were healthy. They also claimed that

> Such an arrangement would quite suit the Egyptian authorities as it would effectually prevent our vessels competing with the Khedive's trading steamers, but it would be most prejudicial to our interests, for not only would it exclude us from the Egyptian trade but it would prevent the transhipment of the Peninsular and Oriental Company, the Messagaries and Austrian Lloyds, as has been arranged with these Companies.[51]

In the light of this correspondence, the Secretary of State for India, Lord Salisbury, demanded to know what the Indian government was doing to prevent the importation of plague. The government replied that there was little danger of this happening and that it did 'not consider it expedient to impose quarantine in any Indian port on the outbreak of plague in Turkish Arabia'. It was, however, prepared to allow ports to carry out fumigation of vessels if the Ottoman Board of Health considered it necessary.[52]

In the meantime, the Bombay government continued to insist on the need for quarantine in its territories and in October 1875 it sent the second draft of its quarantine bill for the approval of the Government of India. It argued that the new bill was substantially more liberal as it proposed to detain only those persons who were unhealthy. But even with these modifications, the bill was still objectionable to the Government of India's Sanitary Commissioner. 'Such measures,' insisted Cuningham, 'cannot be accepted on purely theoretical considerations without one word in their defence.'[53] The Secretary to the Government of India, Arthur Howell, agreed entirely with Cuningham, as did the viceroy in council. Howell thus informed Bombay that 'The Government of India are of opinion that even as the rules now stand they are calculated to interfere seriously with public comfort and convenience besides entailing considerable expense on ship owners; and yet the good to be derived from them will be extremely doubtful.'⁵⁴ But the government in London was becoming increasingly concerned about India's uncompromising stance, which left Britain isolated internationally. The British Foreign Secretary had nothing left to bargain with and believed that it was necessary to make some concessions to other nations in order to obtain relative freedom of navigation in the Mediterranean and the Red Sea.

The Government of Bombay had come to the same conclusion. E.W. Ravenscroft, the Acting Chief Secretary to the Government of Bombay, spelt this out clearly in a telegram to his opposite number in the Government of India's Home Department. 'Quarantine may be established for two objects,' he reminded him. 'One, the protection of the people of the port at which it is imposed: the other, as has been mainly the case in the present instance, that of accommodation to the views of the authorities beyond [our] control, at foreign ports with which it is of importance to preserve uninterrupted communication.' To satisfy the Egyptian Board of Health, he proposed new regulations, not far removed from the original ones: 'it hardly seems probable that the Egyptian authorities, bearing in mind the ravages caused by plague in recent years, will feel any disposition to incur the slightest risk for the purpose of meeting our convenience, and it is for this purpose, that the rules herewith submitted, 15 clear days from arrival are provided for in cases in which there has been actual disease on board'. Ravenscroft pointed out that Egyptian ports had, until recently, been placed in quarantine every summer by the Mediterranean powers. This was not done because it was considered necessary from a medical point of view, but because it was the only means of preserving free communication with French, Italian and Spanish ports.⁵⁵

The actions taken by the Egyptian Board of Health in the summer of 1876 seemed to justify these claims. Plague outbreaks had been reported along the River Euphrates since March that year and ports along the Persian Gulf and throughout the Indian Ocean kept a close watch on vessels from the region. The Government of Bombay was no exception and instructed officers in its ports to take such measures as might be necessary 'to prevent infected vessels entering the port'.⁵⁶ But these measures do not appear to have satisfied the authorities in Alexandria; perhaps because they lacked any solid legal basis. Thus, in June of that year, following further reports of plague in Mesopotamia, the Board once again threatened Indian ports with severe restrictions, resulting in protests from the British India Steam Navigation Company among others. Cuningham echoed their concerns:

> what the Egyptian authorities really threaten is that unless Kurrachee, Bombay and Aden place a quarantine of fifteen days on all arrivals from the Persian Gulf then the authorities at Suez will enforce fifteen days' strict quarantine on *all* arrivals from these Indian ports, no matter whether they have come from the Persian Gulf originally or not.... [T]he proposal is monstrous. India is as free from plague as Egypt is, and it would be just as sensible, so far as plague is concerned, for India to quarantine vessels from Egypt as for Egypt to quarantine vessels from India.[57]

Cuningham deprecated any further sanitary restrictions but this time was overruled by his political masters. A member of the viceroy's council recorded that 'It was agreed in Council that some form of quarantine sufficient to satisfy the Egyptian Government should be established at Bombay, Aden and Kurrachee for all vessels coming to those ports from the Persian Gulf.'[58]

Cuningham therefore suffered the indignity of being asked to draw up regulations for all Indian ports similar to those drafted by Bombay which he had formerly opposed. He protested that such an action was contrary to the views of the previous viceroy and that 'They will concede to the Egyptian authorities what was refused to the Bombay authorities; they will involve fifteen days' quarantine for every vessel, *even when no plague exists on board*, and they imply that if a single case of plague has occurred even the mails, besides the passengers, may be detained for an indefinite period.' He added: 'The results at Bombay, Kurrachee and Aden will be so very serious...'.[59] But Cuningham's intransigence was threatening the Government of India's prized independence from London. As E.C. Bayley, legal member of the viceroy's council explained:

> The question was ... discussed simply on the basis of giving way to the pressure of the Egyptian Government, or of asking the English Foreign Office to interfere. It was decided to adopt the former alternative. Dr. Cuningham's rules, as he proposed them, provided for no quarantine at all. The emendations I have made do provide a quarantine, though, as I have shown, of the mildest kind, unless plague actually occurs.[60]

The Persian Gulf

The viceroy agreed with Bayley's assessment and sanctioned the new regulations but they do not appear to have been drafted, as pressure was relieved by the lifting of restrictions imposed by the Alexandria Board of Health. For the next eighteen months the issue lay dormant but there was a related matter with which the British and Indian governments were becoming increasingly concerned. This was the question of quarantine in relations between the Ottoman Empire, particularly the province of Mesopotamia, and its eastern neighbour, Persia (Qajar Iran). Growing commercial and political involvement in the Gulf placed the British in a difficult position. From 1821, the Ottoman Empire had experienced a series of devastating cholera epidemics which had led many Turks to demand a sanitary administration on the lines of those in modern European countries. However, the Ottoman Council of Health was largely a creation of foreign powers and in the 1870s these remained in a position to influence the measures it took. Nevertheless, the Constantinople Council had a degree of autonomy and issued directives which were seen as covert acts of aggression by some of its neighbours. This was certainly true of Persia, whose shipping was often interrupted by quarantine measures imposed at ports in Ottoman provinces abutting the Persian Gulf and Arabian Sea. The British vice-consul at Basra remarked upon this in 1864, when quarantine was imposed in the port of Fao against all vessels leaving the port of Mahamrah without a clean bill of health from the Turkish consul. He described this 'exercise' as one 'of which ... the Persian Government might fairly object'.[61] It was not so much the right of the Ottoman authorities to impose quarantine that was questioned, as the exorbitant charges levied against ships directed to perform it. The vice-consul noted that 'Cases have come under my notice in which six and even nine kerans representing four or five times the proper fee, have been demanded as Quarantine fees from vessels.' 'Trivial as these are,' he continued, 'these fees are looked upon as exactions, and therefore cannot fail to contribute their mite towards preventing the increase in trade and shipping at Bussorah [Basra].'[62] These measures affected British interests directly. The British Agent in Basra complained that all vessels arriving there and at Fao had to perform 'certain formalities' regardless of whether they had a clean bill of health from their last port of call, which was normally in Persia. The port of origin of most of these vessels was Bombay.[63]

Britain's influence in Persia was growing and it was also seeking to expand its trading interests in Mesopotamia, but as relations between the two regional powers deteriorated it risked falling between two stools. The border between Persia and the Ottoman Empire was hotly contested and tension was heightened by the alleged ill-treatment of Shi'a Muslims from Persia on their annual pilgrimage to holy sites such as Karbala and Najef in Mesopotamia. Pilgrims had been prohibited from visiting Baghdad on the grounds that they posed a sanitary threat.[64] The Ottomans also used quarantine to strengthen their borders and their claims to adjacent territory, much as European states had done in earlier times. In the mid-1870s, the Persians began to retaliate, imposing quarantine directed by a newly constituted International Sanitary Consultative Assembly, loosely modelled on that of Constantinople.[65] It was then the turn of the Turks to protest that 'mere rumour' of plague was sufficient to prompt punitive measures against their ports.[66] The dilemma faced by the British was highlighted by the visit of the shah of Persia to Baghdad in 1870. Nasir al-Din Shah undertook the journey partly out of religious devotion (a desire to visit Shi'a shrines) and partly as a stepping stone to Europe, where his modernizing chief minister hoped that he would be impressed by the marvels of Western civilization. The visit provided the Turkish authorities with an occasion to demonstrate some of the public health reforms undertaken by them under the guidance of the British Civil Surgeon of Baghdad, William Colvill. But at the beginning of the shah's journey, the auguries were less auspicious. The Ottoman authorities claimed that the shah's huge entourage of bodyguards and camp followers (around 10,000 in all) was likely to bring cholera into their territory. The consul in Baghdad therefore urged the British legation in Tehran 'to convince the shah that his Majesty must respect quarantine and induce him to reduce his followers to fitting numbers'. He informed them that 'Cholera is in his camp' and that the position was 'serious'.[67] But the shah was already halfway to the Ottoman border and the British consul in Tehran thought that it would have been useless to make any representations on the matter from there. The only hope of averting a major incident was to dispatch an official from the British legation to join the shah's entourage.[68]

A breakdown in relations between Persia and the Ottoman administration in Mesopotamia would have had serious consequences for British interests in the region. But according to the British legation in Tehran, it was a prospect that might easily have been avoided. 'The Turkish authorities must have been aware all along of the crowd of followers which was to accompany the Shah on his journey and the likelihood of cholera breaking out in the camp,' the consul explained to Lord Granville, the British Foreign Secretary. 'These subjects have been frequently discussed between myself and the Turkish Chargé d'Affaires, and there was ample time from the beginning to provide against the danger now apprehended.'[69] The clear implication of this was that the Turks had deliberately manufactured the incident as an insult to the shah. These developments remind us of the essentially political nature of sanitary regulations and of the fact that, despite growing foreign influence, the Ottoman Empire and Persia were still very much independent actors, with the power to alter the dynamics of international relations.[70] Measures taken in Mesopotamia, at ports such as Basra, and along the province's major rivers, significantly affected British trade, which was already constrained by the protectionist measures of the Ottoman administration. The British Tigris and Euphrates Company was permitted only two steamers, for example. Its trade was mostly in silk and tobacco, which the company imported from Persia, and wheat and skins from Mesopotamia which were sold to Russia and Persia, respectively. Deteriorating relations between Persia and the Ottoman Empire had severely damaged this trade, especially prohibitions on the Shi'a pilgrimage, which had always been one of the main occasions for conducting commerce across the border. In view of the restrictions placed on riverine trade, pilgrims from Persia had become the main purveyors of cotton piece-goods manufactured in Britain.[71]

The outbreak of plague in Mesopotamia in 1876 seemed likely to aggravate the situation and it was important for Britain to get its voice heard on the board of health that had been formed in Tehran. This body, which was generally known as the Central Sanitary Council, had been nominally created in 1868 under the presidency of the shah's personal physician, the Franco-Mauritian doctor J.D. Tholozan.[72] The Council existed in name only but was reconstituted in light of the plague epidemic in the adjacent Ottoman provinces. It was now presided over by Prince Ihtizadel Sullana, the Persian Minster for Public Instruction, although

Tholozan was still a member. Unlike the boards in Alexandria and Constantinople, where foreigners predominated, the rest of the Council was comprised mostly of Iranians, including a number of 'native doctors' (practitioners of *tibb*, or 'Islamic' medicine) and the Director of the University of Tehran, at which the meetings were held. Apart from Tholozan, the only foreign members were those of Britain and the Ottoman Empire, the former being Sir J. Dickson, who had represented Britain on the Constantinople Sanitary Council and at the International Sanitary Conference in 1866. Dickson was well known as an opponent of quarantine but he found little support among the Iranian physicians present. Only one of them, a practitioner of traditional medicine known as Malekela Tubbah ('Prince of Physicians') apparently protested against quarantine. The rest of the Iranian delegates supported some form of cordon along the land frontier and similar measures at the ports. If Dickson reported correctly, it would seem that Tholozan, the most distinguished Western medical practitioner in the shah's service, had not voiced his earlier scepticism about the practicality of land cordons as recommended for the region by the Constantinople Conference.[73] Indeed, it seems that Tholozan's former opinions had led to his removal as head of the Sanitary Council. The assembly had no such qualms and decided that a quarantine of observation of fifteen days against arrivals from Mesopotamia should be imposed at Kasr Shireen on the frontier; orders were accordingly sent to all governors to prevent suspected cases from entering Persian territory. The island of Kasrack, near Bushire, was also selected as a quarantine station for arrivals from Mesopotamia by sea. This measure was probably popular with locals, who had formed their own quarantine against ships from Basra when cholera was reported there in 1865.[74]

However, the government had neither the resources nor the expertise to maintain these facilities and looked to Britain to help out. As most of the trade between the Persian Gulf, Basra and Indian ports was controlled by the British, the Qajar authorities regarded maritime quarantine as Britain's responsibility. Following reports of plague in Mesopotamia in March and April 1876, the Persian government imposed a fifteen-day sanitary cordon along the Turkish frontier and maritime quarantine for vessels arriving from Basra. The British consul in Tehran, W. Taylour Thomson, then informed the viceroy of India that

The Persian Government not having the means of enforcing quarantine regulations at sea, ask for the co-operation of Her Majesty's Indian Government through their Agents and Naval Officers in the Gulf to enforce the inspection of vessels and render the quarantine effective.⁷⁵

In view of its outward show of resistance to quarantine, it is rather surprising that the Government of India replied positively to Persia's request. Indeed, naval officers were soon dispatched to assist the authorities in Bushire and other ports.⁷⁶ This was a pragmatic decision that went against the general line of the Government of India's policy and the opinions of its Sanitary Commissioner. No doubt, the compromise was made with a view to keeping good relations with Tehran and reducing disruption to British shipping in the region. Having control over the maritime quarantine also enabled the British to exercise restrictions with a light touch. But this was not so with land quarantine and it soon became evident that the arrangements made along the border with the Ottoman Empire left much to be desired. Unscrupulous officials extorted money from those crossing the border and many people were obliged to undergo quarantine twice: once on entry into the country and, again, as they attempted to leave.⁷⁷

The plague outbreaks in Mesopotamia during 1876 brought a new issue to the fore. Mesopotamia was an important source of wool exports, including some rare varieties such as mohair, which was used in the British textile industry. Mohair was essential to the economic vitality of cities such as Bradford, or 'Worstedopolis' as it was affectionately known. But wool had long been regarded as a material capable of carrying plague and some other diseases such as anthrax. Indeed, Mesopotamia was regarded by many as the source of the outbreaks of anthrax that occurred in British manufacturing towns including Bradford and Kidderminster, the latter being renowned for its carpets.⁷⁸ However, after the appearance of plague in 1876, it was that disease which was causing most concern, especially to the delegates of the Alexandria Board of Health. In June the Board demanded careful supervision of all vessels leaving the Gulf and ports regularly in contact with it, such as Bombay and Aden. Aden went beyond its usual rules and imposed a quarantine of fifteen days against vessels from the Gulf and also from Karachi which had transshipped cargo from the Gulf. As the Political Resident at Aden explained, 'it becomes necessary to comply strictly with the Board's regulations on pain of execution of threats for non-compliance which would become a matter of serious inconvenience to a port like Aden in the matter of universal trade'.[79] It was not only the severity of the measures which was unprecedented but the expectation that wool would be fumigated before it left the region to pass through the Suez Canal. The British pressed for the Turkish authorities to perform disinfection prior to shipment but this was refused. As a result, transhipment ports such as Bombay had to set up disinfection stations from scratch. As the Commissioner of Customs in Bombay conceded, the whole operation was 'entirely illegal' but, despite protests from shipping companies like the British India Steam Navigation Company, the measures received the sanction of the Bombay government in advance of approval from Calcutta.[80]

The issue of the sanitary regulation of vessels in the Persian Gulf was becoming more serious and it came to a head at the beginning of 1877, when the Ottoman delegate to the Tehran board, Dr Castaldi, recommended a permanent system of quarantine for Persia, together with a permanent sanitary service. However, Dr Tholozan restated his government's belief that maritime quarantine could not be made effective without the support of Britain.[81] Although the Government of India had earlier offered support, the British were reluctant to assume responsibility for quarantine on a permanent basis. The last thing they wanted was to assist in the creation of another powerful sanitary body such as those which existed in Constantinople and Alexandria. Lt-Col. Prideaux, Officiating Political Resident in the Persian Gulf, believed that it would be better to have sanitary measures organized locally, rather than to permit them to be subject to the dictates of a central board. 'Whilst feeling a personal interest in the prevention of disease,' he argued, 'a local body would be less likely than a central committee to overlook important commercial considerations involved in the maintenance of quarantine restrictions.'[82] The London government was similarly inclined and flatly resisted any attempt to turn the Council at Tehran into a troublesome institution like those in Constantinople and Alexandria. But the Austro-Hungarians were pushing hard in this direction and were courting foreign ministries throughout Europe as well as European delegates to the Constantinople Council.

Outbreaks of plague in Mesopotamia aroused old fears about the spread of disease from the Ottoman dominions into the Hapsburg Empire's eastern provinces. Despite demands for the liberalization of land and river quarantine in the middle of the century, most Austrians still regarded the sanitary cordon along their eastern border as indispensable and credited it with having protected the Empire for nearly two centuries. In Constantinople, the Austrian delegate, with the support of his French and Italian counterparts, proposed a veritable stranglehold on the region. The measures recommended included the tightening of quarantine restrictions at Basra; the establishment of quarantine at Aden and the Persian Gulf for all British vessels coming from the Shatt-al Arab (or quarantine in Egypt if they refused); the imposition of quarantine by Russia along the Persian border; the establishment in Persia of quarantines at Bandar Abbas and Bushire against all ships from Basra, as well as the interdiction of pilgrims en route to Shi'a shrines in Mesopotamia.[83]

During 1877, the Austro-Hungarian government increased pressure on Britain, proposing that medical commissions be sent to Mesopotamia and Persia to study plague and that the temporary consultative board of health at Tehran be made permanent.[84] The British Foreign Secretary, Lord Derby, informed the Austrian ambassador that the British government had had frequent cause to complain about the workings of the boards of health in Alexandria and Tehran and that it was firmly opposed to a new one. He would only support a move to make the Tehran assembly permanent if its sole task was to investigate 'the best means of securing the adoption of enlightened sanitary measures'. Derby opposed anything that strengthened existing land quarantines or established new ones along the Turco–Persian frontier. He noted that the subject had been discussed at the Vienna Sanitary Conference and that the conclusion reached was that sanitary cordons were unenforceable and detrimental to international commerce.[85] Dr Colvill, who was attached to the British consulate in Baghdad, lent Derby his support. He told the Foreign Secretary that 'The whole gist of the letters of the Austrian Government, and of the Proceedings of the Sanitary Commission at Constantinople . . . is not only to have more quarantines, but to imprison, if allowed, this earth in a network of them, and the question therefore becomes as much political as medical . . .'.[86]

The proceedings of the Constantinople Council show that the position adopted by Colvill, who denied that plague was a contagious disease, and that of the British government, was rejected by all the other delegates. Derby's views on the matter of quarantine were based chiefly on those of

Colvill and the Secretary of State for India, but he claimed that the stance taken by his government conformed to scientific opinion. The Austro-Hungarian delegate expressed scepticism about this and declared that the British government had always sought to reduce quarantine for commercial reasons. The French delegate also pointed out that Mediterranean countries could not afford to see quarantine relaxed, unlike Britain, which was much further from the source of infection.[87] Britain was left diplomatically isolated and was blamed for paralysing the Austrian initiative with respect to Persia; indeed, its position had hardened considerably since the conference in Constantinople.[88] Nevertheless, Derby stuck to his guns and instructed the legation in Tehran to concentrate on improving sanitation rather than establishing quarantine. This became all the more important after plague broke out on the Persian side of the Mesopotamian border in the province of Resht. The Persian authorities in Tehran found it difficult to compel the provincial government to take the action it required, including the proper burial of the dead and the draining of cesspools. The local government seemed powerless to resist mass defiance of regulations concerning burials and had not instituted a permanent body to carry out sanitary works. The weakness of the Persian state therefore made it impossible to deflect international attention from the issue of quarantine in the Gulf and along Persia's terrestrial borders.[89]

'Meeting the prejudices of other states'

The outbreak of plague in Mesopotamia and its spread to Persia again raised the question of what were the best means of preventing plague from reaching Egypt – the gateway to Europe. The most controversial measures were the quarantine regulations which most countries wanted to see established at ports controlled by the Government of India. Many observers claimed that plague was just as likely to spread via Bombay or Karachi as directly from the Gulf. But while such measures were still anathema to the Government of India, there were British officials who wished to see sanitary restrictions implemented in the Gulf in order to deter the introduction of harsher measures in the Red Sea. Col. J.P. Nixon, Political Agent in Turkish Arabia, suggested that future mail contracts with the British India and Steam Navigation Company would require its steamers to abstain from carrying passengers from Bombay to Basra, which they had declined to do when plague broke out again in Mesopotamia in 1877. But if this were impracticable, he suggested, all steamers trading between Mesopotamia and India should be subjected to a short quarantine of around five days at Karachi, at times when plague was 'active'. He recognized that

> There is little doubt that when legislating on this subject and introducing quarantine into India, Government must be prepared to encounter some odium in doing so, for certainly it is a great clog to trade, but at the same time, if legalized, it is a most efficient method of checking the spread of disease, and when its authority is exercised, under the careful supervision of our Government, the many irritating causes that are so painfully felt in this country would not be so grievous.[90]

Support for some form of maritime quarantine came from the unlikely person of Dr Colvill who favoured it for entirely pragmatic reasons, despite having opposed the imposition of quarantine within Mesopotamia.[91] He proposed that a quarantine station be established at the mouth of the Gulf and that all British vessels bound for India or Europe be detained there before making their journey; the object being to allow ships to pass through the Suez Canal without hindrance. The thinking behind these and similar proposals was that it was better for vessels to undergo quarantine under a more liberal regime controlled by Britain than to be subjected to lengthy delays at Suez and to extortion by Egyptian officials.[92] Colvill's suggestion may have been prompted by the opposition he had faced from the Constantinople Council, which had rejected his non-contagionist account of the plague in the Levant. Indeed, the Council claimed that British opposition to contagion and quarantine was conceived with the sole object of reducing impediments to commerce.[93] But the Government of India believed that 'Such a measure would greatly impede commerce and inflict very considerable loss and hardship on the mercantile community.' It proposed to rely on the measures recommended earlier by Bayley; namely, medical inspection at Indian ports rather than quarantine per se.[94] 'For further preventive measures', it insisted, 'we must rely on keeping our port towns free from those insanitary conditions which render populations susceptible to infection, and without which it is believed that diseases of the nature of plague will not spread'.[95]

This combination of medical inspection and sanitary improvement was in line with the approach taken in Britain with respect to cholera from 1872,[96] but there plague still came under the auspices of the 1825 Quarantine Act. Again, the British position seemed to be inconsistent, which contributed to the isolation of its delegates on the sanitary boards of Alexandria and Constantinople.[97] Having made little headway in Europe, the British launched a diplomatic offensive in Egypt, where they hoped to be able to exert pressure on the Khedive. Since the establishment of a committee overseeing the repayment of Egypt's public debt, Britain and France had considerable control over the country's internal affairs and British influence, in particular, was growing. Derby planned to exert pressure on the Alexandria Board of Health through the Egyptian government, with a view to limiting any restrictions that might be imposed against British vessels in the absence of stringent measures in Indian ports.[98] This initiative proved successful and at a meeting of the Board at the beginning of November 1877, Colucci Pasha – a medical doctor who had formerly been head of the Board of Health – read a proposal to modify existing sanitary regulations for vessels arriving from plague-infected countries. Colucci Pasha claimed that the emergence in Mesopotamia and Libya of what appeared to be a 'milder type of plague', which did not seem to be easily communicable, permitted relaxation of the normal fifteen-day quarantine. Henry Calvert, the British vice-consul at Alexandria, reported approvingly that

> Colucci Pasha's regulations would press less heavily on maritime interests, as he proposes that vessels arriving from plague-suspected ports, and having been five days on the voyage without there having been any compromising illness on board, should be allowed to pass through the canal in quarantine, and should not be required to perform quarantine in Egyptian ports. It would in fact be only when suspicious illness had prevailed on board that the vessel would be required to ride out her quarantine at Tor [the Egyptian quarantine station at El Tor in the Red Sea]; and as a voyage of 15 days would reckon as days of quarantine, vessels with passengers or goods for Egypt (provided the public health on board were satisfactory) would be admitted to free 'pratique' after 24 hours' observation.[99]

No less important was the proposal to exempt mail from quarantine, provided that it passed through the canal in unopened boxes.[100]

Welcome though it was, the easement applied only to plague. Vessels on which cases of cholera had occurred or sailing from cholera-infected ports were subject to the same measures as before. The main sanitary impediments to British imperial trade thus remained intact; British companies and private citizens also complained of 'grave abuses' committed in the administration of quarantine in Egypt such as extortion and bribery, and of similar indignities suffered by pilgrims at the port of Jeddah.[101] Partly in retaliation and partly in an attempt to placate the Egyptian authorities, ports such as Aden took measures of dubious legality in order to place vessels in quarantine.[102] In August 1878, these arrangements came to the attention of the Secretary of State and the Government of India, following revelations that a European man and woman, who had arrived at Aden from Jeddah, had been compelled to perform six days' quarantine in the harbour in an open boat, the huts that had formerly been used for the purpose having been pulled down. The two protested against their situation and particularly their lack of privacy but the Resident later declared that their plight was not as hard as they made out and that they had only themselves to blame by attempting to go to Suez via Aden. However, he as good as admitted that the whole procedure was illegal, as no quarantine had ever been sanctioned for Aden.[103] J.M. Cuningham, who was still the Government of India's Sanitary Commissioner, declared that arrangements at Aden were 'worthy of the dark ages' and continued to insist that government ought not to countenance any form of quarantine in its ports.[104]

But there was a new determination in London to prevent the abuses reported by the British consulates at Jeddah and Suez. The Foreign Office instructed the British consul in Alexandria to

> address a strong remonstrance to the Board of Health and to the Egyptian Government against the disorders, corruption and cruelty revealed in these reports, stating that Her Majesty's Government can no longer acquiesce in a state of things which causes misery, crime and disease, which inflicts needless losses upon legitimate traders, which is worse than useless as a protection against infection, and which appears

to serve no better purpose than the enriching of a set of incompetent and corrupt officials.[105]

The India Office was similarly inclined and declared that 'The abuses brought to light in these and other papers make it imperative that some radical remedy should be applied.' It supported the suggestion of Mr Vivian, the British consul in Alexandria, to separate completely the Egyptian government from the Alexandria Board of Health and to protest against the Board's right to detain vessels passing through the canal if they did not intend to dock in Egypt.[106] Representations were made to the Board of Health but it soon became evident that they were unlikely to secure the desired result. It was now obvious that the Government of India's intransigence was standing in the way of a deal between Britain, Egypt and the other Mediterranean powers. In September 1878, the British government ran out of patience. Lord Salisbury, the Secretary of State for India, ordered the Indian government to draw up quarantine rules applicable to all its ports, pointing out the importance of 'meeting the prejudices of other States as far as possible, so as to avoid the risk of restrictions being imposed on vessels passing through the Red Sea'.[107]

Salisbury thus concluded the tense and protracted debates between the home government and the governments of India and Bombay over whether or not to establish sanitary regulations in Indian ports. All that remained to be decided was what form the rules should take and whether they would be acceptable to all the parties concerned. The Indian government laid down two basic principles for provincial governments to follow when drafting their rules: that a system of medical inspection ought to be introduced for vessels arriving with persons suffering from diseases endemic to India, and that quarantine should be reserved for occasions when there was a danger of alien diseases being imported, principally plague and yellow fever. However, the draft rules sent to Calcutta by the provincial governments were inconsistent and, in February 1879 the Secretary of State for India urged the Indian government to rectify the situation, as the 'present state of affairs materially weakens the hands of Her Majesty's Government in protesting against bad quarantine arrangements at places belonging to foreign powers'.[108]

The Indian government's Sanitary Commissioner J.M. Cuningham was therefore charged with the task of drafting clearer rules on quarantine for the guidance of provincial administrations and, rather ironically, these were based on the measures originally devised by the Government of Bombay in 1877. However, the rules for medical inspection were to be introduced as executive orders of government, rather than as pieces of legislation, despite some doubt as to the legality of this procedure.[109] The draft rules effectively brought practices in India into line with arrangements in Britain, it being intended that medical inspection should entail minimal inconvenience to trade or passengers. Vessels with just one or two cases of diseases such as smallpox or cholera were thus permitted to take up anchorage as normal and there was no need to detain healthy passengers pending inspection. All that was required was for the master of the ship to ensure that no passengers suspected of having disease were allowed to depart before the vessel was inspected. Passengers were to remain on the ship only if two or more were suffering from disease and only then would the vessel be cleansed and fumigated. Even the quarantine regulations were liberal. They permitted vessels from infected ports, without cases on board, to proceed straight to the normal anchorage.[110] But the fact that India was forced to have such a system at all shows that no state, however powerful, could afford to ignore international opinion.

The growing frequency of international sanitary conferences and the existence of international sanitary bodies at Constantinople and Alexandria made it difficult for any government to depart far from prevailing opinion. To do so was to invite coordinated retaliation of a kind that was particularly damaging to a nation like Britain which depended upon maritime trade for its wealth and power. As the Secretary to the British Local Government Board put it when summarizing the quarantine situation in the Red Sea, 'the action of Egypt herself seems necessarily to have become directed by the views of European Powers about quarantine, not only through those views being represented on the International Council, but also through the practice of the various quarantining powers on the Mediterranean littoral'. Egypt was thus obliged to 'bring her own quarantine arrangements up to a high standard of strictness, and to take not only such precautions as [are] in her own interests ... but also every sort of precaution that the apprehensions of the Mediterranean Powers compel her to take'.[111] Sanitary regulations therefore continued to reflect the expectations of other powers as much as perceived threats to health domestically – a statement which might apply to any nation, even today.

But while the international sanitary conferences had made the management of epidemic disease more coherent, the limitations of the system were revealed by the fractured responses to crises such as plague in the Levant. The most serious flaw of late nineteenth-century internationalism was not so much its Eurocentrism,[112] but the fact that competition between states was increasing, undermining whatever ideological affinity had existed in the middle of the century. In the 1830s and '40s sanitary internationalism had been sustained by the spirit of free trade but during the 1870s rival nations began to portray the doctrine as nothing more than a cloak to conceal the special interests of Britain.[113] This continued to be the case in respect of sanitary regulations through the 1880s, with Britain and India remaining isolated diplomatically.[114] While agreements on such issues as quarantine were theoretically desirable, they were not allowed to stand in the way of national interests and, in many cases, these interests were protected by tariffs and other restraints on trade.

By the early 1890s, the situation had changed again. In part, this was a reflection of new scientific knowledge about cholera. The bacterium responsible was discovered by the German scientist Robert Koch in 1883 in a reservoir of water in Calcutta and this confirmed that drinking water was the chief medium through which the disease was transmitted. However, Koch's discovery remained controversial for some years and once accepted did not diminish the necessity of containing infected persons. A more important stimulus to change in the international sanitary arena was the growing volume of shipping through the Suez Canal; not least that of Germany, which had recently acquired colonies in East Africa. Germany was able to use its influence over its partners in the Triple Alliance of 1882 – Austria-Hungary and Italy – to gain their support for the liberalization of quarantine at Suez. The change in outlook was already noticeable at the international sanitary conference held in Rome in 1885, at which the Italian Foreign Minister condemned the 'anarchy' of quarantine arrangements in the Mediterranean and suggested the formation of an international commission to devise standardized quarantine regulations for all states.[115]

The time was not yet ripe for such a body but there was growing support for greater standardization and liberalization. At the international sanitary conferences at Venice and Dresden in 1892 and 1893 there was some relaxation of arrangements at Suez, for example, though not at the pilgrim ports in the Red Sea. For Germany, and especially for Italy and Austria-Hungary, which were now Britain's chief supporters in demanding reform, this represented a remarkable change of heart, these countries having formerly been staunch defenders of quarantine. The sanitary stranglehold which most of the European powers had attempted to impose on the East – and on British power in the region – had been significantly relaxed. But France, whose imperial rivalry with Britain had intensified, stood fast on many issues, though it now found itself in the minority.[116] This was a dramatic transformation and one that owed far more to diplomacy and calculations of national interest than to any change in scientific opinion. But these agreements applied to cholera only and there remained great potential for the abuse of sanitary regulations in relation to plague, the nature of which had not been considered seriously since the Paris conference of 1851. When outbreaks of plague had occurred in the 1870s and 1880s, most states had simply reverted to rigid forms of quarantine, which proved awkward in an age of steamships, railways and global commerce. It was not until the world was shaken by another, far more serious epidemic in the 1890s that anything like a coherent international response to plague was worked out; and then, only after a great deal of disruption to commercial and political life.

CHAPTER 7

Plague and the global economy

In the last chapter we saw that there were occasional outbreaks of plague in China and the Middle East through the middle of the nineteenth century, some serious enough to disrupt long-distance commerce. But for most people, plague seemed like yesterday's disease – a fearful reminder of an age before science and sanitation. In 1890, however, an outbreak of plague in southern China mushroomed into a full-blown pandemic. And, whereas earlier waves of plague had affected Eurasia and North Africa primarily, this one circumnavigated the globe, reaching every inhabited continent. Plague spread rapidly along the arteries of a mature global economy, its path eased by modern transportation. Steam navigation and advances in maritime technology had reduced transit times to a fraction of those a few decades earlier and many large railway projects had been completed or were well under way. New methods of conveyance gave this antique disease a distinctly modern gloss but plague was 'modern' in other senses, too. The Third Pandemic was the first such crisis to which there was something like a coordinated response. Although the first reaction to plague was panic, the chaos that ensued prompted a series of international agreements which were ratified by most of the major powers. This fragile consensus prevented plague from destroying the international economy and provided the foundations for a new sanitary order.

Painful lessons

China's Yunnan province had been troubled by plague for many years but during the 1850s outbreaks became more frequent and spread occasionally into the neighbouring provinces of Guangxi and Guangdong.[1] Rebellions may have been the initial cause but the spread of the disease to other provinces was probably enabled by the lucrative trade in opium, which now linked this upland reservoir of infection to commercial centres to the south and east.[2] Dates for the beginning of what subsequently came to be known as the Third Plague Pandemic vary but in 1890 the disease reportedly spread from Mentze, one of the principal trading towns in south-east Yunnan, to some of the settlements along the Canton River. In previous years the disease had burned out after leaving Yunnan but after 1890 it continued to spread and by 1894 it raged throughout Guangdong. It was then only a matter of time before it reached the populous city of Canton (Guangzhou), the province's capital and chief commercial centre. As bubonic plague thrives in warm weather, the unusually severe heat in Guangdong province during the early part of 1894 most likely favoured its spread.[3]

By May, plague was already at the coast, having arrived at the British treaty port of Hong Kong.[4] Lying at the mouth of the Pearl River and less than eighty miles from Canton, Hong Kong was extremely vulnerable to infection. There was frequent commerce with ports further upriver by junk and steamer and many so-called coolie labourers came annually from Guangdong to Hong Kong in the hope of finding work overseas.[5] Appeals to remain at home were ignored and thousands of labourers entered the city without hindrance. It seems that the population of Guangdong was not particularly alarmed by reports of plague, for many Chinese did not regard the disease as contagious and sought merely to purify the air around their dwellings and businesses.[6] No city in the region employed quarantine in any form and sanitary conditions were poor despite long-standing proposals for reform. The British authorities had connived in the resistance of many Chinese landlords to these reforms, on the grounds that restrictions on overcrowding would diminish the income they could enjoy from rents.[7]

In Hong Kong, plague made rapid progress through the city's over-crowded and insanitary streets. By the middle of June, it had claimed nearly two thousand lives and it was clear that the outbreak was of a different order to earlier ones. Complacency soon turned to panic, with as many as 80,000 people fleeing the city. As a result, commerce was 'seriously affected' and labour was scarce.[8] If the commercial life of the city was to be revived, the confidence of both the local population and the international community had to be restored. In the opinion of most British officials and doctors, this meant demonstrating that the poorest Chinese residents – many of whom were immigrants from plague-affected areas – had been prevented from spreading the disease further. The most impoverished sections of the Chinese population were deemed especially susceptible to infection on account of their squalid living conditions.[9] But early efforts did little to improve sanitary conditions, for they concentrated on forcible removal of plague victims and suspect cases to hospital. The heavy-handed approach taken by the authorities – which entailed house-to-house searches by British troops – had the opposite effect to that intended. It prompted a violent backlash and the flight of thousands more people from the city, worsening its economic situation and spreading plague more widely throughout southern China.[10] As the port's Governor, Sir William Robinson put it, the effect of plague was 'felt in every branch of business, and the loss to the public revenue, to bankers, merchants, shipping companies, the sugar-refining industry, traders, shop-keepers, owners of property and the labouring classes can never be accurately determined'.[11]

Of these groups, it was the shipping interests that were most severely affected. The arrival of plague coincided with the peak of the emigrant labour season when normally thousands of Chinese would have been transported from Hong Kong to the Straits Settlements (Malaya) or further afield. The epidemic caused alarm internationally, too, and Pacific mail steamers leaving for Vancouver refused to take Chinese passengers, as did those owned by British, French and German companies. French and German ships also refused to carry cargoes from Hong Kong and even declined to enter the harbour. This caution was induced not simply by fear of the disease but by the certain knowledge that any vessel that had docked at Hong Kong would be subject to rigorous quarantine at most of the world's ports. Robinson was probably not exaggerating when he declared that, 'so far as trade and commerce are concerned, the plague has assumed the importance of an unexampled calamity'.[12] In his opinion, there was no alternative but to rush ahead with long overdue sanitary reforms, even though it 'might result in the destruction and re-building of one tenth part of Hong Kong' and lead to a massive increase in rents.[13] There would have to be improvements to the water supply and other sanitary works, which had not been much in evidence in this or in any other treaty port until now. The announcement of the discovery of the plague bacillus, independently, by the Japanese researcher Professor Shibasaburo Kitasato and the Swiss-French scientist Alexandre Yersin, did not detract from the perceived need for such reforms. These men – who were both present in Hong Kong – insisted that the habits of the Chinese and the insanitary state of their dwellings allowed the germs to thrive.[14]

Plague prevention in Hong Kong was orchestrated by a committee comprising Dr James Lowson (a doctor at the Civil Hospital), Dr Penny (a naval physician) and Surgeon-Major James (an army medical officer). This committee framed by-laws and other regulations which were subsequently confirmed by Hong Kong's executive and legislative councils.[15] All the while the death toll mounted and by early July 1894 there had been 2,363 deaths and a further 2,000 admissions to hospital.[16] When the disease subsided towards the end of the year, it was probably due to the onset of cooler weather and the exodus of people from infected areas. No further cases were reported the following year but in January 1896 plague broke out again. This time, the reaction of the authorities was immediate: plague victims were removed to isolation hospitals and contacts were segregated on junks moored in the harbour. But the sheer number of victims and their contacts meant that the boats soon became overcrowded and segregation had to be abandoned. The only practicable course of action was to allow temporary residents to return to their families in Canton, providing they had been disinfected. Disinfection was unpopular but after the flight from the city had ceased, the practice of segregation resumed with little opposition, this time in large purpose-built sheds rather than on junks.[17]

There was still no sign of the ambitious programme of urban reform proposed by Robinson two years before, nor had anything been done to bolster Hong Kong's defences against further incursions of plague. Although the Governor was anxious to demonstrate that the authorities could bring plague under control, the colony was not prepared to raise taxes to pay for such measures or to do anything which might interfere with inland trade. This failure to act may have been in deference to the Qing regime in Canton, which was distinctly uncomfortable with the prospect of quarantine and of foreigners inspecting and detaining Chinese.[18] But there was also great reluctance on the part of overseas commercial interests to do anything that would break with the laissez-faire arrangements which had characterized the governance of treaty ports. Up until 1894 few Chinese ports had done much to deal with the threat from epidemic diseases, and medical provisions were rudimentary.[19]

The great exceptions to this rule were the ports of Shanghai and Amoy (Xiamen), which periodically imposed quarantine against ships from India when cholera was reported there. The former took the lead in 1894 and imposed quarantine on ships from southern China when plague appeared at Hong Kong.[20] Immediately after the disease reached the southern ports, the foreign-run Municipal Council in Shanghai instituted the medical inspection of all vessels arriving from Hong Kong and ports in Guangdong. No ship was allowed into the harbour until it had been examined and these precautions remained in place until the middle of September. At the same time, the Municipal Council set about cleaning the settlement and erected temporary hospitals and fumigation stations in case they were required. In 1896 these measures were resorted to once again when plague was detected in other southern ports and, two years later, following further outbreaks in Canton and Hong Kong, a permanent quarantine station was established. It was used frequently in the coming years when plague spread from south China to Amoy, Taiwan, Manila, Swatow (Shantou), Manchuria and Japan. These measures were deeply unpopular with many Chinese but they seemed to keep plague at bay while doing remarkably little harm to business. The volume of trade passing through Shanghai rose steadily from 1894, apart from 1897–8 when commerce was interrupted because of flooding, civil unrest and the Spanish-American War.[21]

While Shanghai continued to benefit from the expansion of trade, Hong Kong suffered severely. From 1898 to 1903, when the British doctor William Simpson made his report on plague in the colony, it had not been free of the disease for more than a month at a time. Plague was either going unreported or the colony was being continually re-infected from the mainland. As Simpson pointed out, 'Nothing is done in China to mitigate the effects of any infectious disease and plague is no exception to the rule.' But the port's authorities had been negligent, too, even though the cost of a having a proper sanitary establishment paled into insignificance when compared with the expenses incurred at foreign ports when vessels from Hong Kong were detained in quarantine. What the city needed, in Simpson's view, was better intelligence about the spread of disease in

China and a system of prevention based upon it. 'This protection may be afforded, not by quarantine,' he argued,

> but by a supervision at the most dangerous period of the year over the junks and steamers trading with infected districts, and insisting on the larger steamers, native and European, which carry hundreds of passengers daily to and between Canton and Hong Kong, having a surgeon on board at the companies' expense to report those that are sick with plague and other infectious diseases, the alternative being a medical inspection before the steamer communicates with the shore.[22]

Simpson's recommendation fell short of the measures that had been introduced by the customs service in Shanghai and did little to reassure those responsible for the safety of ports trading regularly with Hong Kong. Foreign powers were growing insistent that Hong Kong and other Chinese ports introduce quarantine or similar measures to control the spread of plague.[23] But the Qing regime remained opposed to quarantines which it associated with foreign control. Yet the only hope of preventing further interference in matters of trade was for China to adopt its own quarantine system in parallel to that working in the treaty ports. In 1902 the Beiyang Epidemic Prevention Office was created and four years later this and other local initiatives were placed under imperial control.[24] These sanitary institutions were intended largely to allay international concerns but they also allowed the Qing administration to assert its authority in the ports: Chinese personnel would now have the sole power to search and inspect Chinese bodies.[25] But the measures did not have the desired effect and the next few decades would bring further foreign intervention in China's sanitary affairs.

A bridge to the West

Plague's first port of call outside China was Bombay, where it arrived in 1896. Although there were competing theories as to how and from where the port was infected, most observers believed the disease came directly from Hong Kong.[26] Having arrived, it spread quickly through warehouses in the docks and then throughout the city and surrounding region. As in Hong Kong, plague found an agreeable home in the slums and alleys of a burgeoning city. Unknown at the time, the most efficient arthropod vector of the disease – the rat flea, *X. cheopis* – was common in Bombay, probably having been introduced in bales of cotton imported from Persia or Egypt.[27]

The spread of plague in Bombay highlighted the inadequacies of the city's sanitary precautions. Despite several decades of reform, India's greatest port had not shaken off the pariah status it had acquired during the cholera epidemics of the 1860s and '70s. The Municipal Commission had begun tentatively to clean up the city but vested interests acted as a drag on reform. Although Indian men had the right to vote and stand for election, they could do so only on the basis of a property qualification.[28] As a result, the Council was dominated by slum landlords who opposed any regulations which affected the property they owned or increases in taxes to pay for sanitary improvements.[29] The other problem was the light-touch quarantine regime maintained by the health authorities with the connivance of local merchants and the provincial government. A system of quarantine and medical inspection had existed since the 1870s but the measures represented the bare minimum needed to placate international opinion. Thus, when quarantine was imposed in 1894 against arrivals from Canton and Hong Kong,[30] it is likely to have been enforced with less rigour than in many other ports.

Plague probably arrived in Bombay in or around May 1896, when some Indian practitioners began to treat cases of fever with glandular swellings. Apparently, such symptoms were not uncommon in Bombay but the unusually high number reported at this time suggests that plague was present before the disease was officially announced on 23 September. There was great reluctance to diagnose a disease whose presence would surely bring commerce to a halt, and rumours of plague were consequently denounced as 'scaremongering' by the *Bombay Gazette*, which represented the interests of the city's European merchants and shipping lines.[31] At the end of September, officials were still reluctant to confirm the presence of the disease but some precautionary measures were now being taken elsewhere. Quarantine was imposed at Karachi and Calcutta against arrivals from Bombay and Bombay's Health Department began to disinfect and limewash all buildings deemed likely to be infected.[32] In the coming months, the campaign against plague gathered steam and the disinfection of drains and gullies commenced on an industrial scale, together with enforced hospitalization of persons suspected of having plague and segregation of their contacts. These measures produced what the Municipal Commissioner P.C.H. Snow described as 'a wild unreasoning panic' among the city's inhabitants. From October 1896 to February the following year, his office was 'besieged every day by natives of all classes imploring that nothing too drastic should be done, and every consideration should be shown to the people with a view of keeping them in the City and preventing absolute stoppage of trade'.[33] In Hong Kong there had been no such attempt to prevent the population from leaving and the policy adopted in Bombay created unnecessary antagonism. Some of those who had earlier offered their services to the British now became their vociferous opponents. The city's powerful Indian mercantile groups, which had a vested interest in getting the plague under control, also urged moderation. A sizeable and influential minority of these merchants were Parsis – a community which had been at the forefront of medical philanthropy in the city and which had led sanitary initiatives through the Municipal Corporation. Men such as Dinshaw Edulji Wacha, a prominent Parsi politician and later President of the Indian Merchants' Association, were prepared to support anti-plague measures but not if they terrified the population.[34]

The extent of the problem became evident as early as October 1896 when workers from the city's cotton mills gathered outside the Arthur Road hospital which was being used to house plague victims. They threatened to destroy it and attack its employees. On the 29th of that month, a gang of nearly a thousand mill-hands returned and carried out these threats, damaging the building and assaulting some of the staff. The crowd was dispersed by a posse of police constables but it now became clear that the threat posed by plague had to be set against the potential for mass disorder and flight. The Municipal Commissioner therefore decided to modify segregation, allowing some people to remain within their homes. These actions calmed the situation and, from February 1897, many of those who had fled the city began to return.[35] But plague was spreading from Bombay into the surrounding area and the relaxation of measures was roundly criticized by many of the city's European inhabitants who favoured more decisive intervention. Ironically, the loudest criticism came from the *Bombay Gazette*, which had earlier been reluctant to acknowledge the presence of the disease. Its opinions were shared by many senior figures in the British administration who were concerned about the restrictions which had been placed upon Indian shipping.

Within a few days of plague being reported in Bombay, quarantine was imposed at Suez and other foreign ports against vessels sailing from India. The quarantine at Suez was much less of an obstacle than formerly because the Venice International Sanitary Conference of 1892 had instituted a more liberal regime. Now, the greatest impediment to trade was not so much the Board of Health at Alexandria as the punitive measures taken by some nations on an individual basis. The most severe restrictions were imposed by France, and in Marseilles steamers from Bombay were prevented from landing their passengers.[36] Stringent restrictions on Indian shipping were also imposed in Australia, the USA, Brazil and the Ottoman Empire.[37] But perhaps the most damaging restrictions were the bans imposed by France, Italy and Germany on the importation of raw hides and skins from India, on the grounds that these articles were capable of carrying plague. As these three countries together received over 40 per cent of India's exports of hides and skins (India's sixth largest export commodity), the embargo had a disastrous effect; particularly upon Bengal, from which most of the hides were obtained.[38] At the insistence of Calcutta merchants involved in the trade, the Bengal Chamber of Commerce complained vigorously about the restrictions, arguing that such a ban was medically unnecessary. All hides were exported dry and treated with arsenic, which many believed would kill plague germs; moreover, Bengal at this time appeared to be free from the disease.[39]

Although the action taken by the Municipal Commission in Bombay had been accompanied by the imposition of quarantine at Bombay, Aden and Karachi, it was insufficient to allay the anxieties of foreign trading partners and imperial rivals. More drastic action was required. The Government of Bombay's first step was to engage the services of James Lowson, who was appointed Plague Commissioner for Bombay. On the basis of his experience in Hong Kong, Lowson recommended that municipal plague committees should be replaced with smaller bodies headed by military men, containing only one medical practitioner. These new plague committees were given enhanced powers under a new piece of legislation which was intended to signal the government's resolve to check the spread of plague. The Epidemic Diseases Act of 1897 empowered plague committees to segregate contacts in special camps according to caste and religion and to inspect and, if necessary, detain railway passengers suspected of having plague. Europeans, however, were exempt from such provisions.[40]

New legislation enabled sanitary intervention on a scale unprecedented in India even during the worst ravages of cholera. As power passed to soldiers and administrators, the medical establishment was effectively sidelined. Senior figures in the Indian Medical Service normally tended to favour a gradual approach to sanitary reform, which did not alienate influential sections of the Indian population. But some IMS officers, especially younger men, felt frustrated at the lack of sanitary progress in India and welcomed the new legislation as an opportunity to advance a more radical agenda. Many of these younger officers set about segregation and medical inspection with gusto. Some were also enthusiasts of the science of bacteriology and saw the new political framework as an opportunity to introduce methods of prevention such as inoculation against plague.[41] This blend of science and coercion elicited a backlash of fearful proportions. In many Indian towns and cities such measures or the prospect of their implementation led thousands to flee, while in some places opposition to plague committees provided a stimulus to nationalist politics of a more virulent kind than the moderate opposition the British had hitherto faced from the Indian National Congress. In the city of Poona (Pune), not far from Bombay, the British plague commissioner W.C. Rand was assassinated, and in other places across western and northern India there were riots and other violent incidents.[42] Many British inhabitants feared an uprising on the scale of the Mutiny and Rebellion of 1857.[43] The damage done to commerce by population flight and industrial strikes was also very serious. In January 1897 the *Bombay Gazette* reported that the city's stock exchange had closed and that business was at a standstill.[44]

The attempt to control plague using drastic measures thus proved disastrous and the British were forced to enter into dialogue with community leaders. Following these talks, the draconian measures of 1896–7 gave way to voluntary segregation which paid more attention to the requirements of caste and religion. Women were appointed to inspect female passengers at railway stations. An explosive situation was thereby diffused and business in Bombay returned to something approaching normality. Action against plague was also being reformulated in line with evolving medical opinion about how the disease was spread. Two or three years after the initial outbreak in India, there seemed to be little evidence that plague – or at least the kind seen in India – was easily communicable. The Indian Plague Commission, and the foreign plague commissions which descended on India to examine the disease, began to see rats as the most likely disseminators. The theory that the disease was spread by fleas feeding on rats (first proposed by P.L. Simond in 1898) was not confirmed to most people's satisfaction until 1906. However, the observation that mass mortality among rats usually preceded human deaths and that the first cases often occurred, as in Bombay, in rat-infested warehouses, justified the shift of sanitary measures away from the human body to rodents. Henceforth, rat destruction was to be central to the strategy against plague, however ineffectual it proved to be.[45]

These developments were closely bound up with changes in the diplomatic arena, especially negotiations at the International Sanitary Conference in Venice, in 1897. This conference was crucial in deciding India's fate because the resulting convention relaxed restrictions on Indian shipping and put in place a more liberal sanitary regime. Although it did not immediately prevent harsh unilateral actions against infected countries, it made them more difficult to justify. One of the crucial changes as far as India's economy was concerned was the lifting of the ban on articles deemed susceptible of carrying plague. Thanks to the researches of the foreign plague commissions, most medical scientists were now satisfied that hides and skins did not disseminate the disease.[46] Indeed, the role of most merchandise (except grain and rags) in the spread of plague was said to be minimal. The measures agreed at the Venice conference were liberal in other respects, too. Outward-bound sea traffic was to be inspected thoroughly and vessels containing suspect cases were prevented from sailing; measures which greatly reduced the need for quarantine at ports of destination. In the same vein, quarantine was declared unnecessary for overland passengers, the inspection of travellers from infected localities being considered sufficient.[47]

The thrust of the Venice Convention was to bring the sanitary regulation of commerce into line with scientific knowledge and to minimize disruption internationally. To this end, plague was made an internationally notifiable disease, as in the past poor intelligence had induced panic and caused unnecessary disruption of commerce. When plague reached Bombay, most observers had seemed to think it would be a matter of time before other places were infected and states wanted to ensure that they would not be harshly treated.[48] It was therefore in the interests of most nations to ensure that quarantine, medical inspection and other such measures were standardized so that no country would be disadvantaged. As one Spanish physician put it, the Convention represented not simply a compromise, like previous conferences, but a convergence of national interests.[49] While the degree of consensus ought not to be exaggerated,[50] the Venice Convention marked a significant improvement on what had existed before. Even in Britain, a reluctant signatory to previous conventions, there was a feeling that the Venice document was compatible with the needs of commerce. The President of the Epidemiological Society, Professor J. Lane Notter spoke of the Convention as a 'great advance on the part of the nationalities toward a truly *liberal* and truly *scientific* conception of the means to be adopted by respective governments for the prevention and control of infective diseases'.[51] But confidence in this new regime could only be sustained by more exacting sanitary standards and a more rigorous system of surveillance.

Plague becomes pandemic

In the first year after the Venice conference, plague remained more or less confined to India and southern China but in the next decade it spread to most other parts of the world. Of all the outbreaks that occurred as it made its way beyond Asia, the one which caused most consternation was in Egypt. Although the country had been free from plague since 1844, it was still regarded as a staging post from which 'Asiatic' diseases could easily reach the West. It was for this reason that an international body had been created at Alexandria many years before and, in 1899, the Egyptian Board of Health was still by far the most important sanitary authority in the eastern Mediterranean. The Sanitary, Maritime and Quarantine Board of Egypt, as it was now termed, was composed of medically qualified delegates from most European countries as well as the Ottoman Empire. The Egyptian government was also represented on the Board and its delegate had three votes, as opposed to one per delegate from the other powers. The Egyptian Board was notoriously political and swayed with the winds of diplomacy but from 1897 there was more agreement between the delegates than at any time in its history and their decisions were in accordance with the Venice Convention.[52]

However, arrangements in Egypt were insufficient to prevent the arrival of plague, the first case of which was reported in Alexandria in May 1899. By mid-June there were forty-three cases, although the true number was probably higher due to misdiagnosis and concealment. It was widely believed that plague had been imported from Jeddah, where cases had been reported since March. Although the harbour authorities at Alexandria had declared arrivals from Jeddah subject to quarantine, it was an open secret that ships from infected ports had been admitted in the past.[53] On this occasion, there were rumours that plague had slipped into the country with Greek merchants who had become adept at evading Egypt's sanitary restrictions. Yet some felt that the problem lay not with the Sanitary Board but with the Venice Convention, which prevented effective measures from being taken. The Convention permitted the imposition of quarantine against infected ports such as Jeddah but expressly forbade similar measures against nearby ports in which no disease had been reported. It was therefore conceivable that plague had been imported into Egypt from Red Sea ports which had not been officially declared infected. For implacable opponents of quarantine, however, the failure to prevent plague spreading through the Suez Canal revealed its inherent flaws. As one British correspondent put it, Egypt had based its sanitary security on an 'elaborate system of leakiness', the leakage having been very apparent of late, allowing not only plague but cholera to enter the country to devastating effect. It was now idle, in his view, to speculate whether quarantine should have been more effectively enforced, or whether the provisions of the Venice Convention were adequate: 'once more it ha[d] been proved that the most elaborate quarantine arrangements fail to protect a country'.[54]

In its defence, the Egyptian Sanitary Board maintained that the disease had been introduced into Egypt by rail, but this version of events failed to convince many people as there were no outbreaks along any of the main lines during 1899. It was only once the disease had become established in Alexandria and in Port Said (in 1900) that plague spread inland along trade routes: by river-craft, camel caravans and railways.[55] Despite the pronouncements made in Venice, the word on the street was that plague was carried in all manner of infected merchandise and not only in officially suspect commodities such as rags and grain. The close association of the disease with commerce may have owed something to the fact that the first reported outbreaks in Alexandria occurred in groceries and bakeries owned by Greek merchants; an observation which may have been the source of rumours about its importation.[56] Once the disease was announced, the maritime Sanitary Board attempted to prevent its spread from Alexandria to other Egyptian ports in order to minimize the damage inflicted by foreign quarantines. By the end of May 1899, Athens had already imposed quarantine on vessels arriving from Alexandria and the government was concerned lest all its ports face similar restrictions.[57] Action against Alexandria was also becoming more widespread. When the Austro-Hungarian port of Trieste and the remaining Greek ports followed Athens in imposing ten- to twelve-day quarantines in early June, it seemed likely that the whole European Mediterranean would join them.[58]

In order to stave off further restrictions, the Egyptian government was urged to take action on the merest suspicion of plague and it was expected that such action should be drastic and immediate.[59] Responsibility for the prevention of plague in Egypt now shifted from the maritime Sanitary Board to the Public Health Department of the British-led administration. The measures taken in Egypt were similar to those recommended in the report of the Bombay Plague Commission in 1897 but an Act of 1895 had already given the Egyptian Health Department effective powers of notification, isolation and disinfection in the event of epidemics. The Egyptian people were also said to be more receptive to sanitation than the inhabitants of India and Hong Kong.[60] As one observer put it, there had been an 'immense change in the customs, mode of life and general culture of the people'.[61] But others disputed this assertion, claiming that 'As usual, in all semi-civilised communities, the natives resist the authorities in taking sanitary precautions'.[62] In the majority of cases, however, resistance was passive rather than violent as it had been in Bombay or indeed in some other Middle Eastern cities – such as the Persian port of Bushire – where plague-workers, telegraph cabins and other symbols of foreign authority were attacked.[63]

It is certainly true that Egypt avoided the mass fatalities and civil unrest that had afflicted countries further to the east. As well as having a clear mandate under existing legislation, the Public Health Department was assisted in its task by the swift decision by Egypt's Public Debt Commissioners to advance £20,000 to the government solely for the purpose of combating plague.[64] But the Department had also learned a vital lesson from India: that plague measures were unlikely to be effective unless the authorities made an effort to work with the population. To this end, the Health Department and municipal authorities in Alexandria co-opted community leaders, who were rewarded for acts such as the notification of suspicious cases.[65] Although other countries remained alarmed by events in Egypt, the sophisticated approach of the government provided some reassurance and by the middle of September the epidemic in Alexandria had virtually ceased.[66] Restrictions on Egyptian shipping began to be lifted and the situation never became as serious as many had feared. Sporadic outbreaks continued in Egyptian cities such as Alexandria and Port Said but they were easily contained and swiftly stamped out. Better sanitary regulation of trade routes within Egypt and the formation of new agencies such as the Special Plague Service also enabled the government to deal effectively with plague beyond the main coastal and canal settlements.[67]

When plague had arrived in Egypt the prospect of it spreading to southern Europe seemed likely but in only one city did the disease extend significantly beyond infected ships to townsfolk. That place was the Atlantic city of Oporto – an important commercial and industrial centre of some 150,000 inhabitants lying on the River Douro in northern Portugal.[68] In June 1899 the first death from plague (of a Spanish stevedore) was reported. As was so often the case, the early intimations of plague met with denial and disbelief. The physician making the announcement – Dr Ricardo Jorge, Oporto's chief medical officer – faced hostility not only from commercial interests but from most of the city's inhabitants. As one correspondent put it, 'It is no light matter to proclaim the existence of plague in a sea-port town dependent for its prosperity on maritime commerce.'[69] Those who accepted its presence immediately began to point the finger at a British vessel, recalling allegations made during the Lisbon yellow fever epidemic of 1857. Rumours circulated to the effect that the disease had been imported from India or Egypt by the *City of Cork*, which regularly sailed between Lisbon and London. The ship carried goods of diverse origins (tea from China, rice from Burma, tapioca from Ceylon and jute from India) which had been transhipped via London. Its last call at Lisbon was on 5 June, which coincided with the first recorded death from plague.[70]

It was not unreasonable to suppose that plague had reached Portugal from the East but this possibility was soon discounted by a delegation from the Parisian Institut Pasteur, which arrived at Oporto at the beginning of September. The Portuguese had previously sought advice from French experts when faced with serious epidemics in their ports, but the team led by A. Calmette and A.-T. Salimberi typified the scientific internationalism which had been evident since the arrival of plague commissions in India and the passage of the Venice Convention. Although their main aim was to develop a bacteriological (serum) therapy for plague, the team made it its business to look into all features of the outbreak, including its causation and prevention.

Calmette and his colleagues concluded that the disease had probably arrived from Paraguay or Argentina as early as March or April, some months before those countries were officially declared infected. Two crew members of a ship which had touched at Buenos Aires en route from Paraguay had died soon after arriving at Lisbon and a third was found to have buboes in his groin. However, this information seems to have been kept under wraps by the port authorities, allowing the disease to spread throughout the docks before it was officially reported at the beginning of June.[71] Even after plague had been confirmed, the natives of Oporto – and of Portugal generally – remained sceptical of its presence.

But neighbouring countries were nervous. The Spanish imposed quarantine on all arrivals from Portugal and began the construction of five lazarettos along the land frontier. Until these were ready and a system of medical inspection could be established, all communication across the border was prohibited.[72] Land quarantines were still permitted under the Venice Convention – at the insistence of powers such as Austria-Hungary and Russia – but to countries which increasingly relied upon medical inspection they smacked of an older and more insular era. Even Spanish physicians condemned the land cordon as outdated, ineffectual and against the spirit, if not the letter, of the Venice agreement. But in the view of Dr José Verdes Montenegro, the Portuguese had only themselves to blame. Had they not attempted to conceal the outbreak, he argued, Spain would not have reacted as harshly as it did. In his view, Portugal's reluctance to report the disease amounted to 'a case of international irresponsibility'.[73] Montenegro was pleased when Spain decided in 1900 to reopen the frontier and allow travellers through providing that they and their goods were inspected and treated with chemicals if necessary. This, he felt, was more in keeping with the spirit of international sanitary arrangements which had shifted their emphasis from sequestration to disinfection.[74] He also praised the fact that the French had relied on maritime measures rather than the sanitary cordons of old.[75] Apart from quarantine, France warned its fishermen not to land in Portugal or ports in northern Spain in case the disease had spread unnoticed across the border.[76] Other governments – including those of some of Portugal's colonies and former colonies – took a tougher stance. In August 1899, the Azores closed their ports to all vessels from Portugal, and Madeira soon followed suit. In September 1899, the Brazilian authorities also imposed a harsh quarantine against ships sailing from anywhere in Portugal.[77]

Portugal's maritime commerce ground to a halt and, if permanent damage were to be avoided, the federal government believed it had to act decisively to reassure foreign governments. At the end of August 1899, a military cordon composed of eight battalions of troops surrounded Oporto with the object of preventing unauthorized persons from entering or leaving. In taking this step, the Federal Board of Health in Lisbon was flying in the face of the authorities in Oporto and acting in defiance of most of the city's doctors, including foreign observers such as Calmette.[78] But if the drastic measures taken by the Portuguese government were supposed to allay international anxiety they were not working. A number of cases had already appeared in villages beyond the city, leading the *British Medical Journal* to gloat: 'This is the old story: the cordon serves but to spread the disease, and is a form of quarantine which ought to be discredited and forbidden.'[79] Moreover, the cordon was deeply unpopular with the inhabitants of Oporto. By mid-September the ban on exports had thrown 12,000 people out of work and there were violent protests against this and other measures. The medical profession was also blamed for spreading the disease and doctors were attacked as they took victims to the hospital or morgue.[80] These protests led the authorities to allow the dispatch of goods from the city once they had been disinfected and the city's factories were again able to open their gates. People were also permitted to pass through the cordon after medical inspection and disinfection of their baggage.[81]

The relaxation of quarantine was accompanied by an upsurge in cases. During the second week of October – the worst week since the epidemic began – twenty-eight new cases were reported, bringing the total to 148, with forty-four deaths.[82] The immediate response of the city's authorities was to introduce new measures, including a ban on all public entertainments such as theatres and balls. But in the opinion of most foreign observers the real causes of plague in Oporto were not being addressed. Although the upper parts of the city, which rose sharply on a granite cliff above the sea, had delightful villas and parklands, the lower slopes were notorious for their twisting, filthy alleyways and overcrowded tenements. As Dr A. Shadwell, a British doctor in Oporto, remarked in a paper to the Epidemiological Society of London, this, the oldest part of the city, was a 'regular Oriental quarter, consisting of narrow passages and lanes, with lofty houses on both sides, and abounding in Oriental dirt'.[83] According to some estimates, it was not uncommon for fifteen to twenty people to be sharing a single apartment.[84] To make matters worse, the inhabitants of the city seemed to lack the most basic notions of hygiene and appeared to resent any interference in their daily lives. Oporto thus represented a throwback to an older Europe characterized by pestilence, squalor and superstition. Although plague had been absent from Portugal for over two hundred years, some feared that it might even become endemic.[85] Indeed, the episode served as a warning that Portugal was still imperfectly civilized. As one foreign correspondent in Oporto remarked:

> The sanitary officials have much to contend with in face of the prejudice of the populace. Their beliefs indicate an Orientalism which to those who know the Portuguese will seem surprising. The people believe that the disease is the invention of the doctors, and that patients are taken to hospital only to be made away with. We look for such opinions in India or China, but scarcely believe it possible in Western Europe.[86]

Acutely aware of their city's poor reputation and the damage that plague had done to the economic vitality not only of Oporto but of Portugal as a whole, the municipality pledged radical reforms. In late October 1899 it announced that it was trying to raise an enormous loan worth £1 million for sanitary purposes.[87] Like the authorities in Hong Kong and Bombay, the municipality of Oporto came to the conclusion that the city's commercial viability depended on convincing the rest of the world that the Corporation was serious about tackling the disease. Its approach stood in complete contrast to the heavy-handed intervention by the Lisbon government, which aggravated the already tense relationship that existed between the capital and provincial cities.[88] But the most important legacy of the outbreak from an international point of view was that it revealed the fragility of the sanitary arrangements established by the Venice Convention. They were evidently only as good as the system of disease notification upon which they rested and some governments were manifestly not to be trusted. Portugal and Spain – at least initially – had also abrogated the spirit of the Convention by opting for outdated and illiberal measures such as cordons.

As 1899 drew to a close, the epidemic in Oporto subsided with no further outbreaks in Europe. The following year, plague did arrive in some European ports, including Trieste, Glasgow, Naples, Marseilles and Hamburg, but cases were few and there was never any real concern that the disease would take hold as it had in Oporto. Most of these ports had effective systems of sanitary surveillance and competent health authorities which quickly isolated cases and brought the outbreaks to an end. The real concern at the end of 1899, and throughout 1900, was the spread of the disease to countries and continents which had never before been infected. Although epidemics of plague may have occurred in South and Central America during the Spanish and Portuguese conquests, it was generally believed that the disease had never visited the Americas. Thus, when plague was first reported in South America – in Paraguay, in September 1899 – some medical authorities found the news hard to accept, especially as the disease was said to have appeared in Asunción, some 600 miles from the sea.[89] By the end of the year, Paraguay, Brazil and Argentina had all been infected, and in 1900 plague had reached the United States, as well as many parts of Africa and Oceania.

One cannot trace the path of plague with certainty but its movement around the world was closely connected with economic activity. Paraguay and Argentina were probably infected by grain from India imported into the Rio Parana/Rio Paraguay river system, while plague seems to have reached the port of Santos in Brazil on steamers from Oporto.[90] From South America, the disease spread along well-established coastal trading routes, following the path previously taken by yellow fever up to the Caribbean and Central America. International commerce and labour migration were the most likely causes of plague infection in many African colonies, too. Portuguese East Africa, for example, was probably infected as a result of trade with Goa or Madagascar,[91] while British East Africa was probably infected through shipments of grain from India and Indian migrant labourers brought in to work on the Uganda railway. However, in German East Africa and its neighbouring sultanates the most likely source of plague was the return of Muslim pilgrims from the Hejaz.[92] Normal patterns of disease distribution were complicated by the Anglo-Boer War of 1899–1902 which massively increased the demand for certain commodities from plague-infected ports. These two factors combined powerfully in the infection of South Africa. Cape Town and Port Elizabeth were probably infected as a result of shipments of grain from India or Argentina, which were needed to feed British troops, horses and other draught animals.[93]

Plague arrived in Central Asia after moving along caravan routes from India, reaching the ancient Silk Road city of Samarkand and Astrakhan in Asiatic Russia in 1899. The disease moved north into Tibet and Mongolia later the same year, by similar means. These modes of transmission were reminiscent of an era long distant and evoked similarly archaic responses. After plague was reported in Astrakhan, for instance, the Romanian authorities closed the frontier with Russia, which prompted similar measures within Russia in an attempt to contain the outbreak. Trade practically ceased until the Romanians began to allow some travellers from Russia through the border towns, where they were medically inspected. Even then, business remained slow and delays were common because all goods had to be disinfected, not just those deemed susceptible under the Convention.[94]

For the most part, however, plague spread along routes of maritime trade. Invariably, trading vessels – especially those carrying grain – had a large population of rats whose fleas sometimes infected their crews and dock workers in their harbours of destination. By this means, plague spread rapidly from Chinese ports to Japan, for example, infecting Kobe and Osaka in 1899; to Manila, Penang and Hawaii in 1899. From Hawaii, the disease spread west to San Francisco and also possibly south-east to other Pacific islands. The dilemma facing the afflicted countries and their trading partners was how to break these chains of transmission without strangling international trade. Governments also had to square the relatively liberal provisions of the Venice Convention with the expectations of their own populations. In some cases, such as in Oporto, people were slow to acknowledge the presence of plague and hostile to attempts to control it. But the prospect of infection usually produced demands for drastic action to prevent importation or stamp the disease out should it appear. The dilemma was most difficult to resolve in countries which had rarely suffered from contagious diseases in the past.

Nowhere was this more evident than in Australia, which had a strict policy of quarantine against ships carrying convicts and emigrants deemed likely to harbour infections.[95] Although the volume of trade with Asia was not particularly high, the increasing flow of Chinese emigrants to Australia was giving rise to concern that they would introduce 'Asiatic' diseases. Fear of infection mingled with fear of diluting Anglo-Saxon power and led to quarantine becoming an instrument of racial policy.[96] The infectious diseases imported into Australia during the nineteenth century were relatively common ones such as typhus, scarlet fever and smallpox.[97] These often created great alarm, but nothing compared to plague when it arrived in Sydney in 1900.

In January of that year the first official plague victim – who worked in Sydney harbour – was announced. At first, it seemed this might have been an isolated case but in late February more were detected around the wharf. From here, the disease spread rapidly in rat-infested warehouses and overcrowded dwellings close to the harbour, and then to houses in the central ward of the city. The potential for damage to Sydney's trade and that of New South Wales as a whole was enormous. Sydney was by far the most important of Australia's ports, having regular connections with most parts of the world; in 1899 the total tonnage of shipping arriving in the port – some 2,589,457 – dwarfed that of its competitors. Until plague actually arrived, however, there was relatively little anxiety about the disease. Despite improvements in maritime technology, Sydney was still three weeks away from Hong Kong, which Australians regarded as the most likely source of plague. Unknown to them, some smaller ports closer to Australia had recently been infected, one of which was Noumea in New Caledonia. The arrival of plague in Noumena was announced shortly before the disease reached Sydney, which was just three and a half to six days away by steam – significantly less than the incubation period.[98]

As in most other plague-infested towns, the epidemic in Sydney was relatively localized and affected the poorer classes predominantly, particularly those who worked at the docks or in places such as flour mills which were infested with rats. By the time the epidemic ended in August 1900, there had been 303 cases and 103 deaths.[99] The limited geographical and social distribution of plague in Sydney encouraged measures designed to contain the disease within infected localities. Once cases were identified, they and their contacts were quickly dispatched to the quarantine station at Woolloomooloo and thence, by sea, to the quarantine station at North Head. There, plague victims were completely isolated from contacts: the former being sent to hospital, the latter to relatively comfortable pavilion-style houses. By the time the epidemic had ceased, nearly 1,800 people had been placed in quarantine. Accompanying this policy of isolation was one of area quarantine and cleansing: afflicted zones within the city were cordoned off and their residents obliged to remain as gangs of workmen went about the business of disinfecting, limewashing and cleaning. There was nothing very remarkable about these procedures except the fact that they had no basis in law. The Australian colonies had some public health legislation modelled on Britain's but its scope was more limited and local governments had far fewer powers.[100] Although the law allowed the quarantine of immigrants, there were no provisions for the removal of patients and contacts if an epidemic of contagious disease occurred within Australia.[101] The other noteworthy aspect of the response to plague in Sydney was that the government of New South Wales was at odds with its Board of Health over how to tackle the epidemic. The Board argued strongly against the indiscriminate segregation of contacts, maintaining that it was necessary only to remove the sick to hospital. It also advised that merchandise could be removed from some of the areas that had been cordoned off.[102] The government declined this advice and faced an angry response from residents of infected areas and from businessmen who protested against stoppages due to quarantine and damage to goods and premises during clean-up operations.[103]

The tough measures taken in Sydney during the 1900 outbreak seemed to be effective but when plague returned two years later the approach was very different. By that time, plague had broken out in other Australian ports and it may have re-entered Sydney through coastal trade with Queensland and Western Australia rather than commerce with Asia and the Pacific Islands.[104] The presence of plague elsewhere in Australia gave the authorities more room to manoeuvre. This time, there was no knee-jerk reaction but a considered response in keeping with the latest scientific advice. In 1902 there was general agreement that plague was not particularly contagious in its bubonic form and that rats played a key role in the spread of the disease. On this occasion, the Board of Health's advice was accepted and there was no area-quarantine or segregation of contacts. Anti-plague measures concentrated instead on the removal of the sick to hospital, cleansing infected localities, and the destruction of rats. This new approach was more humane, and far more compatible with business interests. As Sydney's Chief Medical Officer, John Ashburton Thompson explained: 'areas which were deemed to be infected were rapidly and thoroughly cleansed, but they were not closed during that operation [and] movement of population and trade were in no way interfered with'.[105]

A fragile consensus

The shift in focus from people to rats was endorsed by the International Sanitary Conference held at Paris in 1903. Even though there was a growing sense that coordination was desirable, the response to plague had been quite fragmented up to this point because many states had been reluctant to relinquish their sovereignty in matters of health.[106] But the damage which hasty and draconian measures had inflicted upon trade galvanized the international community. Moreover, no government wanted to face the embargoes and sanitary restrictions which had brought great port cities like Hong Kong and Bombay to their knees. At Paris, therefore, the delegates strove to bring the control of plague – and other epidemic diseases – into line with both the latest scientific opinion and the needs of international commerce. The Convention signed at this conference – which resulted in the first set of international health regulations – was ratified at Rome in April 1907. It removed most articles from the list of goods formerly deemed susceptible of harbouring plague and gave free passage to vessels on which no cases had appeared within five days of sailing.[107] More emphasis was now placed on the destruction of rats in harbours and ships, as well as on measures to prevent them from boarding.[108] Land quarantines were banned, although persons showing symptoms of plague and cholera could still be detained at the border.[109] The balance also shifted decisively from sanitary action in ports of arrival to measures in ports of embarkation, making it urgently necessary to improve the notification of infectious diseases.[110]

Although major powers such as Britain and France were solidly behind the 1903 Convention, it did not initially cover all of their colonies and dependencies. Nevertheless, when plague struck the British West African colony of the Gold Coast, in January 1908, the administration was determined that it should act in the spirit of the Paris agreement. The outbreak, which occurred in the colonial port and capital of Accra, had probably been under way for some time before it was officially announced on 11 January. Once plague was confirmed, the colonial authorities acted immediately and informed all British and foreign governments in West Africa of the situation and the measures being taken to combat the epidemic. These included the closure of schools, the isolation of the sick, and the destruction of infected dwellings and rats. While these measures were implemented, the Colonial Office dispatched W.J. Simpson, who had experience of plague in Hong Kong. Simpson arrived in Accra in February and took charge of the Committee of Public Health and thus of all preventive measures. Of these, the most notable are those which were designed to improve the salubrity of vessels leaving the port. Cargoes and passengers were fumigated before they could sail, the holds of ships were thoroughly inspected and 'native' crew and passengers were induced to undergo inoculation against plague.[111] Yet, even in sympathetic countries like Britain there were doubts about whether the small medical staff available to Simpson would be sufficient to enforce these measures.[112]

Inasmuch as they emphasized the duties incumbent on ports of embarkation, the precautions taken in Accra were in line with the agreement reached in Paris. However, the price of a more liberal sanitary regime internationally was a tougher one locally. Confidence in the ability of colonial governments to impose sanitary measures effectively was evidently lacking in many quarters and, for this reason, Simpson's committee took what some would have regarded as a retrograde step by imposing a sanitary cordon around Accra. Only those Africans who had been inoculated were permitted to leave but, as many feared, a number of infected persons managed to evade the cordon; some by land, others apparently on fishing vessels. Plague soon appeared at other places along the coast, causing great alarm internationally and the imposition of lengthy quarantines. These smaller outbreaks were quickly brought under control, and in Accra itself the epidemic had abated by October 1908. Its relatively swift end was attributed to the imposition of measures which the Acting Governor admitted were 'drastic'. An Infectious Diseases Ordinance passed in 1908 had greatly enhanced powers over the bodies of infected and suspected persons and over the destruction of property, albeit with provisions for compensation.[113] But those directing plague operations had learned an important lesson from previous outbreaks and worked through local African officials to ensure compliance with controversial measures such as quarantine and mass evacuation.[114]

The response of the British authorities in Accra set the tone for many later outbreaks in the colonies of tropical Africa, in that vigorous action was taken locally in order to reassure other imperial powers.[115] However, the example of the Gold Coast shows that the response of the international community at first fell short of the ideal outlined at Paris in 1903. Many countries still lacked confidence in the ability of governments – particularly colonial administrations – to take adequate measures and report reliably on outbreaks of disease. It was this realization which hastened the creation of the world's second international health organization – the Office Internationale d'Hygiène Publique – in 1909. The creation of such a body (which had been agreed in Paris in 1903) would once have aroused suspicion but most leaders could now see the need for it. As the Office had no executive powers, there was little in the proposal to alarm nations, such as Britain, which were normally sceptical of supranational authority.[116] The main task of the Office was to communicate information about infectious diseases through monthly bulletins, which would also carry occasional articles on epidemiology and disease prevention. The earliest volumes were full of articles on plague and rat destruction, showing that the pandemic had been the main driving force behind the creation of the Office.[117]

The Conference of American Republics had quickly accepted the Paris agreement and it was enshrined in the Washington Sanitary Convention of 1905.[118] However, there were a few changes in the American document, the regulations regarding plague being somewhat stricter than those in force in Europe and its colonies.[119] Mass immigration from Europe and Asia had left many American states – not just the USA – feeling vulnerable to infection. Thus, whereas the Paris conference had settled on a period of five days' surveillance in the case of vessels from ports suspected of having plague, the Washington Convention stipulated a period of seven.[120] These regulations were ratified by Honduras, Guatemala, Ecuador, Mexico, Costa Rica, Peru and El Salvador.[121] Yet there were some individuals who felt the conventions made too many concessions to commercial interests.

Such concerns were voiced most loudly in Cuba, which had a troubled relationship with the USA,[122] but they were shared by many in the United States, too. By 1905 the USA had experienced several outbreaks of plague: in 1899, in Hawaii (which had only recently been incorporated into the Union), in San Francisco (1900), Oakland (1900) and San Diego (1901). These epidemics had resulted in harsh measures directed primarily against Chinese immigrants, who took much of the blame for importing and spreading the disease. Although drastic action was enthusiastically supported by those opposed to the influx of emigrants from Asia, many merchants, together with shipping and labour agents, protested vocally against the disruption of trade and the embargo on migrants needed to work in the sugar plantations, railroads and other concerns.[123] It seemed likely that controls would be weakened further in the interests of big business.

In view of this, the American public health authorities insisted on the need to maintain strict vigilance on traffic from South America, the nearest part of the world in which outbreaks of plague regularly occurred. In 1903 plague was present in many places along the American Pacific coast and it was feared that the disease could easily spread north to California, there being no quarantine station south of San Diego. The surgeon in charge of the San Francisco quarantine station urged the Surgeon-General to begin sanitary inspection of all northbound vessels at Panama.[124] The following year such a procedure was put in place and US public health officials were sent to Colon and other Panamanian ports to inspect vessels, crews and passengers bound for the USA.[125] By 1908 the quarantine authorities at these ports were inspecting over a thousand vessels per year for plague and yellow fever, which resulted in over 5,000 persons being detained in quarantine.[126] The quarantine for ships coming from plague-infected ports was of one week's duration; a period which the US authorities did not think it in their interest to change. However, a few minor concessions were eventually granted in order to free up traffic through the soon-to-be-opened canal. From 1914, passengers and crews of ships from Peru and Chile (countries which were still infected with plague),[127] as well as from Panama, were detained under observation for no longer than five days provided that the vessels had been fumigated at the point of departure. To benefit from this concession, passengers and crew required certificates from a medical officer attesting to their freedom from disease and that their baggage was either considered safe or had been fumigated. If these rules were followed, and if passengers and crew had not resided in a plague-infected area for more than five days before embarkation, they could take advantage of the new regulations. If any of these requirements had not been met, they had to remain under observation for between seven and ten days.[128]

The new regime meant that many quarantine stations were considered obsolete. In 1926 the Chief Quarantine Officer of Hawaii protested against reductions in his staff which made it 'impossible to carry on station activities in an efficient manner'. In his view the cuts were a serious mistake as the Hawaiian quarantine stations were 'the first line of defense between the Orient and the Pacific Coast'.[129] Plague was already present in some of the Hawaiian islands, particularly Maui, where it had become endemic in sugar and pineapple plantations. The US Surgeon-General therefore requested more funds from the Treasury Department for anti-plague works on the island, including facilities for the laboratory examination of rats for plague. He reminded the Treasury that 'The prevention of plague infection in Hawaii gives protection to the United States and the rest of the world and, therefore, is a problem of the Federal Government'.[130] Plague was by no means a distant memory on the US mainland either: in 1919–20 there were outbreaks at New Orleans, Pensacola, Florida, and at Galveston and Beaumont, Texas. No more than twenty cases had occurred in any one of these but they served as a reminder of America's continuing vulnerability.[131]

With this in mind, efforts to prevent plague spreading from Hawaii were stepped up. Far more attention was paid to the control and certification of outgoing freight, which forced sanitary staff to work closely with railway and harbour companies, shipping lines and fruit growers.[132] Medical personnel supervised every aspect of the loading and unloading of cargo on the wharves at Hilo and Kahului, scrutinizing the docks and their surroundings for sanitary defects and interviewing the various persons connected with shipping. New procedures were introduced to prevent the contamination of shipments with rats and fleas, including the packing of sugar, fruit and vegetables. After these initial inspections, there were regular checks to ensure that procedures were being carried out.[133] On the whole it seems that arrangements for the surveillance of produce and transport worked well but occasionally there were infringements of sanitary regulations and disputes between shipping companies about their relative compliance.[134]

The actions taken by the US authorities exemplified a trend which had become increasingly evident after 1903: the move away from a rigid policy of containment to one of risk management based on better intelligence. This was accompanied by a shift in emphasis from quarantine in ports of arrival to preventive measures in ports of departure. Whereas the first few years of the pandemic had resulted in heavy-handed intervention in an effort to appease trading partners, commercially disruptive quarantines had gradually been replaced by more systematic and durable regimes. As in Hawaii and Sydney, these tended to rely more on environmental improvements and the control of rat and flea infestation than on restricting the movement of people and merchandise. But there were some notable exceptions, not least the drastic measures taken to control the spread of plague in northern China in 1910-11 and 1920-1.

The Manchurian epidemic of 1910-11 claimed some 60,000 lives, making it the most lethal outbreak since the one in India which, by that time, had killed millions in the west and north of the subcontinent.[135] Unlike the epidemics in southern China in the 1890s, those in Manchuria followed the railways rather than rivers and coastal trade. Such a scenario had been predicted some years before by the British bacteriologist E.H. Hankin who had witnessed the spread of plague in India. Writing of large railway projects in Asia and Africa, he warned that, 'far from being a disease that promises to become extinct, [plague] threatens to be an increased source of anxiety'.[136] These were prophetic words. Since the 1890s, Manchuria had gradually come under the sway of the Russians and the Japanese, who vied for control of the region. To extend its presence in Manchuria, Russia had built the Chinese Eastern Railway (CER) across northern Manchuria to Vladivostok, which was the final section of the Trans-Siberian Railway connecting Moscow to the Pacific coast.[137] A southern branch line linked the CER to the Russian settlement at Port Arthur but following Russia's defeat by Japan in the war of 1904-5, Russian interests were pushed northwards and the line to Port Arthur fell into Japanese hands, being vested in the semi-governmental South Manchuria Railway Company. When plague struck Manchuria in 1910 it did so by way of this link, which connected plague-endemic areas far inland with important industrial cities such as Harbin and Mukden (Shenyang).[138]

The first reports of plague came from the small railway town of Manzhouli in Inner Mongolia, in October 1910. By the following month the disease had appeared in the Russian-controlled city of Harbin, where two men who had been hunting Siberian marmots (known locally as tarbagans) were diagnosed with plague. Tarbagans provided a wild reservoir of plague infection and cases had occurred among hunters in the past but the new rail link enabled the disease to spread far beyond its usual confines. The railway hub of Harbin soon became a distribution point for plague, which spread rapidly to central and southern Manchuria, and into the adjacent provinces of Chihli and Shantung (Shandong). The lines of infection corresponded closely with the routes taken by labourers returning home for the Chinese New Year.

On this occasion, plague was able to move unusually quickly from its endemic zone because it assumed the virulent pneumonic form, spreading easily in densely packed railway carriages and the overcrowded dwellings of the poor.[139] Plague-control measures in both the Manchurian epidemics were therefore quite different from those in most recent outbreaks in that they concentrated more on people than rats. During the first epidemic, railway carriages were altered into quarantine sheds and in the subsequent epidemic 'rigid measures', including quarantine and burning of infected buildings, were taken at Harbin by the Russians working in conjunction with the Chinese, who had employed a brilliant Malayan-born Cambridge graduate, Dr Wu Liande, to help direct plague operations. In southern Manchuria, the Japanese took similarly decisive action against the plague.[140] Despite or perhaps because of the forceful response, the disruption to commerce in Manchuria was severe. The tarbagan fur trade was banned by both the Russian and Chinese governments in 1911, although it was later permitted to resume under strict regulations drafted by Wu Liande, including the inspection and disinfection of pelts at special stations.[141] More importantly, trade was badly affected by stoppages along the railways and it was estimated that total losses during the epidemic amounted to US $100 million.[142]

Although Manchuria had suffered from plague before, most previous outbreaks had resulted from importation from the south, via coastal trade, rather than contact with indigenous reservoirs: in 1899 and 1907, for example, it was confined to ports such as Niutchwang (Yingkou) and resulted in only a few hundred cases.[143] The plague outbreak of 1910–11 was of a different order and the Chinese government was so alarmed by it that it took the unprecedented step of inviting the Great Powers to send delegates to an international conference which was held in Mukden in April 1911. While the government was genuinely seeking the advice and cooperation of the nations which controlled its treaty ports, it also wanted to demonstrate that it had the ability to keep plague and other epidemic diseases under control. It was this imperative that led to the creation in 1911 of the Manchurian Plague Prevention Service, which worked closely with other powers in the region.[144] During the epidemic of 1920-1, for example, the MPPS took over most of the work in Harbin, leaving the Russian and Japanese authorities to concentrate on other infected areas. As in 1911, all trains entering and leaving Harbin were subject to thorough medical inspection but the city was more badly affected than during the first epidemic: some 3,125 people perished in Harbin in 1920-1 compared to around 1,500 in 1910-11. However, strict surveillance of passengers by the Japanese along the railway to the south meant that there were fewer cases in some of the towns which had been badly hit during the first epidemic. In 1910-11, for example, the city of Mukden experienced around 5,000 deaths from plague but in 1920-1 only four were reported.[145] The sanitary machinery to combat plague was also used to control epidemics of cholera in 1911 and 1926, when the MPPS acted to assist the Japanese South Manchuria Railway authorities in maintaining quarantine against railway passengers.[146]

Such measures did help to prevent plague and cholera from spreading within China, but there remained serious doubts internationally about sanitary precautions in many Chinese ports. When plague appeared in Manchuria in 1910, many southern ports imposed a surveillance period of seven days on vessels from the north but these measures were soon relaxed. Quarantine officers at Shanghai complained that they had been ordered by the port authorities 'to interfere as little as possible with the transport of merchandise'.[147] The main problem at Shanghai and the other treaty ports was that sanitary measures were far more susceptible to commercial interests than in ports controlled directly by the Chinese government. As Dr Andrew, the quarantine officer at Chinwantao (Qinhuangdao) explained:

We had a great deal of trouble and inconvenience in our work, not from the Chinese, who quietly submitted to any regulations we made, but from foreign ship owners. There was one case in which they absolutely refused to comply with quarantine regulations and telegraphed to the consuls and ministers, and it was only after an imperial edict came from Peking that we [were] able to enforce restrictions.[148]

After the international sanitary conference at Mukden, these defects became common knowledge and created alarm in many countries that had regular dealings with China. Nowhere was this more evident than in the USA, whose Pacific territories received many migrants from China as well as many merchant vessels. The gradual liberalization of regulations in American-controlled harbours was tenable only if plague in Chinese and other foreign ports could be more effectively monitored. Representatives of the US Public Health Service were therefore stationed at most of the ports which had regular contact with American territory.[149] These officials were in a position to check local arrangements and, if necessary, to assist in sanitary procedures such as fumigation to prevent rat infestation.[150] In Latin America, the US sanitary presence was already well established but in East Asia arrangements were ad hoc and often ineffective. In 1922 the US government dispatched Dr Victor Heiser of the Public Health Service to China and Japan in an effort to find out what, if anything, was being done to prevent plague from spreading. From 1905 to 1915, Heiser had been Director of Health in the Philippines and he was acutely aware that trade and labour migration from China posed a constant threat to the islands. Plague had entered the Philippines in 1899 from Hong Kong and persisted at a low level until it finally disappeared in 1906.[151] The prospect of re-infection, however, remained great.

The threat came from two principal sources. By far the greater danger came from China, now widely regarded as the 'cradle' of plague. The less obvious threat was posed by trade with Japan, which had been opened up to American commerce since the mission of Commodore Perry in 1853. With its large seaboard, overseas colonies and extensive trading links, Japan was extremely vulnerable to infection. Although the number of plague cases reported there had declined dramatically since the peak of 645 in 1907, they continued to appear sporadically in the major ports, causing foreign powers to remain cautious.[152] Japan had its own quarantine legislation which was very stringent by the standards of the time but sanitary precautions for vessels leaving Japanese ports for American territory were generally arranged by the US Consulate and paid for by local shipping interests, much as they were in China. In 1922, for example, Heiser inspected arrangements in Yokohama, Nagasaki, Shanghai and Hong Kong. In general, it seems that regulations were complied with, but in some ports tensions with shipping lines were evident. At Yokohama, Heiser noted that: 'Hides and similar articles requiring disinfection under the United States quarantine regulations are supposed to be disinfected in accordance therewith. There is much reason to believe, however, that the disinfection is lax.'[153] At Shanghai the American consulate was 'literally besieged on all sides with complaints dealing with the manner in which the quarantine inspection is being conducted'. Ships' captains protested against the 'absurdity of the attempt to require them to fend their vessels six feet off the pontoon wharf which is located in the river in front of Shanghai'. From a marine standpoint, this was dangerous, impracticable and unnecessary.

Shipowners also complained that the American quarantine officer, Dr Ransom, had frequently deputized British doctors in Shanghai to act for him. The latter were allegedly abusive to the crew and steerage passengers, and ungentlemanly in their conduct while aboard. Ransom was also accused of favouritism. His critics claimed that firms which employed him as a sanitary inspector in their factories received preferential treatment during quarantine inspections. The US Consul-General acknowledged that there was intense rivalry among the local doctors and some of the protests against Ransom could be discounted accordingly. But he conceded that 'much of the criticism was justified' and asked for Ransom to be replaced by a regular officer from the US Public Health Service.[154] To end these abuses, Heiser urged that existing officials in the principal ports of China and Japan be replaced with regular medical officers or that a peripatetic inspector should be assigned to visit them, the latter being charged with correcting irregularities and placing sanitary work on a systematic basis. The Chief Quarantine Officer of the Philippines would suffice for this purpose, he believed. The alternative to such a scheme would be to have more stringent arrangements in American ports and to abolish those in Asia. Although Heiser preferred the second option, he saw that it would leave the Philippines open to infection. He also knew that such a move was unrealistic in view of the general drift away from quarantine and because any advantage gained would be insufficient to offset the inevitable 'annoyances to shipping' at home.[155]

Heiser was more attuned than most public health officials to international opinion, being the author of a paper which analysed developments since the sanitary conferences of the nineteenth century.[156] He clearly perceived the trend towards standardization and seems largely to have endorsed it. Indeed, the 1920s brought a stronger drive for uniformity on both sides of the Atlantic. The most recent international sanitary convention – which had been signed in Paris in 1912 – had attempted to reduce periods of quarantine and place more emphasis on surveillance and rat destruction.[157] But after the First World War the Convention stood in need of revision. Some of the nations and empires that had signed it no longer existed, while new ones – like the USSR – had come into being. There were practical problems with the old convention, too. In the Middle East, for example, responsibility for regulation of shipping remained complicated because of the existence of four separate sanitary authorities: the councils of Tehran and Constantinople, the Egyptian Sanitary Maritime and Quarantine Board, and the Tangier International Board of Health. There had long been complaints about the lack of coordination between these bodies and that quarantines were sometimes imposed unnecessarily.

After the First World War there was a strong desire to rectify these problems. Fresh impetus was provided by the formation of the League of Nations – an organization which epitomized a new spirit of international cooperation – and in 1919 the Office International d'Hygiène Publique began to draft regulations on the lines suggested by Britain, which demanded that sanitary arrangements be made more compatible with 'the interests of commerce and traffic'. In the past, such calls would have been dismissed as special pleading but there was now widespread agreement about the desirability of such reforms. The OIHP was not in a position to obtain the information it needed to draft new regulations but the project became feasible when it was endorsed by the League of Nations in 1921.[158] From that time on, the Health Section of the League acted as a kind of clearing house, bringing together national governments to deal more effectively with medical problems. It was not directly involved in the control of epidemics but it helped to coordinate measures internationally and improve intelligence through the collection and dissemination of epidemiological data.[159] Communications technology had also advanced to the point where timely disease reporting had become possible, in theory at least. Networks of electric telegraphy were now extensive and the advent of wireless radio permitted ships to report their health status directly to their intended destination. Ports could also communicate directly with each other, rather than send messages through junctions in the telegraph system.[160]

However, the creation of a reliable and extensive system of epidemiological reporting took time and in the interim the League of Nations Health Committee sought to obtain intelligence about the health status of different parts of the world by dispatching missions of its own. One of the most significant of these was that led by F. Norman White, who was sent by the League in 1922 to investigate the prevalence of epidemic disease in Asia. The primary purpose of White's mission was to examine health in Asian ports, with a view to harmonizing regulations and reducing impediments to navigation. He called at many ports, including Bombay, Colombo, Shanghai, Osaka, Batavia and Singapore. Of the diseases considered by his mission, 'the most important ship-borne disease in the Orient' was thought to be plague. It was now well known that its spread from port to port was due to the movement of 'rat favoured' merchandise, predominantly grain. Instead of quarantine, White recommended a range of measures to prevent rat infestation of ports and ships, pointing to the success that had been achieved in Shanghai, Manila and Formosa (Taiwan) which had been plague-free for some years. These measures included systematic rat examination, the fumigation of small vessels such as junks and lighters (previously neglected), the unloading of suspect merchandise into rat-proof stores, the provision of rat-proof granaries, and close attention to epidemiological intelligence and patterns of spread.[161] By these methods, White reckoned that it would be possible to 'secure absolute immunity to plague ... without vexatious restrictions to the free flow of commerce'.[162]

But some emerging powers like Japan refused to bow to demands from Western countries for greater liberalization. All vessels arriving in Japan from foreign ports were subject to medical inspection under its Port Quarantine Law of 1922 and these procedures were carried out meticulously. Port health authorities were given discretionary powers to determine whatever measures they thought applicable to each case. These regulations could be very tough and, to prevent them from being diluted, the Japanese government refused to sign international sanitary agreements. In its view, the sanitary conventions of 1903 and 1912 were the latest in a long line of 'unequal treaties' which aimed to keep the country subordinate to the West.[163] Japan's tough sanitary regime extended to its colonial dominions, although procedures varied from port to port. In Korean ports such as Busan, for instance, there were periodic examinations of rats for plague infection in the harbour area but no fumigation of vessels, on account of the fact that Korea had never reported cases of plague.[164] But in the colony of Formosa – which had been re-infected by plague as recently as 1917 – systematic rat destruction was carried out in major ports and ships' cargoes were fumigated with sulphur dioxide gas in specially constructed chambers. As in Japan itself, the imperial authorities in Formosa implemented a rigorous inspection regime, examining over 63,000 people in 1922 alone.[165]

Although universal agreement on sanitary measures was out of the question in the early 1920s, the coming years brought closer cooperation in some regions. The Pan American Sanitary Code of 1924, which arose from a sanitary conference held in Havana, was one of several attempts to standardize statistics and sanitary measures and thereby reduce 'unnecessary hindrance to international commerce'.[166] The Convention obliged all the signatory nations immediately to notify the Pan American Sanitary Bureau of the presence of plague, cholera, yellow fever, smallpox and typhus in their countries, as well as of the measures taken to combat these diseases. The Bureau was to supply governments with information on the distribution of disease and to keep them abreast of the latest research. Most importantly, perhaps, it would now have a key role in advising the various sanitary authorities on how to interpret the Code, which meant, in theory, that there would be less disparity in arrangements. Reflecting the advent of air travel, the provisions of the Code were also applied to airports, which were to have facilities set aside for surveillance on the same basis as quarantine anchorages. All South and Central American countries, in addition to the USA, were signatories to the new convention.[167]

Two years later, there was an attempt to draw up a global agreement on similar lines. The main aim of the International Sanitary Conference held at Paris in 1926 was to improve epidemiological intelligence and reduce reliance on quarantine. There were even attempts to limit the damage which chemical fumigation caused to delicate cargoes, and harbour authorities were induced to use other methods to prevent rat infestation. On these principles there was general agreement, but arrangements in Asia and the Americas were covered by separate sections of the treaty and many individual states negotiated opt-outs on specific clauses. The Japanese, for example, were prepared to sign a regional agreement only, although they were now willing for it to be incorporated into the new convention. Although Japan signed up to the idea of creating a regional bureau for epidemic intelligence, it had reservations about reporting even relatively small outbreaks of disease by telegraph, as did some other states, such as China and the USSR. These vast countries – like Japan's more remote Asian territories – lacked the telegraphic infrastructure that would enable them to supply information rapidly to central health authorities.[168] By 1931, however, most major ports in Asia – with the exception of some Chinese ports troubled by political disturbances – were reporting disease outbreaks to the Eastern Bureau on a weekly basis.[169]

Despite the persistence of regional divisions and national opt-outs, the trend in the first three decades of the twentieth century was overwhelmingly towards the reduction and harmonization of sanitary restrictions. This shift from quarantine to surveillance – from containment to risk management – was not due solely to the challenge posed by plague but it was intimately bound up with it. The rapid movement of plague around the world served as a reminder of how intimate humanity had become and few countries seemed to escape its attentions. But plague was still a local problem, too. The disease seemed to take hold wherever squalor and ignorance prevailed, whether in the teeming streets of Hong Kong or the crowded apartments of Oporto. Like cholera in the nineteenth century, plague gave impetus to national governments and municipal authorities which had neglected sanitation for too long. They could no longer afford to ignore such matters if they hoped to remain active players in an expanding and ever more closely integrated global economy.

In the early years of the pandemic, the response to plague was uncoordinated and damaging to trade.[170] By the turn of the century, however, there was general agreement that a different approach was needed; one in which states acted within mutually agreed parameters. There was also a general desire to reduce the frequency and duration of traditional sanitary measures such as quarantine. But a 'light touch' system of sanitary control required two things to operate effectively. The first was reliable information about the prevalence of disease in different parts of the world. The second was confidence in the capacity and resolve of national or colonial authorities to tackle epidemics when they occurred. But whilst the Great

Powers were usually able to monitor events within their domains, they had to rely on international bodies such as the OIHP and the League of Nations to gather intelligence more widely. This remained problematic in view of technical limitations in some parts of the world and the reluctance of states such as Japan to share intelligence or sign international conventions. The USA, which had become a quasi-imperial power in its own right, found a way around the problem by closely monitoring and intervening in the sanitary affairs of other nations.

By the late 1920s, the threat of plague was diminishing and in most parts of the world plague had either disappeared or was in decline. Significant epidemics – those resulting in more than a hundred cases per year – continued to occur in parts of tropical Africa, India and South East Asia but they paled into insignificance when compared to those of a decade or so earlier.[171] How far the measures recommended by international conferences contributed to the decline in plague is unclear but the diminishing threat from this and other epidemic diseases rendered such gatherings redundant. It was fitting that the last of the international sanitary conferences was convened at Paris, in 1938, where they had begun nearly a century earlier. The final meeting had a very limited remit, which was to hand sanitary responsibility for the Suez Canal back to Egypt. Alexandria's sanitary board was thus disbanded and its functions assumed by the national Conseil Sanitaire Maritime et Quarantine, now independent of foreign control.[172] Important agreements, like the sanitary convention on aerial navigation, which came into force in 1935, could now be made without recourse to major gatherings. The Pan American sanitary conferences continued longer than those based in Europe, though their functions were increasingly subordinated to those of the World Health Organization, which was created in 1948.[173] The subject matter of these conferences was different from those of the pre-war era. Cholera, plague and yellow fever had faded into the background, their place taken by a variety of animal diseases and agricultural pests.[174] As we shall see in the coming chapters, the question of how far to regulate trade in order to control these diseases proved just as difficult to resolve.

CHAPTER 8

Protection or protectionism?

In the nineteenth century the world was plunged into a maelstrom of disease. Although its demographic impact was less than the Black Death or the Columbian Exchange, diseases circulated faster and further than before, traversing oceans and land masses along the pathways of global commerce. Potentially the most devastating were those that affected animals and plants. These diseases had the capacity to destroy the agricultural systems on which humans depended and their consequences – in terms of starvation and social collapse – were often more severe than those of epidemics. But while the world became more closely connected by disease, it was far from 'unified', in the famous phrase of the historian Emmanuel Le Roy Ladurie.[1] The more advanced nations now had the resources and knowledge to manage disease successfully, whereas poorer countries – many of them colonies of the former – lacked the means or the will to do so. This was the case not only with epidemics but also with the livestock and crop diseases which struck at the heart of less developed economies.

By the 1900s, the majority of industrialized countries had brought epidemic diseases such as cholera and yellow fever under control or had eradicated them. A fairly sophisticated system had also been established to prevent the spread of infection from Asia, Africa and Latin America to the great ports of the Western world. Disease surveillance and targeted intervention allowed these sanitary safeguards to be maintained without disrupting the flow of the people and merchandise necessary to the health of the global economy. Such a system could only work if there was agreement between the key global players, but this was not reached in the case of the trade in agricultural commodities. In part, this reflected the great diversity of national experiences: some countries had lived with pests for many years, whereas others had only recently encountered them. But the failure to reach agreement on how to regulate the trade in plants and animals was due principally to the different economic interests of exporting and importing countries. The livestock industry played a major part in the economies of the New World, for example, and exports of inexpensive meat from these countries threatened the livelihood of small-scale producers in Europe. Fruit-growers in developed countries similarly faced the prospect of being undercut by producers in poorer ones, where labour and other costs were lower. Sanitary controls offered an ostensibly legitimate means by which these costs could be equalized and domestic industries protected.

Plagues of beasts and men

Prior to the nineteenth century, many parts of Europe and Asia were periodically affected by 'murrains' of cattle. It is not always easy to determine what these were but they usually prevailed in times of hardship and especially in times of war, when the movement of troops and animals spread infection over vast distances. Major outbreaks of 'cattle plague' or 'rinderpest', as it was also known, coincided with the War of the Spanish Succession (1702–13), the War of the Austrian Succession (1740–8) and the Seven Years War (1756–63); the first two diseases seemingly originating in the Russian Steppes and the third in the vicinity of the Black Sea.[2] But, while the circumstances of war provided the conditions necessary for the transcontinental spread of disease, it was assisted locally and regionally by the trade in livestock. Before the mid eighteenth century, most European countries had few if any laws prohibiting the sale of cattle and this allowed disease to circulate for years after it was imported.[3] From the 1740s, however, many European countries passed legislation which provided for the slaughter of infected herds, restrictions on sales and the compulsory disinfection of sheds and farmyards. Most of these Acts were based on rules originally proclaimed by Pope Clement XI (r. 1700–21), who had been urged to take action by his personal physician, Giovanni Maria Lancisi. Lancisi was fascinated by epidemics of fever and saw many similarities between them and the diseases of animals. Like human pestilence, cattle plague seemed to arise from a combination of local infectious influences and transmittable contagion. Lancisi's recommendations for its control therefore looked both to the movement of animals and to their environment, embracing quarantine and the slaughter of plague-ridden beasts, as well the fumigation of their quarters. These measures seemed to bring the outbreak to a halt and other nations took note.

When rinderpest struck Europe again, in the 1740s, many states followed in the footsteps of the Vatican, expecting similar results.[4] Some achieved them, but the majority were less successful. States that were lacking in strong central control, such as Britain, embraced Lancisi's principles tentatively and suffered the consequences. Even when governments were determined to arrest the spread of disease, their capacity to do so was generally limited and smugglers frequently broke through.[5] Part of the problem was opposition from farmers who objected to the destruction of infected herds. Rather than reporting cases of cattle plague, they concealed them and treated their cattle with some of the many dubious substances touted as 'cures'.[6] The sole success story was Prussia, where the sale and advertisement of these remedies were banned, in the belief that they encouraged the evasion of preventive measures. Instead, Prussia maintained a strict military cordon against potentially infected cattle from Eastern Europe and, in some measure, provided a bulwark of infection for other parts of Europe. It was only when Prussia became embroiled in the Napoleonic Wars that its guard dropped, allowing cattle plague to spread across the Continent.[7]

During the mid nineteenth century, animal diseases were on the move once more but this time as a result of long-distance trade. In keeping with the spirit of the times, the removal of old restrictions on the transportation of livestock – tariffs, licences and so forth – allowed European cattle to move more freely than hitherto, while Europe began to receive animals from the New World. The advent of steam navigation was a major boon to this trade as it allowed animals to reach their destination faster and theoretically in better shape than did longer journeys under sail.[8] As the animals crossed borders, so did their diseases, especially as many sanitary controls had been allowed to lapse alongside the tariffs and other impediments to trade. The first indications that something was amiss came in the 1830s, with outbreaks across Europe of a disease called 'lung sickness,' now generally referred to as bovine pleuropneumonia. This highly contagious respiratory infection had been prevalent in parts of Central and Eastern Europe for many years, but in 1833 it arrived in the Netherlands and appeared shortly afterwards in Ireland, England and Scandinavia. In 1843 pleuropneumonia was also reported in the USA, probably having been introduced from Germany. The disease spread around New York City but was stamped out by the slaughter of infected animals. Four years later, it showed up again, this time in New Jersey, in stock imported from England but it was eradicated when the farmer decided selflessly to slaughter his entire herd. Further outbreaks occurred on the eastern seaboard of the USA throughout the 1850s, but each time the disease was eradicated using similar methods. Eventually, though, it gained a foothold. Following the importation of infected cattle from Rotterdam, lung sickness became prevalent in parts of Massachusetts and soon spread to the nearby states of Pennsylvania, Virginia and Maryland.[9]

Because lung sickness has a long incubation period and tends to spread slowly, it was difficult to rouse farmers to united action against it. In Britain, for example, it was allowed to spread unchecked for nearly three decades before legislation was passed to tackle it, costing farmers in the region of £2 million every year.[10] The same was true of foot-and-mouth disease, which was rarely fatal but caused a great deal of distress to the cattle, sheep and pigs it affected. Like lung sickness, foot-and-mouth disease began to spread more widely during the 1830s and 1840s, when it moved across the European continent to infect Britain and the Americas. Generally, it took several major outbreaks before legislation was introduced and in some parts of the world affected animals were simply isolated and nursed back to health.[11] But the resurgence of cattle plague in the 1860s met with a very different response, at least in prosperous countries. Coming at a time when the world was threatened by a resurgence of cholera and yellow fever, a more defensive mindset prevailed. It was as if the world was being scourged by waves of infection in the manner of biblical plagues. This impression was fostered by the nature of the disease itself. Unlike the other animal diseases then in circulation, cattle plague was rapidly fatal, as well as highly contagious. Most cattle died within six to twelve days of the appearance of symptoms, which included fever, nasal discharges and diarrhoea. Rural communities all over the world were devastated.

The 'seat' of rinderpest was widely acknowledged to be in Russia, yet the main focus of attention in the years running up to the pandemic of the 1860s was the Austro-Hungarian Empire. Austria-Hungary received most of the cattle which were exported from Russia, before transhipping them to points further west. Around 100,000 animals passed through the provinces of Galicia and Hungary every year and the long quarantine imposed there incurred great expense and encouraged evasion. Many diseased cattle were smuggled through and rinderpest broke out in Austria-Hungary every six or seven years, often with catastrophic results. In 1849–51, for instance, it attacked 300,000 head of cattle, and in 1863 killed around 14 per cent of stock in the infected provinces.[12] In the majority of cases, the outbreaks were confined to certain provinces of the Hapsburg Empire but, by the late 1850s, the disease seemed likely to spread further. The impending completion of two railways connecting the great ports of Rotterdam and Hamburg with the stockyards at Pesth and Lemberg raised the prospect of rapid and constant infection from the east. Furthermore, quarantine facilities along the Galician railway were rudimentary and unlikely to check the spread of rinderpest from Russia.[13]

This grave threat was the subject of an unprecedented gathering of European veterinary surgeons in Hamburg, in 1863. The brainchild of a British veterinary surgeon, Professor John Gamgee, the conference convened with the object of pooling information from different parts of the Continent and suggesting preventive measures which would be mutually beneficial to European governments. At this conference, there was widespread agreement on the nature of the problem presented by the trade in cattle from Russia and on the need for some kind of quarantine. But delegates could not agree on the duration. The majority favoured a reduction in the length of quarantine – normally twenty-one days in Prussia and Austria-Hungary – to ten days, in the belief that this was sufficient and that a shorter time would provide less incentive for evasion. However, a minority, dominated by the Prussian veterinary surgeons, preferred to retain the longer term.[14] At their next meeting, in Vienna in 1865, a few delegates continued to insist on a quarantine of twenty-one days but most who had formerly opposed it had changed their mind.[15]

But while the veterinary surgeons were moving towards a consensus there was none among the governments they represented. Very few had even thought about the problem. Even Prussia, which had hitherto been vigilant against the disease, had its eyes elsewhere, distracted by war with Denmark and an impending conflict with Austria-Hungary.[16] Britain's island position may have contributed to its complacency but those who placed their faith in the English Channel were soon disappointed. On 24 June 1865, there were reports of a contagious distemper among livestock in Lambeth, in London. Gamgee and other veterinary experts were called to investigate and confirmed that the disease was rinderpest. They concluded that it had probably arrived with Russian cattle imported from the town of Ravel, just before the disease was reported there. These cattle had come into Britain through the north-eastern port of Hull but were sent to London shortly afterwards, where they probably infected domestic beasts.

Stung by criticism of its tardy response to recent outbreaks of yellow fever, the Privy Council acted decisively against this new threat and in advance of public opinion. It immediately ordered the inspection of herds and the slaughter of all those which had been infected, causing great hardship to farmers, who received no compensation.[17] These drastic measures added to the misery of the plague and conjured images of Old Testament misfortunes. With cholera once again on the march, many people began to wonder whether these diseases had been sent to punish a world that had grown wicked and materialistic. The Archbishop of Canterbury called for a 'Day of Humiliation' to be observed across the country on 9 March 1866 to atone for the sins of the nation. Fiery sermons poured forth calling on the country to reflect on its failure to observe the Sabbath and to repent of its sins.

Sermons of a similar kind had often been delivered during epidemics of cholera and plague but some of those concerning the cattle plague had novel features such as their preoccupation with 'science' or, more precisely, with the fact that the public was apt to place more faith in it than in revealed religion. In the 1860s the Church was fighting a battle, not only against apostles of Darwin's theory of natural selection, but against the more general belief that the world could be understood solely through knowledge of the physical laws which governed it. This anxiety was expressed in sermons on the cattle plague and in condemnation of the government's policy of stamping it out, which was widely seen as an inhumane act dictated by scientific reasoning.[18] It was also evident in the conviction that cattle plague had developed not from contagion but from the transgression of moral laws which had been divinely ordained.[19] One had to look no further than the conditions in which animals were transported to find the seeds of the pestilence, claimed Rev. William McCall:

> Some persons have thought that the cattle plague is owing mainly to the cruel way in which foreign cattle are treated while on their way to us; and many who have seen the poor animals on their voyage, have told pitiable tales of the heedless cruelty which has been practised. There may be other sources of the disease, but surely this is a sin, a national sin which requires punishment.[20]

The Day of Humiliation reflected public disquiet about the trade in livestock and the measures taken to control the cattle plague; both of which resolved into a disinclination to accept that the plague was contagious. Such notions were to be found equally among men of faith and of science: the physician Charles Bell was far from alone in his belief that rinderpest arose from an 'epidemic influence' already existing in Britain rather than from some imported infection. In his view, recently imported cattle were usually the first to succumb because they had been exhausted by a long journey often made in appalling conditions. These opinions were not so very different from those expressed in some of the sermons but Bell placed more faith in down-to-earth remedies. He believed that the only way to prevent cattle plague was through the efforts of farmers, not the measures devised by government. Each farmer ought to be induced to

> convert his farm-steading into a sanitarium in which his cattle will have their constitution improved by suitable means, and their condition carefully examined every day, so as to ascertain the earliest indication of disease, in order that proper remedies may be had recourse to at a period when they are most likely to be effectual.[21]

These recommendations were compatible with those of some of the more moderate advocates of contagion theory, such as the Edinburgh chemist Lyon Playfair, who acknowledged that poor ventilation and hygiene on farms and in railway trucks and steamers had played their part in the recent epizootic. While he denounced as 'fanatics' those who thought the cattle plague a punishment for whichever sins they most despised, Playfair believed that God had made laws to govern the welfare of his creatures and that transgression of these would necessarily bring penalties. The task of scientists, he believed, was to comprehend the wisdom of the Creator as expressed through his laws and to ensure that society arranged itself in harmony with them.[22]

It is conceivable that popular support could have been mobilized around a campaign to eradicate cattle plague; one that embraced both quarantine and sanitary improvement. But the policy of slaughter overshadowed everything. One of its most vigorous and influential exponents was Gamgee, who had based his recommendations on the practices he had observed in continental Europe. In 1866 he moved from his professorship at Edinburgh University to become Principal of the Albert Veterinary College in London and from this elevated position he espoused quarantine and slaughter in the full knowledge that it went against the grain of public opinion.[23] He was resolutely opposed to anything that detracted from this simple message and thought the British government should emulate Prussia by declaring all attempts to treat or advertise remedies for the cattle plague illegal.[24] He was similarly opposed to preventive measures that removed the focus from slaughter, including attempts to inoculate cattle against rinderpest using material from infected beasts. This method had long been used by farmers in continental Europe but it was often deadly and would have been slow to implement nationally. For the same reason, he deprecated the conclusions of the Royal Commission, which reported to Parliament three times in 1865–6, but failed to reach a clear and unanimous agreement on preventive measures. While he had no objection to such measures in principle, the disease had penetrated too far and wide for these to be practicable. As he put it, 'It is not expedient nor economical to adopt any other method of prevention when the cattle plague enters a country such as England, except that of extinguishing all centres of infection with promptitude and determination.'[25] But to make slaughter more palatable, he recommended compensation for the farmers affected, based in the longer term on a system of national insurance.[26] Aware, too, of the substantial opposition that existed to quarantine in Britain, Gamgee made the case that such interruptions were normally short-lived and the inconvenience caused by them was incomparable to the havoc created by the disease when it became established.[27] It was easier to enforce quarantine of animals than of humans, he argued, because the former travelled in large numbers under supervision and epizootics such as cattle plague were often spread by contagion pure and simple, with few contingencies of note.[28]

Although the government wavered in the face of opposition from farmers and those who opposed the slaughter on humanitarian grounds, it was moving tentatively in the direction advocated by Gamgee. The Cattle Diseases Act of 1866, which permitted the wholesale slaughter of infected herds, was the first of several pieces of legislation which conferred powers on local authorities to deal with all animal diseases, including pleuropneumonia and foot-and-mouth.[29] The policy of 'stamping out' had edged alternative methods out of the frame. Indeed, such measures enjoyed near-unanimous support among veterinarians and were becoming more popular with the public.[30] But there remained implacable and vocal sceptics. One of these, George Foggo, a former member of the Legislative Council of Bombay, drew an analogy between cattle plague and cholera, and wondered how reliable experts were when they had changed their mind about both so rapidly.[31] Not so long ago, he recalled, bodies such as the French Academy of Medicine had pronounced against contagion and France had abandoned quarantine on vessels sailing to and from Algeria. As late as 1865, the British Privy Council, too, had declared quarantine against Egypt, where the disease then prevailed, to be unnecessary. What could have changed within the space of just a few years to convince such august bodies to reverse their earlier policies? In Foggo's opinion, their sudden volte-face smacked of panic and cast doubt on both the reliability of experts and the wisdom of government. Moreover, all the evidence seemed to point to the fact that quarantine and similar measures were less effective in preventing animal and human diseases than reducing overcrowding and improving sanitation. The government ought to attend to such matters and have no truck with 'vexatious interference with trade or with individual liberty', he insisted.[32]

But this was Britain, not Prussia, and Britain had a strong tradition of local governance. The passage of legislation in the 1860s and 1870s therefore left considerable scope for interpretation and local authorities varied greatly in the diligence with which they implemented it. However, growing frustration at the apathy of these bodies led to their functions being usurped by central government in the form of the Board of Agriculture. This was an unpopular move at first but it made an enormous difference to the control of rinderpest, which was eradicated in Britain by the end of the century.³³

The state of affairs in Britain's colonies could not have been more different. When rinderpest began to spread in India, the response of the British administration was decidedly muted. During the 1860s, outbreaks of what officials identified as rinderpest were noted in many provinces, causing the death of hundreds of thousands of cattle every year. The rapid spread of the infection was attributed to the commercialization of agriculture and the movement of livestock, yet no legislation was introduced specifically to tackle the disease apart from in the southern province of Madras, which passed a Cattle Diseases Act in 1866. This law was much weaker than its British equivalent, for India's vast expanse and the size of its herds made it impossible to maintain effective surveillance. British officials also expected determined opposition from Hindus to the slaughter of their sacred animals.³⁴ For these reasons, and no doubt because of the expense which would have been necessary to provide inspection and other facilities, the Government of India continued to resist demands by veterinary surgeons for controls such as those which existed in Britain.³⁵

The prevalence of rinderpest in India meant that it was only a matter of time before the disease spread to countries which imported its cattle. India was almost certainly one source of the epizootics which devastated the Philippines and other parts of South East Asia from the late nineteenth century but it may not have been the only one. Southern China was also infected by this time, as were countries such as Siam (Thailand) which abutted British India. These countries had known rinderpest for some years but the commercialization of agriculture, changing consumption patterns and the growth of the long-distance livestock trade meant that the plague became more widely distributed. Dairy cattle and beasts of burden were exported from India to Malaya and Singapore; Siam regularly shipped beef cattle to Singapore and Sumatra; and sheep, goats and cattle were exported from Hong Kong to many parts of the region, including the Philippines. Any or all of these places could have been the source of rinderpest in South East Asia. The region was also particularly vulnerable to infection because most of its ports had an open door policy. Even when the risks became apparent, quarantine, slaughter and other restrictions evolved piecemeal in most of the European colonies. Progress was made in some areas, such as in parts of the Philippines, but it was generally undermined by the difficulty of maintaining effective barriers between infected and non-infected regions.[36]

The situation in Africa was similar. When rinderpest first appeared in the north-east of the continent, in 1889, it made extraordinary progress, reaching the southern tip by 1896. At this time, no colonial governments had legislation relating to animal diseases and none took measures to prevent its spread. As a result, rinderpest brought untold misery to pastoralists, whose principal means of livelihood was destroyed. One consequence was that resistance to European imperialists was undermined, allowing them to establish their authority more easily.[37] After a while, some of these colonial governments did begin to take measures to control cattle plague, for as white farmers began to settle, their livelihoods came to depend on keeping their cattle healthy. Unlike the situation in India, the colonial authorities did not see themselves as bound by indigenous sentiments towards animals and in some areas, such as Cape Colony, quarantine and stamping-out were vigorously pursued.[38] Nevertheless, the disease became endemic in many parts of Africa and occasionally escalated into full-blown epizootics due to uprisings and other upheavals which disrupted controls. Even in times of peace, the difficulty of quarantining herds grazing over vast distances and of controlling disease among wild animals rendered traditional control measures useless. Had the early 1900s not brought a new technique – in the form of safer inoculation – then rinderpest would probably have continued to ravage much of Africa even longer than it did. The new inoculation was developed in 1897 by the German bacteriologist Robert Koch, using attenuated rather than fully potent germs. Although it took many decades for effective implementation strategies to be worked out, inoculation enabled the eradication of rinderpest in Africa and India by the end of the twentieth century.[39] It was a rare success story but one which came very late in the day by comparison with other parts of the world.

Before effective control measures were introduced, rinderpest wreaked havoc in many parts of Europe, Africa and Asia but it was unable to colonize the American continent. North America never experienced an outbreak and, although the disease appeared briefly in Argentina and Brazil during the 1870s, it was localized and quickly stamped out. There are two possible explanations for this. The fact that countries such as Argentina and the USA were net exporters of meat and cattle meant that the probability of infection was lower than in countries which were net importers. The shipping routes between America and Eurasia were also considerably longer than those between India and the South East Asian archipelago, for example. As such, it was more likely that diseased animals would have been noticed en route, especially as the symptoms of rinderpest were far more striking than those of cattle affected by diseases such as foot-and-mouth. However, the American continent had enough problems of its own. As well as having to deal with imported infections like pleuropneumonia and foot-and-mouth, America had native infections. In the mid nineteenth century some of these diseases were able to spread outside the areas in which they were normally confined to affect other parts of the country, aided by a booming trade in livestock and the absence of formal controls. In the 1860s, for instance, the USA began to experience frequent outbreaks of a disease known as 'Spanish' or more commonly as 'Texas' fever. This murrain was endemic in the southern states of the Union and was often fatal, but its long incubation period (30–40 days) made it difficult to detect.

Texas fever had caused many deaths among cattle prior to and during the Civil War, but did not become a major issue until the middle of 1868, when its characteristic symptoms (fever, emaciation and bloody urine) were noticed among cattle from Texas which had been sent north for fattening. Long feared by ranchers in the adjacent states of Missouri and Kansas, Texas fever created greater alarm in 1868 because this was the first time that the disease had appeared in the northern Midwest.[40] The outbreak, which began at Cairo, Illinois, intensified a campaign of violence and intimidation against drovers and stock dealers who handled southern cattle. Writing to a farming magazine, a Missouri rancher warned:

> Talk to a Missourian about moderation, when a drove of Texas cattle is coming, and he will call you a fool, while he coolly loads his gun, and joins his neighbors; and they intend no scare either. They mean to kill, do kill, and will keep killing until the drove takes the back track. . . . No doubt this looks a good deal like border-ruffianism to you, but it is the very way we steer clear of the Texas fever; and my word for it, Illinois will have to do the same thing yet.[41]

Very soon, dealers and farmers who sold or owned southern cattle were being threatened, attacked and called to account for their actions at angry public meetings. Professor Gamgee, who was invited by the US government to investigate the outbreak, was not exaggerating when he declared that 'the prevention of splenic fever', as he termed it, 'implies, in many instances, the prevention of lawlessness and the preservation of the public peace'.[42] Such disorder had a direct impact on trade within the USA but reports of Texas fever and of loose or non-existent sanitary controls in American stockyards were creating panic overseas, too, resulting in temporary bans on livestock and hay imported from the USA.

Gamgee's report for the US Commissioner of Agriculture would clearly have serious implications for commerce and he was well aware of the various interests which feared curtailment of the cattle trade. He knew that the price of meat in many northern cities of the Union was rising and that there was a demand for cheaper food which could only be satisfied by importing cattle from the South. Gamgee was equally aware of the political influence exerted by meat traders in Chicago and those involved in the rearing and shipment of cattle from Texas. Such interests could not be easily overridden. As a foreigner, too, Gamgee was more reticent about making recommendations than he was in the case of rinderpest in Britain, where he had shown no hesitation about entering the political fray. However, some important principles for disease prevention were buried deep in his report. Gamgee went against the received wisdom of Midwestern ranchers and declared the disease to be non-contagious; in his view, it seemed to arise from cattle which had eaten contaminated food in their native southern environment, foodstuffs which might contain poisonous plants or which had been infected with the spores of an organism resembling anthrax. This organism, whatever it might be (it was later found to be a parasite transmitted by a tick), seemed to thrive best in the humid conditions of the American South but it could also propagate in warm weather in pastures further north. In Gamgee's view, it could be passed on from infected cattle through their faeces, which deposited the organism on pastures on which hitherto healthy cattle might graze.[43]

The transmission of the disease therefore seemed to be indirect and required very different measures than a contagious distemper like rinderpest. In the long term, Gamgee concluded, an answer to the problem might be found in the drainage of southern pastures and the extirpation of poisonous plants. But that was a long way off and for the time being the only measures which could conceivably be effective, in his view, were to limit the movement of stock to winter, when imported and native cattle were likely to be kept in sheds, and, if possible, to keep imported cattle in specially designated areas. In effect, this meant creating a vast quarantine belt across the northern states. This proposal was unlikely to be popular with ranchers but without some such demarcation Gamgee believed there was no way of stopping the disease and the economic disruption that attended it.[44] As he suspected, the recommendations were not to the liking of the businesses involved, which at first resisted any suggestions that southern livestock needed to be quarantined. Exporters could, however, draw comfort from Gamgee's reassurance to the British government that there was nothing to fear from hay imported from the USA. As he pointed out, hay exported to Britain was produced in the eastern and western states of the Union, not in the South or Midwest where cases of Texas fever had been discovered.[45]

But European countries remained nervous, if not of American hay, then of livestock entering their ports. American ranchers and exporters protested that concerns about the health of their animals were unfounded and suspected that imports were being restricted to protect farmers in Europe.[46] Nevertheless, in the 1880s, a number of Midwestern states began to introduce controls of the kind suggested by Gamgee. Some, like Oklahoma, began to quarantine Texas cattle on their arrival, while others, like Kansas, banned the trail drives altogether. In the early 1890s, when it was found that Texas fever was a tick-borne disease, farmers also began to control the infestation by dipping their cattle in vats of water and crude oil. But these measures were unpopular with the owners of small herds, who saw them as an expensive nuisance, and there were numerous violent protests against them. Doubts over the efficacy of such measures remained until they were brought under federal control in the early 1900s.[47]

Texas fever was not the only source of tension between American exporters and European importers. The British authorities were also up in arms over the importation of American cattle said to be suffering from lung sickness, formerly a European disease. This malady showed itself several times during the 1870s in shipments of American cattle to the English port of Liverpool and, in 1879, the British government mandated that all cattle from the USA were to be slaughtered within ten days of arrival in British ports. This greatly annoyed American ranchers because the restrictions prevented them from selling properly fattened stock, thereby realizing their commercial potential.[48] Other European countries such as Belgium imposed quarantines on American cattle for the same reason and, in some cases, maintained them for decades.[49]

But the most serious sanitary dispute which erupted during the late nineteenth century was over the export to Europe of American pork. In Germany, Sweden and other European countries hundreds of people sickened and some died after eating pork products, some of which appeared to have been imported from the USA. In 1878, Dr Richard Heschl, Professor of Anatomy at the University of Vienna, made this connection firmer when he claimed that nearly 20 per cent of American hams were infected with the parasite *Trichinella spiralis* which seemed likely to be the cause of the outbreaks. On the strength of this, in 1879 the Italian government decided to place an embargo on imports of US pork and it was quickly followed by Austria-Hungary. In 1880, Germany also banned American pork and France did the same in 1881, with Spain, Greece and Turkey joining in soon afterwards. The combined effect of these measures on the American economy was potentially devastating, for pork then constituted 10 per cent of the country's entire export market and was worth nearly $80 million per year.[50]

Of all the bans imposed against American pork, the German embargo was the most damaging, as the country was the single largest importer of this product. Right from the start, the ban was seen in the USA as thinly veiled protectionism, coinciding as it did with the passage of the Tariff Act in 1879 – a measure vigorously championed by the German Chancellor, Otto von Bismarck.[51] As the German Liberal Party, which opposed the ban, realized, it was far easier to prohibit pork on health grounds than for purely economic reasons.[52] But the German ban provoked retaliation from America and sparked a decade-long trade dispute known as the 'Pork War'. Tariff wars between European states were not uncommon and even in Britain, the traditional champion of free trade, the clamour for protection and 'imperial preference' had been growing as the agricultural sector began to feel the effects of cheap foreign imports. But although there was a 'return to protectionism' in many countries, it should not be overstated, for there was a good deal of grass-roots support for free trade. The key concern of most people – many of them now voters – was to ensure that bread, meat and other basic foodstuffs remained affordable.[53] The extent to which protectionism prevailed in each country therefore depended on the strength of producer over consumer interests. Governments also had to balance the benefits of commercial freedom against the actual risk of infection and in many cases the fear of infection – among both producers and the general public – was very real. If exporting states attempted to allay these anxieties by improving the sanitary standards of their products, they had a reasonable prospect of having any restrictions lifted or reduced. This is precisely what happened in Germany, in 1891, when the government was persuaded to lift its embargo following the passage of legislation improving sanitary safeguards in the USA.[54]

The American authorities were now taking vigorous action to control all kinds of animal disease. Federal authorities were tightening their control over individual states and working more closely with business interests such as importers of hides, wool and other animal products. Legislation passed in 1905 enabled the Secretary of Agriculture to impose a strict quarantine on shipments of cattle between states and across national borders, in much the same way as the Federal Public Health Service acquired control over human quarantine. As one government publication put it, 'Foot-and-mouth disease should be classed as an undesirable alien enemy.'[55] At the same time, greater efforts were made to improve the safety of meat products. An Act of 1902 forbade interstate transport of impure food and drugs and the Meat Inspection Acts of 1906–7 closed interstate and foreign commerce to meat condemned as unfit for use.[56]

Whilst these measures did much to improve the reputation of American meat and livestock at home and overseas, the Pork War and similar disputes had shown the difficulty of resolving disagreements bilaterally. It was becoming increasingly clear that international forums and agreements – similar to those to regulate cholera and yellow fever – were necessary if the traffic in animals and meat products was not to be disrupted continually. It was also recognized that many countries would benefit from internationally agreed standards, sanitary certification and the sharing of scientific information. In the years running up to the First World War such sentiments were often expressed but they were not successfully translated into action. In the aftermath of war, however, the spirit of international cooperation symbolized by the League of Nations aroused hopes for an enduring agreement. In 1920 the French government seized the initiative, announcing that an international conference on the study of epizootics would shortly be held in Paris. Recent outbreaks of cattle plague in neighbouring Belgium meant that the issue was a pressing one from the French point of view, but the economic disruption caused by the war made the reconstruction of trading links a matter of urgency for most nations. The French government intended the conference to be a springboard for cooperation in the study of animal diseases and that any new organization arising from the conference would be based in Paris – like the OIHP – helping France to remain at the centre of international affairs.[57]

At the conference, which began in May 1921, the delegates adopted unanimously most of the resolutions tabled by their hosts. These included the need for more scientific research; the desirability of regular sanitary bulletins; tougher measures to control the spread of disease; standardized certification; and the creation of an international bureau similar to the OIHP. But despite its ambitions to be an international body in the fullest sense, the Office International des Épizooties (OIE) was a predominantly European enterprise. Some Latin American states were enthusiastic supporters of the idea and signed the convention which established the OIE in April 1924, but many non-European countries, including the USA, did not.[58] The governing committee of the Office was therefore dominated by European governments and the only non-European delegates appointed were those from the colonies of European powers rather than independent nations.[59] This narrow membership necessarily limited the effectiveness of the committee, especially as far as trade with the Americas was concerned. There, the USA held sway over animal health much as it did over human health. Nevertheless, the OIE proceeded diligently and energetically, publishing scientific articles on disease prevention in its bulletins and providing reports on disease outbreaks around the world.[60] The Director of the OIE, E. Leclainche, also made an impassioned plea for greater standardization of sanitary reports and certificates to accompany exported animals and meat.[61]

In 1928, however, the direction taken by the OIE was endorsed by the League of Nations, which was becoming increasingly concerned by crypto-protectionist measures that threatened to sow the seeds of discord. In February of that year, the Secretary-General invited member states to appoint veterinarians to a committee of experts which would advise the League's Economic Committee on disease and the trade in animals and meat products. It was hoped that this subcommittee would examine sanitary measures and strike a balance between the interests of exporting and importing nations. The OIE was to have a key role in assisting these experts through the provision of up-to-date information.[62] But the subcommittee was a consultative body and had no powers to force states to take certain measures or otherwise to intervene in their sanitary practices; nor did it have the power to adjudicate on trade disputes involving sanitary questions.[63] It also suffered from the same problem as the OIE because the ten experts were drawn exclusively from Europe. This limitation proved a major handicap in dealing with the Great Depression, during which many countries were inclined to protect their domestic industries.[64]

In an effort to improve the credibility of the veterinary subcommittee, specialists from three countries outside Europe were invited to become members: New Zealand, Argentina and Uruguay.[65] In 1931 the League of Nations also began to press more strongly for an international convention on veterinary matters.[66] Members of the Veterinary Committee and the League's Economic Committee drafted a document which would have made it obligatory for member states to maintain effective disease surveillance, to inspect livestock and meat-processing facilities, and regulate the transportation of meat and live animals. The inspection of exports was to be carried out by a designated veterinary sanitary service in each country, and each nation or colony was to publish a veterinary sanitary bulletin which would be sent regularly to the OIE.[67] But the League still found it difficult to rouse enthusiasm for its venture beyond the European empires. The main stumbling block remained the attitude of the US government, which was keen to maintain preference for its neighbours – Mexico, the Central American republics, Canada and the West Indies – while keeping its blanket quarantine against livestock from elsewhere: fifteen days in the case of pigs, thirty in the case of cattle. Livestock from Mexico and other countries enjoying trading privileges with the USA was merely subject to inspection for disease.[68]

US sanitary policies and the country's refusal to join the OIE protected its ranchers and meat producers from foreign competition, a principle which remained in place for decades to come. However, it was permitted to lapse from time to time in the interests of international relations. In the early 1930s, for example, the administration of Franklin D. Roosevelt sought a tariff agreement with some of the South American republics as part of his 'Good Neighbor Policy', which was designed to limit US intervention in Latin America and contribute to economic and political stability. To this end, the US government sought the removal or relaxation of all restrictions upon trade, and in 1935 it signed a treaty with Argentina – the region's foremost cattle and beef exporter – to limit bans on imports to infected areas only. The treaty was in line with a resolution at the fourth Pan American Commercial Conference of 1931, which declared that sanitary regulations should not serve as a cloak for protectionism. But soon after the treaty was signed, the State Department was inundated with protests from farmers who claimed that their interests had been damaged. The Bureau of Animal Industry refused to support the treaty and the Senate hesitated to ratify it, provoking critical comments in the Argentine press about protectionism. Despite implementing measures designed to reassure US public opinion, the Argentine government was disappointed. Faced with implacable opposition at home, the State Department gave up all hope of ratification by 1939.[69]

Argentina also faced difficulty in exporting its meat to countries which were members of the OIE. It had a particularly important trading relationship with Britain, which imported large quantities of meat from Argentina and exported to Argentina a significant amount of manufactured goods. British investors had built up the meat industry in Argentina, together with much of the country's infrastructure, including the railways that linked the stock-rearing areas of the pampas to the country's ports. Following the discovery of new methods of freezing and refrigerating meat from the 1880s, British businessmen had established slaughterhouses and meat-packing and refrigeration plants on the estuary of the River Plate. Improvements in ship design also enabled the transportation of tens of thousands of live animals from Argentina to Britain every year.[70] Foot-and-mouth disease, which had entered Argentina in the 1860s and '70s with European migrants, occasionally interrupted the shipment of livestock but it was not until the 1920s – after a serious outbreak occurred in Britain – that veterinary officers suspected that the virus causing the disease might also be spread by infected meat. Unlike the US government, however, the British were reluctant to ban imports of Argentine meat, the interests of exporters and investors taking precedence over those of farmers and consumers. Lacking a large livestock industry of its own,

Britain would have been unable to meet demand for meat without foreign imports.

In the mid-1920s, as firm evidence emerged linking outbreaks of foot-and-mouth to meat imports, the British government was forced to introduce restrictions, though not a complete embargo, against Argentine meat. It asked Argentina to implement tougher disease controls but the Argentine government protested that it did not have the resources to do so. Many cattle producers and exporters also claimed that the restrictions were a political act, lacking scientific justification. Protectionist impulses certainly existed in Britain but farmers were more afraid of their own government than of foreign competition. British farmers feared foot-and-mouth primarily because of the devastating stamping-out policies used by the government since the 1860s to combat 'alien' diseases when they invaded British shores. In Argentina, however, foot-and-mouth disease was endemic and occurred in a milder form because of a high degree of natural immunity. But faced with the prospect of a complete ban on its exports, the Argentine government overrode domestic opposition and introduced more stringent controls on the movement and slaughter of infected animals.[71]

The Second World War ushered in a new era in the governance of human health but brought no end to the diverse practices and bilateral arrangements governing the trade in animals and meat products. For a time, even the future of the OIE seemed uncertain. The United Nations, which replaced the League of Nations in 1945, established two specialist health agencies: the Food and Agriculture Organization (FAO) and the World Health Organization. The WHO had responsibility for human health only and thus did not impinge on the OIE, but the FAO did. However, the OIE's member states wished it to continue as an independent organization that would deal specifically with diseases of animals and, in 1952, an agreement was reached with the FAO to establish lines of demarcation and collaboration. In the years ahead, the OIE pursued its traditional mandate of harmonizing the sanitary regulations of member states, focusing in the first instance on the newly formed European Community.[72]

Despite this, relations between Latin American and European countries continued to be plagued by sanitary disputes. In 1967–8, Britain and Argentina were again at odds over the alleged spread of foot-and-mouth disease from Argentina to Britain. There had been over a thousand cases in Britain since the last major outbreak in 1951 but most had been easily contained. Foot-and-mouth disease broke out in Britain in late October 1967 and quickly spread over much of the country, eliciting much the same response from the British government as in earlier outbreaks – mass slaughter. By the time the epizootic ceased, in March 1968, 208,811 cattle had been culled, 100,699 sheep, and 113,423 pigs, at an estimated cost to the British economy of over £100 million, including £35 million paid in compensation to farmers. Vaccination began to be talked about as an alternative to slaughter but was rejected by the government for a variety of reasons, including cost and doubts about its efficacy for all species of animals affected by the disease. Nevertheless, there was still uncertainty about the policy of slaughter and a widespread feeling that farming interests had been sacrificed yet again. Apart from the culling, farmers were angry at the government's reluctance to impose a ban on meat from Argentina, which seemed increasingly to have been the source of the original outbreak. The vast majority of primary cases – 179 in all – were attributed to infected meat, swill, bone offal or meat wrappers shipped from South America, and in most cases from Argentina.[73]

Argentine controls on cattle diseases were therefore regarded as lax but the British government did not introduce a ban until 1 December 1967 and lifted it in April 1968, immediately after the outbreak had finished. This half-hearted response drew protests from the Country Landowners' Association and the National Farmers' Union, which were angered by what they saw as surrender to the interests of the manufacturing sector, which would have lost out if Argentina had carried through its threats of retaliatory action.[74] Suffering from a record trade deficit, the British government calculated that it could not afford an all-out trade war with Argentina and backed down. After the ban was lifted, meat imports from Argentina quickly resumed and were back to normal levels within a year or so. But this was far from a victory for trade liberalization and merely reflected the economic difficulties which Britain was then experiencing. As Argentina quickly found out, the mere threat of commercial retaliation could reap substantial rewards.[75]

Despite frequent disputes of this kind, membership of the OIE was expanding rapidly. On the eve of the Second World War, it had only forty-four members, most of which were European, but by 1964, as it celebrated its fortieth year, it had eighty-one members drawn from four continents.

The global scope of the OIE was now reflected in divisions centred on regional offices, each of which occasionally staged conferences whose proceedings were reported in the OIE's bulletin.[76] With this broadening scope came a stronger commitment to aiding economic development in poorer countries of the world. This meant support for programmes designed to contain animal diseases, such as vaccination, but also a continuing focus on sanitary measures that restricted trade. Again, the aim was to harmonize measures internationally and end their abuse as means of economic protection. But the USA still showed little interest in joining the OIE and while it stood outside the international framework there was no prospect of a meaningful agreement on sanitary regulations governing the trade in animals and meat products. The situation did not change significantly until 2003, when the OIE became the World Organization for Animal Health (although the old name continued to be used) and the USA became a signatory.

The new organization was committed to removing abuses of sanitary regulations, which it recognized affected poor nations disproportionately. As the WOAH clearly stated, 'an unjustified trade disruption along with an important decrease in consumption … affects small farmers and producers all around the world who lose their income resources'.[77] But the recent history of trade disputes shows that the WOAH/OIE was often unable to achieve its goals. Foot-and-mouth disease remained a source of tension between Argentina and the United States, for example, and any sign of a settlement between the two governments was denounced by livestock interests in the USA. To farmers in Argentina, it seemed that the USA was bent on protecting its ranchers at all costs.[78] In the opinion of some observers, regionalization was the only possible solution; that is, permitting different standards of controls in different areas depending on the prevalence of disease and the state of sanitary infrastructure.[79] As the new century dawned, the prospect of global standardization still seemed a distant dream, much as it did for the trade in plants.

The prospect of eternal argument

In 1845–52 the Great Irish Famine, or 'Potato Famine', had resulted in the deaths of around a million people and the emigration of a similar number to other parts of Britain and North America. The immediate cause of the famine was potato blight but it was not simply the shortage of produce that proved so destructive: the government's inadequate response to this agricultural disaster, spuriously justified by contemporary economic theory, was roundly condemned and proved to be a turning point in Irish history.[80] While the famine itself was only partly a result of the destruction of that staple of the Irish peasant diet, the potato, it was an ominous sign of things to come. Just as the march westwards of lung sickness had heralded a resurgence of livestock disease, so the arrival in Europe of an American pest led many to fear ruin in the arable sector.

The disease which brought about the Irish famine was caused by the fungus-like organism *Phytophthora infestans* which most likely reached Ireland from Peru in shipments of guano – a commodity then widely in demand as a fertilizer. If weather conditions were suitable, the spores of this fungus could spread quickly, causing plants to rot before they were capable of bearing fruit. Although no subsequent pest was as devastating as the potato blight, other plant pathogens native to the Americas began to arrive in Europe soon afterwards. One of these was a parasitic fungus or 'mildew' affecting vines. It first appeared up in 1843, in a greenhouse in south-east England and spread from there to France, Italy and Spain. In an attempt to alleviate the problem, European wine producers imported American vines which were known to be resistant to the disease, only to find that they, too, were infected with an even more deadly pest, *Phylloxera vastatrix*. This aphid-like insect was a native of North America and had been discovered as recently as 1856, by the American entomologist Asa Fitch. But in 1864 it appeared in vineyards in southern France, where it was destroying the roots and leaves of grapevines, leading in many cases to secondary infection with fungi. From the south, phylloxera spread slowly to infest all the wine-growing regions of France, as well as those of Portugal, Italy, Switzerland, Austria and Germany.

At first, no measure seemed to control the insect. Wine-growers tried burying roots in sterile sand, spraying their plants with chemicals and even flooding vineyards to protect the roots; all to no avail. Following the precedent set in human and animal diseases, wine-growers and scientists from the most badly affected countries came together to share information and find a way of controlling the pest. This meeting resulted in the Berne Agreement of 1878, which set rules for the notification of phylloxera outbreaks and the restriction of movement of plants and plant materials from infected areas. Information was also exchanged about pest control and the merits and defects of the various methods then used. The Berne Agreement was signed by Germany, Austria-Hungary, Spain, France, Italy, Portugal and Switzerland and was followed by a further document signed in Bordeaux in 1881.[81] These agreements marked the first international attempts to regulate the trade in plants but they proved powerless to stop phylloxera and it infected South Africa in 1886, Peru in 1888, and New Zealand in 1890. The infestation was probably too widely distributed and the measures insufficiently enforced to make any real impact. Its eradication from many vineyards by the 1920s was due not to restrictions on trade but the importation of disease-resistant strains from North America.[82]

The phylloxera conventions may have been ineffectual but they were symptomatic of growing concern about the propensity of long-distance commerce to spread a host of alien diseases and pests. The prospect of wholesale destruction of many species loomed large and, as a result, some countries began to pass legislation banning or regulating the trade in certain plants. In 1873, for example, Germany placed an embargo on the importation of plants and plant-products from the USA in an attempt to stop the spread of the Colorado beetle (*Leptinotarsa decemlineata*), which had infested potato farms in America. Four years later, Britain passed a Destructive Insects Act amidst growing public alarm at the spread of the beetle across North America. Aided by railroads and steamships, it advanced from its home in the west towards the Atlantic at the rate of seventy miles per year. In 1877, it was found amongst plants at Liverpool docks and legislation was quickly passed on the precedent set by the Contagious Diseases (Animals) Acts. The Act empowered the Privy Council to prevent the landing of suspect cargoes, to destroy crops suspected of sheltering the beetle, and to provide compensation to the affected parties.[83]

The Colorado beetle was unable to colonize Britain but the alarm it caused throughout Europe in the 1870s marked the beginning of legislative intervention to regulate the trade in plants. Together with the Berne and Bordeaux agreements on phylloxera, these measures signalled an end to free trade and the emergence of a sanitary regime similar to that for human and animal diseases. This trend, which began in Europe, was soon followed elsewhere. From the 1870s, some American states began to pass legislation designed to prevent infection of valuable crops, and in 1891 a federal plant quarantine station was established in San Pedro, California, to prevent the spread of infections from tropical areas of South America and the Pacific. This was followed in 1912 by a Plant Quarantine Act, aimed at controlling a range of infections and infestations known to be endemic in Europe and Asia.[84] The Act created a Federal Horticultural Board to carry out disease surveillance and administer protective measures. Across the Pacific, too, there was growing concern about the spread of plant pathogens and pests. The Japanese colony of Korea passed a Plant Protection Act in 1912 and the Philippines followed a decade later; the British and Japanese colonies of India and Taiwan made similar provisions in 1914 and 1921 respectively.[85]

These measures were uncontroversial to begin with but some people came to suspect that quarantines were masking protectionism. In the USA, for example, critics of the federal quarantine law alleged that restrictions had been introduced on the merest suspicion that foreign plants might be infected, no reasonable proof being required. While such measures benefited some large domestic producers, others involved in the horticultural business deprecated the practice, claiming that

> It is depriving the public of plant materials it has come to use and desire. It is checking horticultural, educational and investigational work and the initiative of skilled pioneering amateurs. It is tending to lower horticultural standards. It is threatening to disrupt the nation's export trade in important commodities and disturb amicable international relationships in that it jeopardizes the basic principle of international commercial reciprocity.[86]

Those, like the North West Fruit Growers' Association, who opposed the policy of the Board, believed that it would invite retaliation from exporting countries and be detrimental to their interests in the long run. Although the Board had repeatedly denied that it was protecting infant American industries, it was seen to be doing just that.[87] Some countries had already begun to retaliate against the USA: Spain was hesitating to renew its trade treaty on account of an American embargo against certain kinds of grapes, while Britain and Ireland had banned imports of American potatoes for similar reasons.[88] The Board's protectionist instincts were also anathema to many Americans who were fearful of 'big government'. Its opponents claimed it was eroding civil liberties, many of which had already been lost through 'the steady and stealthy encroachment of a silent, invisible government'.[89] As an alternative, the Board's critics suggested that protection could be maintained by careful inspection of plant materials at ports of embarkation, where cargoes would be certified as healthy and clean. It added that this was how the legislation had been implemented from 1912 to 1919 and that it was similar to the kind of controls used in Britain and the 'majority of other civilized countries'.[90]

Disputes of the kind alluded to by critics of the Federal Horticultural Board in the 1920s were common and generally stemmed from efforts to protect fledgling industries or more established interests facing cheap foreign imports and/or domestic over-production. During the same decade, phytosanitary precautions were introduced in several other countries, including Australia which banned imports of pip fruits from New Zealand in 1921, in response to the bacterial disease 'fire-blight'. The disease had appeared in many parts of the world in the previous decades, including the USA and Japan. After arriving in New Zealand, in 1919, it took around ten years to spread through the country's orchards, causing great hardship. The government responded by placing a ban on the importation of nursery trees from the USA, which were believed to be the source of the infection, but this did not satisfy one importer of New Zealand's apples – Australia. The Australian government banned New Zealand apples soon afterwards, for reasons which the New Zealand government regarded as unjust. At the time, both countries were attempting to expand their pip fruit industries as part of their post-war soldier-resettlement programmes. But fruit production in the two countries gradually diverged. New Zealand farmers became proficient at growing new varieties of apple and had an effective marketing strategy. Australian producers, who continued to rely on old methods and traditional varieties of fruit, became comparatively less efficient and their produce was undercut by cheaper alternatives from New Zealand. The Australian ban, which had been introduced with disease prevention as the paramount object, began to seem like a form of protection.[91]

After the General Agreement on Tariff and Trade was concluded in 1947, the dismantling of formal tariffs was offset by the erection of numerous sanitary and other 'technical' barriers to trade. Thus, when

Japan officially opened its apple market in 1971, it was able to rely on a long list of pests to keep cheap foreign imports at bay. Using these methods, fruit from the USA was kept out of Japan until 1995, in which year certain varieties were permitted to enter under strict phytosanitary conditions.[92] Fruit was not the only commodity affected by such measures. In the early 1970s the international trade in wheat was devastated by a disease known as 'dwarf bunts', which damaged crops of winter wheat in the USA. But even after scientists had devised measures to control it, countries such as China and Brazil refused to accept American grain.[93] These measures went against the spirit but not the letter of attempts to lower trade barriers.

As they did not cover non-tariff barriers, the GATT talks could do little to prevent the abuse of phytosanitary legislation by protectionist governments. The only important step towards harmonization at this time was the International Plant Protection Convention, signed by member states of the United Nations' Food and Agriculture Organization in 1952. Under the IPPC, which was revised from time to time, signatory countries were encouraged, but not required, to harmonize their measures for preventing the spread of plant diseases. As in the case of animal diseases, any country that chose not to adhere to the international standard was required to base its phytosanitary measures on a risk assessment which could be subject to challenge by any country affected by it. Tougher than normal measures were considered legitimate if countries could demonstrate that they were free from disease and that imports posed a high risk of infection. As with animal diseases, this involved some concession to regional peculiarities but it opened the door to all manner of dubious practices and did nothing to halt long-running trade disputes such as that between Australia and New Zealand. Despite numerous challenges to the ban, the Australians continued to maintain their embargo on the grounds that New Zealand apples were infected with fire-blight, despite New Zealand's insistence that the risk was minimal.[94]

The formation of the World Trade Organization in 1995 raised hopes that such disputes would be resolved swiftly and equitably. Founded to replace the General Agreement on Tariffs and Trade, the WTO came to preside over most aspects of international trade, from tariffs to safety requirements. It is in the latter capacity that it became involved in matters of health, drafting guidelines which aimed to prevent or reduce the threat from disease in the shipment of various commodities. Just as the WHO drafted International Health Regulations (IHRs) which were binding on signatory states, the WTO devised Sanitary and Phyto-Sanitary (SPS) Regulations which aimed to reduce the risk of disease transmission in agricultural commerce.[95] The SPS Agreement stipulated that all such regulations should be 'based on science' and that they should not arbitrarily or unjustifiably discriminate between countries. But although member states were encouraged to use international standards, they were permitted to raise them if there was scientific justification for doing so.[96] This left the door open to a bewildering variety of measures; some seemingly imposed more for protectionist purposes than to deal with genuine health risks.

If any state's SPS measures were challenged, the WTO was supposed to play a mediating role in the first instance but if the disputing parties were unable to resolve their differences, the complainant could submit a document to a panel of the SPS Committee – chosen in consultation with the disputing parties – which would consider the evidence and invite oral testimony at hearings, together with additional evidence from experts in the field. After making an interim report, to which the parties would have the opportunity to respond, the panel would submit its final report to the WTO's Dispute Settlement Board (DSB), which was the Council of the WTO in another guise. The panel's report was theoretically advisory, but the DSB would be able to dissent from its findings only if there were a consensus among its members that it should do so. Parties would also be permitted to appeal on points of law and the appeal would be heard by a seven-member appellate body established by the DSB.[97]

The key test of any phytosanitary measure was whether it was based on a relevant international standard; that is, one set by the IPPC. If not, the measures were to be subjected to a risk assessment. If the panel deemed the measure to be in violation of the SPS Agreement, the offending government would have the choice of changing it or keeping it and compensating the complaining party. If this did not happen, the complainant would have the right to suspend a trade concession of equivalent value to its lost exports.[98] However, it could take up to two years for a deadline for implementation to be agreed and during that time, the complaining party might have incurred serious losses.[99] The WTO had the power to 'retaliate' if its directives were not complied with, but the sanctions it could impose could not be applied retrospectively. The

23 Disinfecting apparatus such as this came to be used widely throughout the world to combat yellow fever, once the insect vector of the disease had been discovered. The Clayton disinfector was also used in a more general way in an effort to destroy plague germs during the epidemics of the 1890s and early 1900s.

24 Panama was on the front-line in a war against yellow fever. The American administration of the Canal Zone insisted on strict vigilance of shipping passing through the Canal when it opened in 1914 but allowed free passage to vessels which had been certified as free from yellow fever when leaving their port of embarkation. Such measures reduced the need for quarantine and the staff of quarantine establishments was reduced considerably in the coming years.

25 The dramatic symptoms of cholera and its sudden onset made it one of the most feared diseases of the nineteenth century. After violent cramps and profuse diarrhea, victims typically suffered a state of nervous collapse before dying – the characteristic 'blue' stage of cholera pictured here.

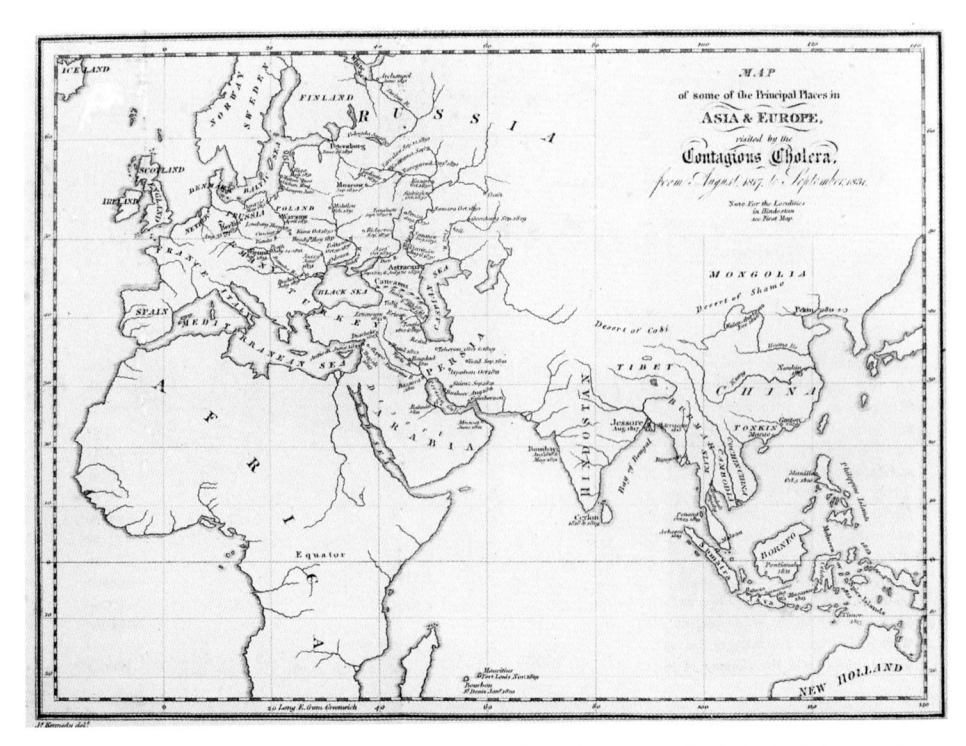

26 This map indicates the areas afflicted by cholera after it became epidemic in the deltaic region of Bengal in 1817. In the coming decade it spread to other parts of Asia and by the early 1830s it had appeared in Europe. The disease was spread by trade, pilgrimage and labour migration but at the time controversy raged over whether it was actually contagious.

MISTAKING CAUSE FOR EFFECT.

Boy. "I SAY, TOMMY, I'M BLOW'D IF THERE ISN'T A MAN A TURNING ON THE CHOLERA."

27 This sketch parodies the view, newly expressed, that cholera was spread in contaminated drinking water. John Snow later became famous for establishing this connection in his epidemiological study of the cholera in London, in 1854, but even then it took some years for the water-borne theory of cholera transmission to be generally accepted.

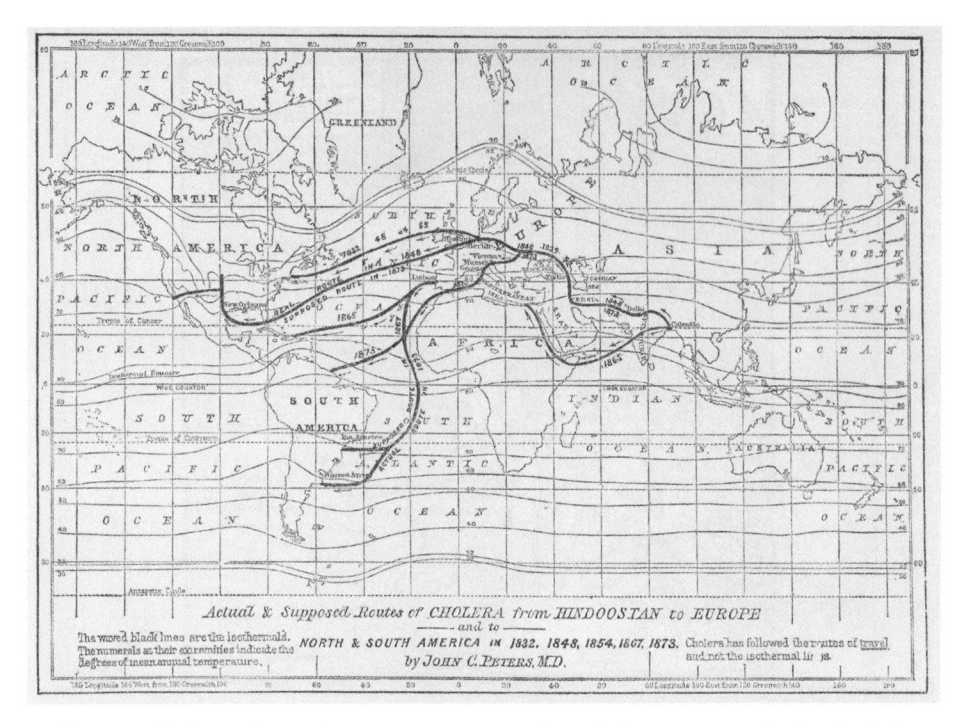

28 This map alludes to the continuing controversy over how cholera was transmitted and whether it was in any sense contagious. In particular, it refers to the debate between advocates of contagion and those who continued to claim that the spread of cholera was governed by meteorological conditions. Which theory one adhered to had implications for preventive measures and ultimately for disruption to commerce.

THE CHOLERA IN EGYPT: QUARANTINE EXAMINATION AT BRINDISI.

29 The medical inspection of passengers at Mediterranean ports became a regular feature of navigation through the Suez Canal in the 1880s and 1890s. Such measures were deeply resented but they reduced to some extent the need for general quarantine restrictions which posed obstacles to trade and communication between Europe and Asia.

30 This device for perforating and disinfecting letters was one of many which came into use during the nineteenth century in an attempt to reduce damage to mail which was often opened in the process of fumigation. Correspondence containing potentially important political or commercial information was often opened on the pretext of sanitary vigilance.

31 The attempt to combat plague in Hong Kong used draconian methods, including wholesale destruction of property in plague-affected areas. Such measures were a desperate attempt to calm fears that the plague might spread to other countries, and which had resulted in the suspension of trade with this major imperial port.

32 The picture shows an affluent family forced to leave their home due to plague in their neighbourhood, sitting outside temporary huts in a segregation camp. Such indignities caused rioting and mass flight from Bombay and other plague-afflicted cities, bringing business to a standstill. The British authorities were soon forced to come to terms with community leaders and relax their coercive measures.

33 This drawing depicts popular fears about the spread of plague to the United States. It shows the country being assailed from East and West although the spotlight was very much on East Asia and Latin American countries. The USA maintained strict vigilance in Asian and Latin American ports and sometimes intervened in their sanitary affairs.

34 The continuing danger posed by ship-borne trade is highlighted in this drawing by the American artist Albert Lloyd Tarter. By the early 1900s, the connection between rats, fleas and bubonic plague was generally accepted and anti-plague measures shifted away from restrictions on human movement to the prevention of rat infestation. This entailed new responsibilities for ship-owners but reduced the necessity for quarantine and other impediments to trade.

35 This drawing shows that plague was perceived as a global problem, even though it highlights the danger to America. The notion of 'pandemic diseases' – diseases which had the potential to spread around the whole world – was a relatively new one and was shaped to a significant extent by the experience of the Third Plague Pandemic, beginning in the 1890s.

INSPECTION OF FOREIGN CATTLE AT THE METROPOLITAN CATTLE MARKET.

36 During the 1860s there was increasing concern that livestock diseases would spread quickly from Russia by way of newly constructed railways which linked stock-yards in Hungary to ports in Northern and Western Europe. Quarantine and other sanitary arrangements at the stockyards and on some of the railways were lax and were often evaded completely.

```
     O Lord God Almighty, whose are the cattle
on a thousand hills, and in whose hand is the
breath of every living thing, look down we
pray Thee, in compassion upon us, Thy servants,
whom thou has visited with a grevious murrain
among our herds and flocks. We acknowledge
our transgressions, which worthily deserve Thy
chastement, and our sin is ever before us; and
in humble penitence we come to seek Thy aid.
In the midst of judgment, do thou, O Lord, re-
member mercy - stay, we pray Thee, this plague
by Thy word of power, and save that provision
which Thou hast in Thy goodness granted for
our sustenance. Defend us, also, gracious
Lord, from the pestilence with which many
foreign lands have been smitten; keep it, we
beseech Thee, far away from our borders, and
shield our homes from its ravages; so shall we
ever offer unto Thee our sacrifice of praise
and thanksgiving, for these Thy acts of pro-
vidence over us, through Jesus Christ Our
Lord. Amen.
```

37 The spread of cattle plague (rinderpest) during the 1860s elicited a powerful response in many European countries. Being reminiscent of Biblical plagues and coinciding with the global march of cholera, it resulted in a good deal of soul-searching. This prayer was typical of many penitential offerings in Britain after the arrival of rinderpest in 1865; all of which attributed the plague to the wickedness of humanity.

CATTLE PLAGUE.

NOTICE.

NOTICE IS HEREBY GIVEN, That persons who are not employed on the Farm of LANGRIG, in the Parish of Eccles and County of Berwick, are prohibited from entering any Building or Enclosed Place on said Farm, without my permission in writing.

JOHN DOVE.

LANGRIG,
19th November, 1867.

NOTICE IS FURTHER GIVEN, That any person contravening this Order is liable, under the Act of Parliament 29 Vic. cap. 15, to a Penalty of £5, or imprisonment for each offence, and the Police are authorised to apprehend offenders, or report them for prosecution.

GEORGE H. LIST,
Chief Constable.

COUNTY POLICE OFFICE,
Dunse, 19th November, 1867.

J. M. WILKIE, PRINTER DUNSE.

38 This notice was one of many issued during the outbreak of cattle plague in Britain. It signifies the growing involvement of the state in the suppression of animal diseases and the perceived need for restrictions of movement to achieve this. Legislation such as the Contagious Diseases (Animals) Act of 1866 reversed a trend towards the liberalization of animal exports and the neglect of sanitary precautions.

39 There was great unease over the official policy of 'stamping out' rinderpest by slaughtering whole herds of cattle. As shown here, the government's heavy-handed actions were regarded by many as a greater menace than the disease itself. Although many farmers continued to complain about the policy and the lack of compensation, the public came to support it.

PUNCH, OR THE LONDON CHARIVARI.—November 18, 1865.

BEEF?
FOURTEEN PENCE
A POUND
Ha Ha!

THE DEMON BUTCHER, OR THE REAL RINDERPEST.

40 In 2008 protests against the importation of American beef sparked huge demonstrations in the South Korean capital of Seoul. On this occasion, the potential threat of infection with BSE/*v*CJD served to crystallize more general grievances against the government and American influence.

41 SARS was the first pandemic of the twenty-first century. Although the numbers who contracted the infection were relatively small, the disease sent shock-waves through a world already nervous about terrorism and the vulnerability of the global economy. The response to SARS produced demands for greater cross-border security against infection.

42 South Korea was one of many countries affected by outbreaks of 'avian' influenza during the 1990s and 2000s. The disease generated fears that a new pandemic would arise, perhaps as deadly as the influenza of 1918–19. H5N1 did not spread from human to human but it caused considerable damage to farms and generated a bitter debate over biosecurity and animal welfare on intensive poultry farms.

43 The use of pesticides on an industrial scale to prevent plant disease was one response to attempts to block exports of tropical fruits from the Philippines on phytosanitary grounds. However, such methods generated great controversy on account of the alleged damage caused by these chemicals to human health.

deadline for implementation was also quite flexible, because each country was allowed a 'reasonable time' to alter its sanitary practices. Although this was not normally to be longer than fifteen months, it could be subject to negotiation. In the interim, producers and exporters might suffer major hardship and, in some cases, go out of business.[100]

These problems were evident in many of the SPS cases considered by the WTO during its first fifteen years. Ten years after the SPS Agreement was signed, the Australian and New Zealand governments were still far from resolving their long-running dispute. In 2005, New Zealand apple-growers staged a protest at the Australian High Commission in Wellington by bowling apples underarm beneath its gates, alluding to the infamous incident in the 1981 cricket test match when the Australian captain, Greg Chappell, ordered the bowler to bowl underarm to avoid the possibility of the New Zealand batsman scoring a 'six' to tie the match. The action was widely condemned as unsporting and New Zealand apple-growers believed that the Australian ban was similarly unfair. The New Zealand government therefore took Australia to the DSB over the risks posed by fire-blight disease if imports of New Zealand apples were permitted. The DSB's panel of experts ruled that there was 'a negligible risk' of the transmission of fire-blight through apples but Australia maintained its ban nonetheless, under considerable pressure from Australian farmers to do so. Australia's refusal to heed the WTO ruling prompted an angry response from New Zealand's Minister of Agriculture, Jim Sutton, who protested that 'Australians cheat in matters of biosecurity, and the concept of honest science has no meaning to them'.[101]

In June 2005 the New Zealand government lodged another complaint against Australia with the WTO and demanded that Australia produce a scientific risk assessment. In November the following year, Biosecurity Australia, the agency responsible for risk assessment, responded with a report which recommended that imports should be permitted subject to a range of phytosanitary measures. It was endorsed in March 2007 by Australia's Director of Animal and Plant Quarantine. However, the conditions attached were stringent and, in the view of the New Zealand government, lacked scientific justification.[102] It insisted that there was no evidence to show that fire-blight disease could be transmitted through commercially traded apples. In August 2007, New Zealand requested consultations with Australia under Article 11 of the SPS Agreement and Article 4 of the

WTO's Understanding on Rules and Procedures Governing the Settlement of Disputes. Consultations took place in Geneva on 4 October 2007 but they were unable to resolve anything. These bilateral talks having failed, the New Zealand government requested the establishment of a WTO panel. Australia did its best to prevent the WTO from investigating the legality of its ban on New Zealand apples and used its veto in an attempt to block the formation of a panel. But this was a move which Australia could only use once under the WTO's rules and when it met again in January 2008, Australia was unable to prevent a panel from being established.[103] The composition of the panel was decided in March but the Australian government again lodged an objection, claiming that New Zealand did not have grounds to request the formation of a panel, and this resulted in claim and counterclaim until the panel issued a preliminary ruling in June 2008 to the effect that it would continue to consider the points raised by New Zealand.

In the coming months, both parties made their written submissions in accordance with the WTO's dispute resolution procedures. Hearings also took place in September 2008 and June–July 2009.[104] After nearly ninety years of what the New Zealand government and apple-growers regarded as an unjustified ban, it seemed as if they might finally see the removal or at least the easing of restrictions, which, on a rough calculation, had cost NZ $20 million a year.[105] The final WTO ruling came in August 2010, when it called upon Australia to amend its import restrictions after finding its safety checks to be 'unscientific' and unnecessarily disruptive of trade.[106] This example shows that even once dispute resolution had been initiated it could take a long time to reach a conclusion.

Australia's geographical isolation had enabled it to remain relatively free from human, animal and plant diseases and its natural advantages had been enhanced by strict vigilance at its ports.[107] Quarantine was therefore popular in Australia and was seen by both farmers and the government as an important resource. SPS restrictions either banned or increased the cost of imported goods, while helping to maintain the confidence of countries which imported produce from Australia. A joint statement by the federal ministers for agriculture and trade in August 2001 explained: 'Most people think quarantine is just about keeping things out of Australia, but it is also one of the unsung strengths of Australia's export performance.' 'We have used the WTO's quarantine rules to achieve more than 240 market access gains for Australian exporters over the past five years,' they boasted, stressing that such measures were 'based on science and not on industry protection'. While a 'zero-risk' approach was incompatible with WTO rules, Australia took an unabashedly conservative view of quarantine and risk assessment. As the ministers' bulletin was at pains to emphasize, such an approach was fully within the bounds of the SPS Agreement.[108]

Other countries, such as New Zealand, clearly thought differently, as did the Philippines, which wanted Australia to open its doors to imports of tropical fruits. In 2003 the Philippine government announced that it was going to initiate settlement proceedings against Australia over its allegedly unfair use of quarantine to keep out imports of papayas and bananas. The two countries had been in talks for some years regarding the Australian ban, which was in keeping with the country's long-established policy of prohibiting all imports of fresh fruits and vegetables except under licence. There had been frequent requests from some Australian importers to source papayas from the Philippines but in 1995 additional requests were received for licences to import plantains and bananas. These were not granted and bilateral discussions ensued under the auspices of the WTO. Australia claimed that it was unsafe to import these fruits from the Philippines because certain diseases such as black sigatoka (a fungal disease affecting bananas) were endemic there. However, the Philippines claimed that Australia was breaching WTO rules, as its quarantine measures were neither scientifically grounded nor in line with international standards.[109]

A great deal was at stake in this dispute. Filipino producers already exported more than $AU30 million of agricultural products per year to Australia and were regarded as formidable competitors, enjoying a clear cost-of-production advantage over Australian growers. However, Australian farmers were genuinely terrified by the prospect of black sigatoka disease taking hold in Australia: in 2001, cases of the fungal disease were detected in Queensland and some farms lost their entire crop of bananas.[110] The disease cost a great deal to control using fungicides and other measures, adding around 15–20 per cent to the price of bananas.[111] Australian banana-growers put together a 'world class' team of scientists and lawyers to press their case for the ban at the WTO. As 'banana industry hardhead', Ross Boyle, CEO of Bananas NSW, put it: 'We agreed to stick to the science, but we realized that if they [the Filipinos] resorted to politics, then we would have to match them.'¹¹² The Philippine government certainly sought to place the dispute in a wider context. In the wake of the terrorist attacks of 9/11, Australia and other countries had become very concerned about Islamic terror cells operating in the Philippines and the Australian government had committed $AU5 million to fighting extremism in the country. But some politicians in Manila claimed it would be easier to prevent disenchantment with the regime if Australia were to remove the ban on exports, thereby stimulating the economy. Most of the country's bananas were grown on the island of Mindanao where there was chronic poverty and political unrest among the large Muslim population. According to some economists, removing the ban would have allowed Filipino producers to earn up to US $50 million per year.¹¹³ The Philippine government retaliated by imposing a ban on imports of Australian beef after its quarantine officers claimed to have found traces of anthrax contamination in some shipments from Australia.¹¹⁴

The looming trade war between the Philippines and Australia was linked to concern over the inequities of global commerce, which was increasingly coming under scrutiny from anti-globalization and fair-trade groups.¹¹⁵ Writing of the banana ban, one activist claimed that 'protectionist trade policies by powerful nations including the US, EU and ... Australia in areas such as textiles and agriculture where developing countries have an advantage, have caused the Philippine leadership to question the WTO system.'¹¹⁶ The ban was coming under attack in Australia, too. Although Australia's quarantine policies were generally popular, some commentators claimed that the embargo on fruit from the Philippines was detrimental to national interests. The trade and public policy analyst Peter Gallagher believed it likely that Australian quarantine procedures were in breach of WTO rules and would provide ammunition to rival countries and trading blocs. At home, too, quarantine barriers seemed to benefit Australian producers to the detriment of consumers, who were paying higher prices because of the restriction on more competitively priced imports.¹¹⁷

Much depended on the WTO's decision but, again, no resolution was forthcoming. Although the DSB agreed to establish a panel to rule on the Philippines' complaint on 29 August 2003, the panel was not convened until December 2005. This seems to have been partly the fault of the Philippine government, which did not supply the DSB with terms of reference soon after it announced that a panel would be formed. In the meantime, in February 2004, Australia issued a revised report on the importation of bananas from the Philippines which recommended that import restrictions be relaxed to permit the entry of hard green bananas subject to quarantine. The move was welcomed in Manila and presented Australia in a more reasonable light to the WTO.[118] However, it caused outrage among Australian banana-growers. The Australian Banana Growers' Council claimed the report watered down Australia's tough quarantine standards and it staged a protest rally in Cairns on 5 March 2004. By the end of the year, Biosecurity Australia had been asked by the government to review all risk assessments – a process which would clearly take months.[119]

The government of John Howard was anxious to maintain its credibility as a force for multilateral trade liberalization and was sympathetic to the interests of Australian consumers. However, the banana industry carried a great deal of weight in Queensland where the bulk of bananas and other tropical fruits were grown. Howard's government was presented with a dilemma: it faced further protests and strikes if it continued to relax quarantine unilaterally but it risked becoming an international pariah if it did not.[120] In view of these competing pressures, a protracted dispute resolution process may not have been wholly unwelcome to the Australian government. Any decision which might ultimately be taken would also be taken by the WTO rather than the Howard administration alone, thus absolving the latter of blame.[121] However, Howard was keen to put Australia–Philippines relations back on track as part of an ongoing attempt to improve not only commercial links but regional security and defence cooperation. Mutual concerns about Islamic extremism and terrorism now topped the political agenda in the wake of the invasions of Afghanistan and Iraq, to which Australia lent military and other forms of support.[122] Thus, in 2007, John Howard met the Philippines' Agriculture Secretary Arthur Yap and announced that a formal risk assessment for Philippine bananas would be completed by the end of the second quarter of 2008; the export of bananas to Australia was expected to begin, subject to stringent quarantine, in the third quarter of the same year.[123]

In the meantime, banana-growers in the Philippines had been doing their bit to ensure that produce from the country posed as little risk as possible to importing nations. Aerial spraying of fungicides was stepped up in parts of Mindanao, causing concern among inhabitants, who complained of rashes on their skin and of feeling unwell.[124] One of the chemicals used as a fungicide – dithane – had been listed as carcinogenic by the US Environmental Protection Agency, although its use was permitted in the Philippines. Davao City on Mindanao Island therefore decided to ban the aerial spraying of local plantations in February 2007 and, in 2009, there were moves in the Philippine Congress to ban it on all plantations. The Philippine Banana Growers' and Exporters' Association, which included multinational giants such as Del Monte and Dole-Stanfilco, challenged the Davao City ban in the courts but it was upheld.[125] Yet, spraying continued in other areas, eliciting vigorous protests from environmental groups.[126] The Philippine government continued to face a tough choice between its export earnings and further civil unrest.

The developing accord between the Philippines and Australian governments involved difficult decisions for both parties. Producers' interests, regional security and public health considerations had all to be weighed in the balance. In Australia, the declaration by the two governments met with a predictably furious response from the banana lobby which remained concerned about the introduction of pests and more competitively priced fruit. The price of Philippine bananas was then US $5.50 per carton, whereas Australian bananas were priced at $18; it seemed unlikely that Australian producers would be able to compete in the long term, except in organic and other niche markets.[127] Although Biosecurity Australia had issued a revised risk assessment in 2007, recommending that bananas be permitted to enter Australia subject to stringent quarantine, the banana industry was not confident that the Australian Quarantine Inspection Service could adequately protect producers under the terms of the new risk assessment. The Queensland government appealed against the decision in early 2009 but the appeal was rejected; nevertheless, banana-growers retained considerable support. The supermarket chain IGA, for example, announced that it would be banning all Filipino bananas from its 1,270 stores when local products were available.[128] There was still no sign of Philippine bananas entering Australia in large numbers and Filipino producers continued to complain that risk assessments were being used unfairly to target their bananas, even though they had had diseases such as black sigatoka under control since the 1970s. They also pointed out that other countries with stringent safety standards, like Japan and several

Middle Eastern states, had allowed imports of bananas from the Philippines for the last three decades.[129] Banana-growers in the Philippines were fortunate that these markets were buoyant, particularly that in Japan, where imports were up on the previous year by 11.2 million cartons due to a banana-diet craze.[130]

Australia was portrayed as the villain of the piece in several of the disputes brought before the WTO but the Australian government was more flexible than some of its citizens wanted. Perhaps the clearest example of this is the free-trade agreement concluded with the USA in 2004. Up to this point, the US government had complained frequently about Australia's strict quarantines against plant and animal products and American producers and exporters generally welcomed the prospect of an agreement likely to reduce barriers to trade. One extremely burdensome restriction was a measure limiting the export of Californian grapes to Australia. The California Farm Bureau Federation thus supported the proposed agreement but remained cautious, warning that any new arrangement would have to be based on 'sound science'. In its view, the final agreement ought to include a mechanism for regularly addressing new phytosanitary issues as they arose. Unless a truly independent body was created to resolve potential disputes, the CFBF suspected Australian domestic politics would continue to 'cause months of delay and often result in decisions based on considerations outside the risk assessment'.[131] Something resembling this body was actually incorporated into the agreement. Indeed, two committees were established as a result of the free-trade agreement: an SPS committee to enhance understanding of each country's regulations and a standing technical working group on animal and plant health measures to focus on quarantine. Both countries affirmed that any decisions made about quarantine or food safety would be 'based on science'.[132]

In Australia, however, there were fears that the government had caved in to pressure from the USA. A submission to the Senate Select Committee on the Free Trade Agreement protested that it was contrary to the interests of the Australian people and claimed that the Australian government had been in a weak negotiating position because of the relative size of the Australian and US economies. It protested that the Agreement was 'likely to compromise our quarantine standards over time through pressure from skilled US negotiators who have no interest in Australian agriculture or biodiversity'. It also pointed out that the agreement had not explicitly stated that industry representatives or 'puppets' would be excluded from the two technical committees designed to ensure that safety standards were maintained.[133]

Like the disputes that were referred to the WTO in its first fifteen years, the controversy over the Australia–USA agreement demonstrates the difficulty of reaching consensus on risk assessment. It also highlights a fundamental weakness in the SPS Agreement. The WTO's regulations permitted great diversity in the measures taken by signatory states, the only requirement being that they were broadly in line with international standards and 'based on science'. There were good reasons for this: disease environments differed greatly, as did the technical capacities and resources of member states. But apart from regional diversity, the SPS Agreement allowed a fair amount of latitude to members in cases where scientific evidence might be deemed 'insufficient'. Under Article 5, paragraph 2, members were permitted to adopt on a provisional basis measures which could be verified or not at a later date. Whilst they might later have to be dismantled, these temporary constraints on foreign imports proved onerous for many producers, particularly in developing countries. Moreover, under Article 3, paragraph 3, WTO members were permitted to introduce measures which resulted in 'a higher level of sanitary or phytosanitary protection than would be achieved by measures based on the relevant international standards'. They were required to present a 'scientific justification' for doing so but as one commentator put it, 'in the world of science judgement about whether a pest or disease can establish itself offer the prospect for eternal argument'.[134]

CHAPTER 9

Disease and globalization

By the end of the twentieth century, the austere dichotomies of the Cold War had all but melted away. The collapse of communism in Eastern Europe had given rise to feelings of euphoria, of capitalism and liberal democracy spreading triumphantly under the protection of a single hyper-power – the USA.[1] The creation of the World Trade Organization and the removal of many formal tariff barriers also seemed to herald a new era of free trade. But the future was rather different than the prophets of liberal capitalism had foretold. By the 1990s, the rise of Asian 'Tigers' such as Singapore, alongside more established Asian economies like Japan, pointed to an economic future that would no longer be determined by the West. The massive growth experienced in India and China a decade later made this realignment appear irreversible. Lower production costs and technological sophistication gave these countries an advantage in the export market and made them attractive to investors from outside. Indeed, many Western-based concerns began to 'outsource' parts of their operations to Asia, a move enabled by a revolution in electronic communications.

These were just some facets of a new phenomenon which supporters and critics alike dubbed 'globalization'. Global integration was not a new phenomenon but global wars and ideological divisions had meant that it had proceeded fitfully for much of the twentieth century.[2] From the 1990s, however, there were fewer political barriers to hinder the flow of commodities and finance, especially once Marxist ideology in China had been diluted in order to allow capitalism to flourish. In this multipolar world it seemed that individual nations would matter less and that institutions of global governance would grow in stature. But globalization was a more diverse and fractured process than many imagined. Transnational businesses were forced to adapt to local conditions, taking account of different tastes and sensibilities. Globalization also encountered resistance from those who sought to preserve local economies and cultural diversity.[3] By the end of the first decade of the new century, the smooth progress of integration no longer seemed assured and, as a worldwide recession loomed, some commentators feared that protectionism would reassert itself. As some Asian countries began to emerge from recession these warnings seemed prophetic, for allegations of an undervalued currency threatened a trade war between the world's two largest economies – China and the USA.

These tensions are clearly evident in the recent history of commerce and disease. While proponents of globalization have celebrated the potency and diversity of the market,[4] its detractors have often pointed to the risks which globalization poses to health as well as to domestic producers and cultures. In many cases, these fears have combined powerfully in anxieties about the health status of imported goods, especially foodstuffs. Concern has also been expressed over the spread of disease due to the accelerated flow of people and commodities. Global cities have been linked by a chain of infections ranging from multi-drug-resistant tuberculosis to SARS, while the long-distance trade in animals threatens to spread new strains of influenza and other pathogens capable of crossing between species.

These are all legitimate concerns but the most important consequence of globalization from a public health standpoint is not, perhaps, the wider circulation of disease or even the rapidity with which it can spread in an age of cheap international air travel. It is the pressure which globalization places on sanitary regulations. Globalization has exerted enormous pressure on sanitary standards and it has done so unevenly. Powerful countries have sometimes prevailed upon weaker ones to reduce sanitary barriers to trade, but they have also faced fierce competition as a result of the removal of formal tariffs. In these circumstances, states have been increasingly inclined to use non-tariff barriers such as the SPS Agreement as forms of agricultural and industrial protection: a situation which institutions of global governance have been largely powerless to prevent.

Crossing species, crossing borders

Some of the sanitary issues related to globalization began to surface in the late 1980s and 1990s as a new disease attracted the attention of the world's media. Bovine Spongiform Encephalopathy, or BSE as it came to be known, affected cattle in the first instance and then seemed to 'jump' species to affect humans. The disease had its origins in feeding practices established years before it became manifest. Although various theories were advanced to explain its emergence,[5] a scientific consensus formed around the role of infectious agents called 'prions', which are types of malformed protein. These agents probably entered the food chain in the 1970s and 1980s when commercial feeds for cattle started to contain left-over materials from slaughtering, especially nerve tissue and bone-meal. This process had advanced furthest and fastest in Great Britain and it was there that the first cases of BSE were reported in the mid 1980s. This, at any rate, was the conclusion reached in 1988 by the Southwood Committee, which had been established by the British government to look into the causes of BSE. Shortly after the committee issued its report, the government ordered the slaughter of all potentially infected animals and banned feeds derived from animal proteins. But the rot had already set in and the government was forced to slaughter around 4.4 million cattle. These drastic measures helped to bring the disease under control but not quickly enough to prevent a great deal of infected meat entering the food chain. In an attempt to reassure the British public, the government banned beef offal from baby foods in February 1989, following this with a total ban on its use. Alarm was growing internationally, too, and the European Commission prohibited the export of British beef cattle born before July 1988, as well as the use of bovine brain and spinal material for human consumption.

Although these measures were seen by some as excessive, public concern proved to be well founded. In 1995 a British citizen became the first person to die from a new variant of the degenerative brain disorder, Creutzfeldt-Jakob Disease, or vCJD as it became known. This horrific and invariably fatal disease created widespread alarm and the British government was forced to admit that there was probably a connection between it and BSE. Responding to concerns in other European countries, in March 1996 the EC announced a worldwide ban on the export of British beef; a policy which met with howls of protest from the British government and farmers, who claimed that *v*CJD had probably been contracted before the offal ban had been introduced. The British government declared a 'beef war' against its EU partners and embarked on a policy of non-cooperation in an attempt to have the ban lifted. It also stepped up measures at home, introducing, in 1997, an unpopular ban on the sale of beef-on-the-bone, which was often flouted openly. But despite appeals against the legality of the EC's decision, the ban remained in force until the summer of 1999.[6] Even then, France persisted in its embargo, citing continuing uncertainty over the risks of BSE/*v*CJD transmission in its defence, but the British government suspected France of protecting its own producers and the European Court of Justice agreed, ruling against France in December 2001. Its ban was eventually lifted in October 2002.[7]

As the first country to suffer from BSE, Britain's experience of the disease and its human equivalent became the reference point for other countries facing similar dilemmas. After hearing of official incompetence in Britain, the inhabitants of these countries became anxious and wondered whether they were being told the whole truth about their food supply. Many people also felt that the disease was a consequence of modern farming practices, which seemed to fly in the face of nature. Like the cattle plagues of the 1860s, BSE appeared to be a punishment for the sins of humanity.[8] But having seen the devastation caused to British agriculture, farmers and officials in other countries were reluctant to confirm its existence. Amidst widespread panic, proponents of economic protection saw that they could play upon public anxiety in order to block imports of foreign meat.

These fractures in the global economy began to show themselves shortly after BSE was reported outside Britain. By the late 1990s, BSE was found among cattle in several European countries, including France, and before long it appeared on the other side of the Atlantic, in Canada and the USA. In most cases, the disease was probably transmitted in contaminated feed traded on the world market.[9] But whether it spread through feed or infected animals, BSE had become widely dispersed and many countries began to fear that their domestic herds and their citizens would become infected. The importation of animals and meat thus became a focal point for concerns about globalization.

Nowhere was this more evident than in South Korea, where BSE/*v*CJD came to define the country's ambivalent relationship with the USA. BSE first became a hot topic in South Korea, in December 2003, shortly after a case of 'mad cow disease' was identified in America. The USA had hitherto been the source of most of South Korea's beef imports but confidence in its safety soon evaporated. Other major importing countries such as Japan and Mexico also banned American beef and, within days, the USA lost 90 per cent of its export market.[10] The US government moved quickly in an attempt to reassure consumers. There was already a surveillance system which tested cattle displaying signs of neurological problems at slaughter but the testing of apparently healthy animals was limited. After the first BSE case, additional precautions were taken, including the removal of certain high-risk materials like bone and nerve tissue from the food chain and a ban on mechanically recovered meat. A thorough investigation was also made of the circumstances in which the first BSE case arose and it was claimed that the animal had been imported from Canada, where it had been born before a ban on feeding meat and bone meal came into operation.[11] In other words, neither farming practices nor feedstuffs were to blame.

These findings, and the additional precautions taken to safeguard beef, enabled the US government to argue that embargoes against American meat were unjustified and throughout 2005 and 2006 it exerted great pressure bilaterally and through the WTO to have the bans lifted.[12] South Korea and Japan – the two countries which had held out longest – finally relented. At the end of 2005, Japan decided to remove the ban but it was reimposed in January 2006 following revelations that banned beef products had entered the country despite an agreement to exclude meat from cattle over twenty-one months old, as well as high-risk material such as nerve tissue.[13] A new agreement was reached in July 2006, following an inspection conducted in the USA by Japanese officials, but consumer confidence in American beef remained low.[14]

Exactly the same thing happened when South Korea decided to resume imports in 2006. By October 2007 the ban was reintroduced when proscribed animal parts were found in cargoes of American beef and it remained solid until the newly elected President, Lee Myung-bak, began to re-examine the issue in April 2008.[15] The first conservative president in a decade, Lee had been elected with a landslide majority. A former senior executive of the Hyundai engineering and construction company, 'the bulldozer', as he was known, aroused great hopes in those disappointed by his left-leaning predecessor Roh Moo-hyun. Although it had enjoyed success as an export-driven economy (then the thirteenth largest in the world), South Korea was facing difficult economic problems. Inflation was increasing rapidly, growth was faltering, and the pensions system was in need of reform. Lee promised to tackle these issues with the same determination he had shown in business. Among other things, he aimed to privatize state-owned sectors of the economy and challenge the power of the unions. Only then, in his view, would South Korea be able to take full advantage of the expanding global economy. Many South Koreans welcomed these reforms but Lee's abrasive style was beginning to alienate former supporters even before the beef scandal erupted.[16] At a deeper level, many Koreans were profoundly ambivalent about the effects of globalization upon their national culture as well as their economy. But until Lee flew to Washington, DC to discuss the beef ban in April 2008, opposition parties and other interest groups had been unable to tap these feelings of unease and channel them into a political movement.

The beef ban rose to the top of Lee's agenda in April because American politicians had been threatening to scupper a trade agreement with South Korea which Lee considered vital. South Korea had formerly been a major market for American beef and it seemed as if the resumption of imports would be a condition for any agreement on trade with the USA. Had Lee canvassed widely before his visit, there might have been a chance of enlisting public support, but he gave the go-ahead to beef imports without consultation. Many people believed that Lee had bowed to American pressure, making him seem weak as well as arrogant. The agreement also aroused long-standing fears about South Korea's independence. Tens of thousands of US service personnel had been based in the country since the war of 1950–3 and large military bases in the heart of the capital were a constant reminder of their presence. For many South Koreans, the issues of food sovereignty and political sovereignty were inseparable. On 10 June 2008, nearly 100,000 people gathered in the centre of Seoul to protest against Lee's decision to resume imports of American beef. The timing of the demonstration was highly symbolic: it was the twenty-first anniversary of the massive street protests which had turned South Korea away from authoritarian rule and towards democracy. Now, after what seemed to be a unilateral and secretive decision by the President, the country was poised to take a step back in time. But the protests in 2008 were not openly about a return to the past or even the prospect of cheap foreign meat flooding the domestic market. What united the protestors who came on to the streets throughout June and July was their fear of 'mad cow disease' and its human equivalent. Although the number of deaths from vCJD in other countries had not reached the proportions which some had feared, a steady stream of cases around the world was sufficient to cause alarm among South Koreans. Their President was asking them to accept beef from a country which had reported cases of BSE in the past and which had allegedly flouted food safety regulations in exports to a number of countries, including South Korea.

The massive protests in Seoul forced Lee to delay the resumption of imports for a few weeks. Together with his agriculture minister, Chung Woon-chan, Lee attempted to allay public fears and pointed to the fact that the World Organization for Animal Health had declared the USA a 'controlled BSE risk'. But South Koreans were far from reassured, for the agreement with the USA appeared to place too much faith in vested interests to police the trade. Lee considered America's BSE inspection procedures adequate, even though they examined only 1 per cent of the country's cattle. Nor were Korean consumers reassured by the fact that 18 per cent of cattle slaughtered for consumption in the USA were older than thirty months.[17] The USA's track record in beef exports to Asia and the misleading information issued by the British government to its own people in the 1980s and '90s led many to doubt the reliability of official sources. Indeed, throughout the summer of 2008, there was intense interest in South Korea in what had happened in Britain, particularly the government's ill-informed reassurances with respect to the danger of BSE crossing the species barrier.

With trust in the government running low, thousands of protestors marched in candlelit processions through the streets of Seoul. The nightly demonstrations were organized online, making it difficult for the government to deal with them. There were no obvious leaders at first but soon figures emerged to voice the concerns of protestors and orchestrate demonstrations. Most of these protests were peaceful and included many people who were not normally to be found on demonstrations, such as young mothers and children. The symbolism of vulnerable people threatened by the actions of a 'cavalier' government and an 'uncaring' foreign power was potent. One schoolgirl aged thirteen years told an American reporter, 'I am afraid of American beef. I could study hard in school. I could get a good job and I could eat beef and just die.'[18] The protests continued throughout June, attracting a disparate band of parents, students, office workers, trade unionists and reserve soldiers.[19] Many hoped that the delay would lead to a resumption of the ban but they were disappointed by the announcement of a new deal with Washington on 21 June, which released beef on to the market within days. Lee's apology and further guarantees of the safety of American beef had evidently failed to convince.

Although the protests had no fixed leadership, the first demonstrations were instigated by the main opposition party, the United Democratic Party. More radical elements saw popular outrage against the importation of American beef as an opportunity to mount a broad-based opposition to Lee's government. One of these was Kim Kwang-il, a member of the South Korean Socialist Workers' Party who hoped the demonstrations would provide a platform for opposition to the ruling Grand National Party's neoliberal, 'anti-democratic and imperialist' policies. Kim believed that the candlelit marches embodied the principle of 'People before Profit' and that this united Korean protestors with struggles against neoliberalism all over the world.[20] But political parties had to share the limelight with other organizations. At the end of June, 200 Roman Catholic priests and around 8,000 others attended mass at the oval plaza in front of Seoul City Hall and marched through neighbouring streets for about an hour. Afterwards, the priests urged most of the protestors to go home but ten priests began a hunger strike, urging the government to renegotiate the deal with Washington. These actions spurred another round of protests based on public masses. Other religious groups entered the fray. At the beginning of July, Buddhists and some Protestant churches declared that they would be joining the protests, condemning brutality by the police towards some of the protestors.[21]

Up to this point, the protests in Seoul were seldom violent. There had been a few minor confrontations between protestors and riot police but no determined use of force by either side. But in the second half of June tempers were becoming frayed. On the 26th, thousands of protestors clashed with police in the capital and there were signs that trade unions were becoming militant: 300 members of the Korean Confederation of Trade Unions, one of the country's largest umbrella groups, with more than half a million members, blocked roads leading to eighteen warehouses in which American beef was stored. The KCTU simultaneously announced an indefinite strike against the beef deal,[22] affecting some of South Korea's most important export industries. A two-hour walkout by 55,000 workers on 2 July cost the Hyundai motor company an estimated US$ 28.8 million.[23] Within a few days, the Korean Teachers' and Education Workers' Union joined the protests and hung banners at schools, opposing the resumption of beef imports.[24] As tension mounted, the government decided to crack down hard on the protestors. By the end of June, 746 had been taken into police custody and six were arrested. On 1 July police also raided the homes of seven leaders of the People's Conference Against Mad Cow Disease and confiscated computers and documents; arrest warrants were issued for eight persons who had organized the rallies.[25]

The government's tough stance provoked accusations of a return to undemocratic rule but it was not without support. Owners of restaurants and other businesses badly affected by the demonstrations organized their own protests against the People's Conference. Others denounced the growing use of violence by protestors and opposed the involvement of teaching unions and religious groups in political rallies.[26] The pent-up demand for American beef also became clear when the ban was lifted and huge stockpiles of meat were released from the customs sheds. Many retailers and restaurant owners were eager to get their hands on the beef and wholesale companies such as 'A-Meat' based in Seoul struggled to keep up with demand.[27] Nevertheless, the Korea Chamber of Commerce and Industry thought it wise to reassure the public with a beef-eating event involving doctors and executives who were determined to show there was no risk of infection.[28] Koreans normally resident in America also staged demonstrations in Seoul in support of the resumption of imports. Korean-Americans were in a difficult position and were eager to show loyalty to their new homeland, yet they strove to reassure South Koreans that American beef was safe.[29] Americans in Korea were in a similar predicament, calling for resolution of the dispute by more conventional means.[30] Despite these gestures of support, the Lee administration was isolated and weak. Even those who welcomed the resumption of imports, or at least who could not see any significant health risk, believed that the President had unnecessarily antagonized moderates. There were repeated calls for Lee to embrace the opinions of others and seek a meaningful dialogue.³¹ But the administration alienated potential supporters by continuing to repress dissent. In early July, the President ordered the police to intensify action against protestors in the light of a fast-deteriorating economic situation. As the stock market plummeted following news of a huge trade deficit, Lee declared the stoppages in car factories 'illegal' and threatened arrests if workers failed to turn up because of an ostensibly political strike.³² The announcement met with defiance. Tens of thousands of supporters, including religious and labour groups, gathered outside the City Hall in Seoul chanting slogans such as 'Lee Myung-bak out' and 'The people will triumph'.³³ Similar protests occurred in the weeks ahead, some of which were dispersed by riot police.³⁴ The government's grip intensified over the coming months and the number of regular protestors began to dwindle. Much of the discontent expressed during the crisis was forced beneath the surface.

BSE/vCJD was quite different from most of the diseases which had hitherto been linked to trade. It was not simply that it was a new disease, capable of moving between species, but it called into question the reliability of governments and the experts they employed. After the BSE scandal in Britain, confidence that governments would act in the interest of public health was shattered. The same mistrust was even more evident in South Korea, where BSE crystallized latent political tensions, not the least of which was the country's ambivalent relationship to the United States. The potential of BSE to spread through imported meats was also viewed through the lens of globalization, the relaxation of the embargo on American beef illustrating the vulnerability of even a medium-sized economy in the face of an economic giant such as the USA. Few countries, it seemed, were in a position effectively to protect their own industries and the health of their people. Those that were able to do so were to be found among a handful of powerful nations and members of trading blocs such as NAFTA and the EU. As we learned in the last chapter, they were often able to exploit loopholes in international agreements and erect numerous impediments to delay the process of dispute resolution. But when the world faced a new threat, in the form of SARS, the principal issue was not protection but rather the propensity of disease – or fear of disease – to undermine the confidence of investors. Its potential to destabilize the global economy meant that SARS assumed an importance that far exceeded its impact in epidemiological terms.

SARS, 'security', and the limits of free trade

In November 2002, a government official in southern China fell ill with a bad case of pneumonia which did not respond to the usual forms of treatment. By the middle of February, over 300 cases of this 'acute respiratory syndrome' had been reported in Guangdong province, with five deaths. A medical team was sent from Beijing to investigate the outbreak and to meet Dr Zhong Nanshan, the head of the Guangdong Institute for Respiratory Disease, who had done more than anyone to understand the nature of the new infection. At this point, the outbreak had not been announced officially but concern was growing locally and pharmacies were mobbed as people tried to get hold of any drug they thought useful. Dr Zhong and his colleagues had little success in treating the disease but he discovered that it was spread by droplet infection and attempted to prevent it by placing his patients in quarantine. These findings were overlooked by the government's health officials, but by mid-March the problem could no longer be denied.[35] There had been outbreaks of the disease in the commercial and financial centre of Hong Kong and from that point on, Severe Acute Respiratory Syndrome, as it became known, was no longer a domestic issue but a global concern. On 15 March the WHO confirmed that SARS was a 'worldwide health threat', reporting that cases had already been identified in Canada, Thailand and Vietnam. By 19 March suspect cases had appeared in the USA, Spain, Germany, Slovenia and the United Kingdom.[36]

The speed with which the disease travelled was an unsettling reminder of how small the world had become. Inevitably, SARS soon came to be associated with the phenomenon of globalization: both in the physical sense – as the disease spread along networks of commerce and tourism – and in more subtle ways, as the lightning-fast transmission of information enabled by electronic media created a virtual pandemic. Febrile speculation, rumours and deliberate falsehoods spread unchecked, undermining business confidence and magnifying the instability inherent in commodity and currency markets. Even those who openly praised globalization acknowledged that the new economy would be a less predictable creature than the one it was replacing. In the late 1990s, a financial crisis had brought to an abrupt end the rapid rise of South East Asia's Tiger economies but on that occasion the 'contagion' of financial meltdown had been contained. The future was far less certain, for the world economy seemed increasingly vulnerable, not only to financial contagion but to the real thing.

Between 2000 and 2010 or so, the emergence and spread of so-called 'Hot Zone' diseases such as Ebola, together with a resurgence of better-known infections such as yellow and dengue fevers, reminded the world of its continuing vulnerability to disease. Ebola and haemorrhagic fevers rapidly burned themselves out and did not constitute a major problem for the global economy. Yellow fever, dengue and some other mosquito-borne infections posed a greater problem because they were more widely distributed. Indeed, their geographical reach appeared to be spreading because of unplanned urban development, climate change and the extensive trade in used tyres, which easily harboured mosquitoes capable of carrying dengue, for instance. But the greatest challenge was posed by the emergence of HIV/AIDS in the 1980s. Although AIDS was rapidly becoming a chronic disease in the West because of the availability of antiretroviral drugs,[37] it remained a geopolitical concern because it threatened to destabilize those parts of the world in which the disease was progressing, particularly sub-Saharan Africa.[38] And yet, not even HIV/AIDS posed an immediate threat to the global economy: it was not a disease of trade in the usual sense and lacked the shock potential of a fast-spreading, rapidly fatal epidemic. But SARS fitted the bill perfectly and its potential to play havoc with business was soon recognized, coming as it did in the wake of the Asian banking crisis and attacks on New York's World Trade Center.[39] Soon after the WHO made its first announcement about SARS, concern was expressed about the potential of the disease to halt South East Asia's fragile recovery and plunge the world into a major recession. SARS was important for another reason, too. Coming at a critical time for the sub-discipline of public health, the disease provided a means by which public health workers could demonstrate the relevance of their field. Combined with the accelerated dissemination of news and rumours through the internet, these imperatives ensured that the disease would receive undivided attention.

By April 2003 there was no sign of the disease abating and SARS was present in several populous nations, including India. However, it was still chiefly prevalent in southern China and on 2 April the WHO recommended the postponement of all non-essential travel to Hong Kong and

Guangdong. Embarrassed by this unwelcome attention, the Chinese government issued an apology for having responded so slowly to the epidemic. The Prime Minister Wen Jiabao and the new President, Hu Jintao, mobilized the country and removed the officials who had concealed the outbreak. These measures were designed to reassure international opinion but outside China there were grave doubts about the sincerity of the government. The Chinese authorities needed to do more and they decided to take drastic action, including placing 10,000 people in quarantine, in the eastern city of Nanjing. Spitting in public was banned and those who broke quarantine were threatened with life imprisonment or execution. These measures caused a certain amount of unrest in China, especially in rural districts, but they conveyed the message which the rest of the world wanted to hear: that the Chinese government was no longer in denial and was taking its responsibilities seriously.[40]

As the Chinese geared up for an offensive against SARS, commercial and financial pundits began to speculate about the consequences of a pandemic for the region's economy. Even if SARS did not affect the rest of Asia as badly as China, China was the world's fastest growing economy and was involved in one in five commercial transactions globally. The WTO was therefore concerned that the disease would have serious repercussions, especially when major investment banks such as J.P. Morgan reduced their forecasts for growth in other Asian economies.[41] In April and May 2003, such anxieties were understandable in view of performance indicators from several sectors of the economy. The Hong Kong Retail Management Association believed that it would take months for consumer confidence to recover from SARS, with many shoppers staying at home. The tourist industry had also been badly affected and there was a 10 per cent fall in visitors to Hong Kong in the second half of March, compared with the previous year. The volume of traffic at the former colony's airport was down 25 per cent, as tourists and business travellers stayed away. Transnational firms such as Nestlé cancelled commercial events in East Asian cities, while sanitary restrictions prevented or hindered many business people from travelling. Firms were in a no-win situation: if they decided to postpone travel they could lose valuable business but if they sent representatives to infected countries they might have to quarantine them for up to ten days (the apparent incubation time for SARS) on their return. SARS also hit trade conventions which had become a vital part of the international economy. Hong Kong's watch industry, for example, was forced to withdraw from a trade fair in Switzerland because of concerns that representatives might be carrying the disease. The Federation of Hong Kong Watch Trades and Industries estimated that the ban would lose its members the equivalent of US $1.8 billion.[42]

The situation was not yet catastrophic but the cancellation of package tours and business trips was sapping economic vitality and chipping away at investor confidence. The appearance of around a hundred suspected SARS cases in Singapore, for example, led banks to slash their growth forecasts for that country, calculating that it could lose around US $13 million a week in revenue from tourism. The World Bank estimated that SARS could knock 0.3 per cent off of Asian economic growth in the course of 2003.[43] But fear of SARS was now hurting economies further afield. The decline in travel to Asia was beginning to be felt in the USA, in particular, with flights from San Francisco and Dallas being badly hit. Revenue from passenger fares was down 10 per cent on the previous year and it seemed likely that the airline industry would be one of the biggest casualties of the pandemic.[44] It also seemed that SARS might have an economic impact beyond restrictions on business travel and tourism. Before SARS was reported, there had been growing pressure on many Asian countries to revalue their currencies, which, in the opinion of the US government, were set too low in relation to the dollar. But fears about the effects of SARS on commerce in Asia meant that China, Japan, South Korea and Taiwan were more inclined to resist revaluation. Each country was reluctant to increase the value of its currency, lest other Asian countries gain a comparative advantage. The *Herald Tribune* warned that these countries were laying themselves open to retaliation that would be devastating to their global trading prospects. Some feared that Asian nations were creating conditions for 1930s-style competitive devaluation and protectionism.[45]

The potential of SARS to wreak havoc on the global economy depended more upon perception than reality, however. Media images of Asians donning white face masks – then something of a novelty in the West – served to increase fearfulness and magnify the economic fallout from the disease. On television and the internet, scare stories and half-baked news items abounded. Western governments also ramped up the pressure on infected states by treating SARS as if it were a threat to national security. In the USA, for example, President George W. Bush announced that military quarantines could be used to contain SARS in the event of large outbreaks in the country.[46] These powers were never used – the US health authorities relying on isolation of infected individuals rather than general quarantine – but talking tough was probably intended to allay public fears raised by alarmist reports of the prospect of mass mortality.[47] Writing of the economic implications of these reports, Tim Harcourt, Chief Economist of the Australian Trade Commission observed, 'Governments are scared that people's reactions to SARS will cause more damage to the economy than the virus could itself – no matter what progress is made on the medical side of the issue.' However, he believed that the pandemic was likely to postpone business rather than destroy it and that SARS would hit Hong Kong and Singapore most on account of their reliance on tourism and professional service industries. Although Australia would suffer from any fall in tourist revenue, its economy would weather the storm, he concluded, because it derived most of its income from the export of commodities which were not directly affected by SARS.[48]

By the end of May, some experts in China reckoned that the impact of SARS upon their country's economy would be limited by the government's determined efforts to tackle the outbreak.[49] But other prominent economists sounded a note of caution. Professor Jong-Wha Lee and Warwick J. McKibbin of the USA's Brookings Institution pointed out that it was very difficult to model the economic effects of a disease such as SARS and that simply adding up the amount of money lost from cancelled flights and so forth was unlikely to produce an accurate figure. They argued that the economic costs of SARS went beyond direct damage to the most badly affected sectors and that economic shock was likely to spread from the hardest-hit countries to others through its effects on trade and capital flows. 'As the world becomes more integrated,' they warned, 'the global cost of a communicable disease like SARS is expected to rise.' Added to this was uncertainty over the fate of the disease itself: would it fade away or mutate and become more virulent? The effects of this uncertainty on investor confidence were potentially catastrophic: if investors believed that Asian countries were likely to be affected by SARS on a regular basis, they would demand greater risk-premiums from investing there, which would have an immediate impact upon business confidence in those countries.[50]

The most badly affected nations therefore had little option but to take firm action in tackling the disease, their primary aim being to restore the confidence of investors and the international community. In this respect, the response to SARS was similar to the response to plague in the 1890s, denial having given way to determined intervention in order to reassure foreign governments and businesses. Most countries again reverted to quarantine in an attempt to combat the disease and in Hong Kong an old colonial law, passed following the plague epidemics a century earlier, was invoked to enforce isolation and other methods of containment; schools and daycare centres were closed and flights carrying passengers suspected of having the disease were suspended.[51] These measures were implemented with little opposition and retained public confidence, for responsible media coverage consolidated public opinion and brought it into harmony with official action.[52] But in Taiwan, where the government attempted to introduce similar measures, public opinion was already divided by long-running debates over the island's relationship to the People's Republic. Right from the start, the ruling, anti-PRC party, the Democratic Party, portrayed SARS as a Chinese disease and used the opportunity to cool cross-Strait business fever, which it regarded as a threat to Taiwanese independence. The opposition National Party countered with its own proposal to allow direct flights to China avoiding Hong Kong. As a result, there was no agreement on how to combat the disease and attempts to impose quarantine and other restrictions came too late and were widely ignored.[53] The response to SARS in Taiwan had the hallmarks of a political crisis rather than a scientifically calculated intervention.[54]

However, Taiwan was very much the exception amongst states afflicted with SARS. Using screening at airports, contact-tracing and enforcement of home quarantine, Singapore brought its outbreak under control within weeks, and was declared free from SARS by the WHO on 31 May 2003.[55] Its handling of the outbreak was said by the WHO to be 'exemplary'.[56] After Canada was made the subject of a WHO health advisory because of an outbreak of SARS in Toronto, the Canadian government also sought to allay anxiety by imposing restrictions on movement, including a ten-day quarantine on those who might have come into contact with infected persons. The impact of these measures was softened by offers of compensation for any income lost as a result.[57] Toronto was declared SARS-free on 2 July after twenty consecutive days without new cases and China was removed from the WHO's list on 23 June. Both countries had imposed quarantines and travel restrictions of varying degrees of severity, so the lesson seemed to be clear: quarantine, contact-tracing and isolation had brought the epidemics under control.[58] As Georges C. Benjamin, Executive Director of the American Public Health Association later put it: 'In the end, it was old-fashioned epidemiology and disease control that saved the day.'[59] The underlying message was that public health was as relevant as it had ever been; perhaps more so. According to many experts, globalization had powered ahead, leaving a yawning gap between profit-seeking and public health. A measure of this neglect was the fact that the WTO still lacked a committee to consider the implications of trade for human health (as opposed to the SPS Committee, which regulated trade), despite having working groups on the environment and other matters of contemporary concern. The most urgent task of public health workers was to bridge this gap and restore the world's 'health security'.[60]

But the lessons of SARS were far from clear. Despite its ineffective response to the outbreak, the disease quickly fizzled out in Taiwan and it was removed from the WHO's list of infected countries on 5 July 2003, only three days after Canada. A question-mark therefore remains as to whether the disease died out naturally in other countries, too, but the more important question is why did SARS assume such prominence in international health? The answer lies partly in the novelty of SARS and the peculiar fears it engendered. Some scientists and public health workers speculated that it might produce mortality on a scale not seen since the terrible influenza epidemic of 1918–19, when at least 25 million people died.[61] In April 2003, for example, one American public health expert predicted that 60,000 Americans would die from the virus.[62] The world was also a very different place than it had been in the 1960s and '70s, when pandemic respiratory diseases had last circulated the globe. Although the Cold War had ended, there was a pervasive feeling of insecurity in the wake of the terrorist attack of 11 September 2001 and the rise of Islamic fundamentalism. The demise of the Soviet Union also brought with it the prospect of bioterrorism on a scale seldom imagined, it being feared that former Soviet scientists would sell their expertise or deadly microbes to terrorist organizations. Added to this were concerns about the loss of sovereignty and the shift of economic power from the West to Asia: processes related to globalization. There were obvious parallels with the past – with nineteenth-century fears about the spread of cholera and yellow fever via international trade, for example – but economic interdependence now transcended the exchange of commodities. The flow of capital between nations was much greater than before and happened far more quickly. The prevalence of visual news media and the internet meant that people were informed about epidemics and the actions of governments more frequently and vividly than in the age of print and telegraphy. Although many people were very sophisticated in the way in which they viewed the media and its news reports, the net effect of these new technologies was probably to engender panic and increase their sense of vulnerability.[63]

Asian governments and business leaders were well aware of the propensity of the internet to generate panic and were concerned lest SARS be used to justify restrictive measures against their countries. They also feared that capital flight might follow scare stories calculated maliciously to damage business confidence: the tough response to SARS in countries such as Singapore was primarily intended to ensure that this did not happen. With this in mind, it is important to note that some of the key initiatives in tackling SARS came not just from scientists and public health workers but from the business community, which had long been alert to the potential consequences of the disease. International coordination in the Asia-Pacific region was championed by the trade ministers involved in Asia-Pacific Economic Cooperation (APEC), which had set up an Emerging Infections Network in 1996 to consider 'trade-related infections'.[64] When they met in Thailand in June 2003, the trade ministers of APEC therefore resolved to take all necessary action against SARS in cooperation with international organizations such as the WHO. They saw it as essential that measures in APEC member states should be harmonized and that screening measures be calculated to maintain business confidence without interfering unduly with mobility. Above all, they resolved that 'fear of SARS shall not be used as a pretext to protectionism or raising non-tariff barriers that restrict the movement of people, goods and capital'.[65] Recent controversies over BSE may have primed APEC's response to SARS but, unlike BSE, there is little evidence that SARS was used to justify protectionism. It was a very different kind of disease and the only commodity trade adversely affected by it was that in certain animal species – palm civets, raccoon dogs, and Chinese ferret badgers – which were suspected of being reservoirs of infection.[66] The dangers which SARS posed to commerce were, rather, indirect: the impact of control measures on business mobility and the propensity of media reporting to diminish business confidence in infected countries.

Ultimately, the economic impact of SARS proved to be less than many had feared, as both commerce and tourism in places such as Hong Kong recovered relatively quickly.[67] Nevertheless, SARS revealed the potential of pandemic disease to upset the global economy and the response to it set the tone for international public health over the next decade. During the SARS outbreak, pandemic preparedness was conceived largely in terms of disease containment; a focus which some public health workers thought too narrow. It certainly disappointed those who had entertained hopes that primary health care would again rise to the top of the international agenda, after years in which it had been downgraded under the World Bank's structural adjustment programmes.[68] In the context of SARS, public health seemed to be shorn of its humanitarian associations and became an aspect of national security.[69] Writing about SARS in the *National Review* in May 2003, David Gratzer, a Toronto physician and senior fellow at the Manhattan Institute, wrote portentously that 'This flu-like virus is a warning. September 11 shows us the consequences of unheeded warnings.' Like states which sponsored or sheltered terrorists, Gratzer argued, those which failed to notify disease or take measures to prevent it from spreading should be treated as 'international pariahs'.[70]

In international circles the security agenda was not so obvious but bodies such as the WHO were nonetheless affected by it and to some extent adopted its rhetoric. In 2005, for example, the WHO issued new International Health Regulations with the intention of dealing more effectively with 'emerging diseases'; a category dear to the hearts of security theorists who tend to locate disease threats in marginal populations or other nations.[71] The new regulations were welcomed in some quarters as a 'landmark event' which elevated public health to its rightful place in international politics.[72] As one enthusiastic commentator put it, 'germ governance' had become synonymous with 'good governance'.[73] But while the international community seemed to be taking public health more seriously, anxiety over pandemic preparedness had unintended consequences. It became easier for national governments to take actions which exceeded international norms and to justify patently protectionist measures as legitimate responses to disease.[74]

Panic and protectionism

Despite the dire projections of some public health officials, SARS came and went relatively quickly, leaving remarkably few casualties. At the end of the pandemic, only 8,446 people were recorded as having contracted the disease, 876 of whom died. But SARS had been a 'wake-up call' which renewed vigilance over the spread of disease and led to an increasing focus on its containment. Quarantine was the key to this approach and Asian governments did not hesitate to reintroduce it when SARS reappeared briefly at the beginning of 2004.[75] Although some health workers believed that such techniques required further evaluation, the strategy of containment went largely unquestioned.[76] Yet this was not the only legacy of the pandemic. SARS reconstituted social relations in a way that reinforced the new state-led emphasis on 'germ governance'. The use of face masks became more common in outbreaks of viral diseases, even in the West, where such practices had been largely unknown outside Asian communities; temperature screens were erected in airports and railway stations; and organizations drafted disaster plans to deal with pandemic threats. In civil society, as in political life, a security mindset prevailed. As in former times when there had been intense concern about the spread of infectious diseases, the penalties for stepping outside the consensus could be severe: states, individuals and various private and public bodies risked ostracism and ruin if they bucked the trend.

The first major challenge to global health in the wake of SARS came in the form of 'bird flu': the disease caused by the influenza virus H5N1. As its popular name suggests, this was largely a disease of poultry, although it had the potential to infect and kill human beings. Avian influenza was not a new disease, and there had been serious outbreaks as recently as the 1990s, mostly in Hong Kong, where thousands of poultry had died, as well as a few humans who had been in close contact with infected birds. Fearful of the consequences for trade and tourism, the Hong Kong government culled around 1.5 million hens. But the spotlight on avian influenza intensified after SARS. Each new outbreak of the disease was minutely scrutinized for cases of human-to-human transmission and the virus was continually examined for evidence of mutation. In December 2003 a new strain of the H5N1 virus did emerge – initially in South Korea, on a poultry farm near the capital, and within the space of a month in Vietnam,

Hong Kong and Thailand. By the end of the year the same strain of the disease had been reported in Europe. A number of human deaths were recorded but there was no evidence of transmission between humans. Nevertheless, the spread of this new and highly lethal strain of H5N1 caused great alarm globally. The losses incurred by farmers in affected areas were catastrophic. In South Korea, for example, over 1.3 million chickens and ducks were either killed by the virus or destroyed within the first two weeks.[77]

H5N1 had the potential to ruin poultry farming and the trade in birds and meat products. As a result, some of the worst-affected regions began to tighten their biosecurity. In Hong Kong there was a concerted effort to improve infection control in poultry farms with the aim of maintaining confidence in the export market, despite the higher costs that such measures entailed. But elsewhere, in the Guangdong province of China, for example, comparatively little was done despite a major outbreak of the disease. Nearby centres of the poultry trade, notably Hong Kong, thus remained vulnerable to infection. The experience of SARS and 'bird flu' seemed to point in the same direction – towards better surveillance and containment. As the Assistant Director-General of the WHO put it, 'The possibility, opened by SARS, that emerging diseases might be stopped has given the roles of national and international disease surveillance for epidemic-prone diseases even greater importance.' In his view, avian influenza had had a similar effect, having 'encouraged governments to develop or strengthen pandemic preparedness plans and to find ways to use the unprecedented international collaboration seen during SARS to protect the world against other threats.'[78]

In the wake of SARS, H5N1 was portrayed as a threat of catastrophic proportions and a pandemic resulting from a mutated, more easily transmissible strain of this virus was said to be inevitable. In 2005, the WHO claimed that it would probably kill between one and seven million people worldwide.[79] Over the next few years, projections of deaths became rather more circumspect but the security mindset continued to shape the language and practices of public health, at least in relation to influenza. Following precedents set during the SARS pandemic, airline companies made emergency plans to deal with bird flu and governments prepared to place restrictions on incoming travellers. In the USA there was talk of using quarantines and of giving the military the task of dealing with an epidemic emergency in light of the failure of civilian agencies to deal adequately with Hurricane Katrina.[80]

Unlike SARS, avian influenza had a more direct impact on trade and its effects on the international economy were potentially more severe. The first outbreaks of the disease in Asia seemed to be linked to intensive farming of poultry and the response in most cases was to cull infected flocks. In Vietnam 17 per cent of poultry were killed, causing losses equivalent to US $83.3 million. When the disease spread across Asia and into Europe the losses were nearly as great. When H5N1 was reported in Europe in 2004, Belgium destroyed 2.3 million birds and the Netherlands 2.6 million. This old-style culling or stamping-out strategy was reinforced by biosecurity guidelines framed by the European Commission; guidelines which also advised bans on the movement of live poultry and restrictions on the movement of people on poultry farms. Around the world, methods which might have lessened the need for such measures were either ignored or expressly ruled out by many governments. For example, after trials of a vaccine against H5N1, the government of Thailand banned the practice because it feared that farmers would lie about vaccination to avoid having their birds slaughtered.[81]

In the worst-affected areas of Asia, these drastic measures were easier to justify than in Western Europe where the number of H5N1 cases among poultry was significantly lower and the risks to human health negligible. In 2006 there were some cases of avian influenza which seemed to be linked to trade between infected farms in Hungary and the United Kingdom but such cases were rare.[82] In view of the much lower density of poultry and human populations in Europe, Alex Schudel of the WOAH's Scientific and Technical Department declared that 'There is no way this [the H5N1 outbreak] could turn into a human virus like SARS'.[83] One can only conclude that other considerations weighed on the minds of the European Commissioners and the national and regional governments concerned; not least, the fear of appearing weak or negligent by comparison with nations that were taking more robust action. As in the past, governments were fearful of the sanctions which might be imposed against them if they stepped outside international norms. Disease prevention was guided more by fear of retaliation than by realistic perceptions of the risks posed by H5N1 to either animal or human health.

The faith displayed in this combination of culling and biosecurity was fundamentally misplaced, according to some critics. As early as 2007, a report by the campaign group Compassion in World Farming pointed out that measures designed to restrict the movement of vehicles, persons and materials between farms had not worked and were almost impossible to enforce. The thrust of the report was that biosecurity detracted from the real problem, which was the intensive rearing of poultry on 'factory farms'. It was this, according to CWF, which explained the alarming rise of avian influenza and its recent appearance in countries hitherto unaffected by it.[84] The surge in cases of H5N1 coincided with a massive expansion of poultry production – 300 per cent over the previous two decades – and this explained why the disease had recently appeared in Indonesia, a country which had begun to shift from small-scale to intensive rearing of poultry.[85] Up to this point, wild birds had largely been blamed for infecting farmed poultry but CWF pointed its finger at the international trade in poultry and meat products. Recent molecular and phylogenetic studies had found different regional subspecies of virus, whereas if migrating birds had been the culprits there would have been fewer such differences. This led CWF to conclude that the poultry trade – both legal and illegal – was the vector for the spread of H5N1; it cited several established cases in which the cross-border trade in birds and meat had spread infection. The report concluded that intensive farming provides the ideal environment for generation of highly virulent avian flu strains and that 'the frequent flow of goods within and between countries' meant that the 'potential for disease spread is high'.[86]

The same issues arose even more starkly in 2009, in response to a strain of influenza which had allegedly spread from pigs to humans in Mexico. The origins of what was to become the century's second new pandemic were quickly traced to the village of La Gloria, where there had been cases of a severe respiratory illness since March 2009. By 21 March, around 60 per cent of the village's inhabitants were said to be ill with influenza-like symptoms: many attributed the disease to infection from pigs reared on a nearby farm owned by the giant corporation Smithfield Foods. Although any link was later denied by Smithfield and the Mexican government, the name 'swine flu' stuck to H1N1 and this materially affected the way in which the world responded to it.[87]

By the beginning of April 2009 there had been around 1,800 cases of what appeared to be H1N1 in Mexico and several deaths; the numbers climbed to 1,995 suspected cases and 149 deaths by the month's end, by which time it was clear that the disease had spread to other countries, apparently as a result of tourists and business people returning from Mexico. By 28 April there were forty confirmed cases in the USA, six in Canada, two in the United Kingdom, one in Spain, plus suspected cases in several other countries.[88] Unlike 'bird flu', the H1N1 virus could be transmitted easily between humans and its pandemic potential was therefore greater. Coupled with what, at first, appeared to be a high death rate, this was enough to prompt calls for emergency measures to prevent travel from infected regions. The EU's Commissioner for Health, Androulla Vassiliou, caused a stir when she advised Europeans to avoid travelling to Mexico and the USA. The Commissioner's remarks were immediately denounced by European diplomats, business people, and some medical experts, who warned that such advice was unnecessary and likely to damage the lucrative transatlantic travel industry.[89] National health ministers and the WHO urged calm and did their best to dampen public fears.[90] But the messages from scientific experts and health officials were mixed. While the WHO endeavoured to appear measured in its public pronouncements, the impression given was that a pandemic was 'inevitable' and 'imminent'.[91] The WHO's pandemic alert levels generated intense media coverage suggesting a dramatic unfolding of events; as if the world were heading inexorably towards disaster. By 27 April Phase 4 had been announced, with human-to-human transmission of influenza in community-level outbreaks reported; three days later Phase 5 was reached after regional-level outbreaks had occurred in at least two countries. All that was necessary for a pandemic (Phase 6) to be declared was evidence that the disease was spreading in two or more continents. This level was reached in June and a pandemic was officially announced on the 11th of that month.[92]

The tension generated by these announcements was perhaps unavoidable given the extent of media interest in swine flu. But the more controversial issue was whether the WHO had deliberately exaggerated the severity of the pandemic. The organization later claimed in its defence that it had not done so, pointing out that most cases were likely to be mild; however, it had also held open the possibility that a mutation of the virus could enhance its lethality. In all probability, the WHO, like other authorities, was erring on the side of caution.[93] Yet some national health officials did make announcements which may, in retrospect, appear alarmist. In the United Kingdom, the Chief Medical Officer Sir Liam Donaldson estimated that the number of deaths would probably be between 19,000 and 65,000, the most optimistic assessment being that only 3,100 would die. In July, the Cabinet emergency committee, Cobra, met three times a week to ensure that the country was fully prepared for the expected onslaught.[94] There were regular updates on the progress of the disease and employers in both the private and the public sectors were informed of the steps that needed to be taken in the event of a pandemic being declared. Already, schools and other public bodies in which there had been outbreaks were closing, following confirmed cases among pupils and employees. In the northern hemisphere, however, the declaration of a pandemic coincided with the summer and shortly afterwards the number of cases began to fall. Interest in 'swine flu' flagged as it became apparent that the mortality rate outside Mexico was no higher than for ordinary seasonal influenza; less so, in fact.

By December 2009, estimated deaths from swine flu in the United Kingdom had fallen to fewer than 1,000, which was considerably lower than the annual average of 21,000 deaths from seasonal influenza. Sir Liam Donaldson resigned amidst criticism that he and other public health officials – including those at the WHO – had overreacted.[95] But Donaldson and some other authorities still expressed concern that the disease could mutate into a more deadly form, possibly mixing with H5N1 in the intensive pig- and poultry-rearing areas of East and South East Asia. In Europe and North America there were fears that the disease would return in the winter, more virulent than before, just like the influenza pandemic of 1918–19.[96] Although many remained sceptical and suspected health officials of exaggerating the problem, the media retained an interest in flu-related stories.

In China the boot, for once, was on the other foot. Many feared that the region would become a mixing bowl for different strains of the influenza virus, but in mid-2009 it was less badly affected by H1N1 than many countries in the West. China lost no time in employing the battery of measures which had seemingly proved successful in controlling SARS and implemented stringent quarantine against all persons suspected of being infected. Screening at airports detected incoming foreign nationals with raised temperatures and they and their contacts were placed in quarantine.[97] Chinese nationals who had visited countries considered severely affected were placed in quarantine for ten days regardless of their state of health. To some travellers from other countries, these measures appeared drastic and unnecessary but Western governments could hardly complain when they had urged China to take decisive action in the past. However, according to some Western commentators, the economic downturn had revived protectionist instincts which had found expression in measures of biosecurity.

In April 2009 several countries announced bans on imports of pork produced in Mexico and the USA. The Philippines temporarily banned all pork from both countries, while China, Russia and Thailand banned pork from Mexico and parts of the USA, including Texas, Kansas and California. The pork industry responded by issuing safety guidelines to producers, and Smithfield Foods, the largest US pork producer, insisted that it had found no evidence of influenza in its herds or among the company's workers in Mexico. The bans were potentially very damaging, for the USA alone exported nearly $5 billion of pork products every year, mostly to Japan, China, Mexico, Canada and Russia.[98] Later in the year, when H1N1 was reported in Canada and several European countries, Russia, China and eight other countries also banned imports from those nations and the EU protested to the WTO that such measures were unjustified and against internationally recognized guidelines.[99] Canadian pork producers, too, feared that thousands of jobs would be lost, as they were facing losses of around $10–20 per animal: a major blow for an industry already troubled by high feed prices and the global recession which began in 2008. The Canadian government denounced the bans and threatened retaliation against China and other countries if they were not lifted.[100]

The WHO, the WTO and the WOAH all denounced boycotts of pork on the grounds that it might transmit the H1N1 virus; in their view, there was no evidence that it could be spread through meat as opposed to live animals. Such a ban appeared to violate both the SPS Agreement and the WHO's International Health Regulations (2005).[101] However, the SPS Agreement was sufficiently ambiguous to enable some countries to persist in their bans. The Chinese government, for example, showed no sign of lifting its embargo on meat products from affected areas. It prohibited pork imports, not only from affected states or provinces of American countries, but also pork that had travelled through them. This took its toll on transhipment hubs such as ports in California, at which numerous small loads were normally combined in order to reduce transportation costs. Having to re-route shipments to gain access to the Chinese market meant massive increases in costs for exporters. This position contrasted markedly with that of Japan, which focused on informing consumers about the safety of both imported and domestic products. Embargoes like those declared by China therefore seemed out of place as well as contrary to the declarations of international bodies. However, the WTO had limited powers to enforce compliance. Critics of China's stance alleged that its powerful position within the WTO meant that the organization had little ability to force its will on the country or, indeed, on other powerful states that had taken similar action.[102] But China claimed that its embargoes were entirely within WTO rules and that the emergency measures it had taken were also in accordance with those of the WOAH. Chinese Foreign Ministry spokesman Ma Zhaoxu explained that international regulations allowed members of the WTO and WOAH to take strict quarantine measures in emergencies such as the H1N1 pandemic.[103] Indeed, while international bodies believed that there was sufficient evidence to exclude pork meat as a possible source of influenza, the case was at least arguable.

Some countries – including South Korea and Russia – lifted their bans before long but not perhaps for medical reasons. Russia began to lift its embargo at the end of July, in the first instance permitting imports of pork from Wisconsin in the USA and Ontario in Canada. Without admitting that they had been at fault scientifically, the Russians justified their decision on the grounds that outbreaks in these areas had stabilized.[104] However, Russia may have been playing politics from the start. Having the largest economy outside the WTO, it had been attempting to join the organization for the last fifteen years and US diplomats suggested that lifting the ban on American pork would boost Russia's chances of success.[105] As in the past, trade embargoes could serve as bargaining chips in international diplomacy.

The presentation of the H1N1 outbreaks as a global security threat conferred legitimacy on what many pork exporters regarded as spurious practices. China claimed that its bans were justified because of its large and vulnerable population, and the potential burden of a H1N1 outbreak upon its public health system. It also claimed that there was a clear link between the virus which had infected humans and that which had infected pigs, which have a similar genetic make-up.[106] Although there was no evidence to support the ban on shipments of pork meat, it was later shown that the virus causing the pandemic was descended from a triple hybrid human–pig–poultry virus which might have been circulating in North America for over a decade.[107] The WTO's right to enforce action against China and other nations maintaining bans was also challenged by international lawyers. Professor Steve Chanovitz of George Washington University accused the WTO of 'mission creep' and questioned whether its joint statement on pork with the WOAH and WHO was constitutional and whether it could be used to preclude a defence on health grounds if the pork ban became subject to the dispute resolution process. The position of the US government and others seeking to rescind the ban was weakened after the respected magazine *New Scientist* carried a piece which suggested that officials at the US Centers for Disease Control had come under pressure from the pork industry to play down any links between human H1N1 and pigs, whereas the author stressed that the links were clearer than ever.[108] In view of this, it is hardly surprising that a meeting of the WTO's SPS Committee which convened in June to discuss trade bans related to H1N1 failed to reach an agreement. The interests of net-exporting and importing countries proved irreconcilable.[109]

The 'swine flu' controversy was complicated by its connection to a trade dispute that had been rumbling on for months. In July 2009, China requested that a WTO panel be established to consider a ban recently imposed by the USA against Chinese poultry. An American law passed in March 2009 contained a section prohibiting the use of funds to facilitate the import of poultry products from China; a provision which the Chinese claimed violated WTO regulations. The American legislation was the latest in a series of spats between the two countries over poultry imports, the USA having proclaimed a similar ban in 2004 following outbreaks of H5N1, which met with a retaliatory embargo by China against US poultry. China lifted its embargo following the establishment of a Sino-American commission on commerce later in 2004 but the USA seemed to have reneged on the agreement.[110] China's ban on pork was therefore seen, in part, as retaliation. The combined effects of the economic recession of 2008–9 and the advent of 'swine flu' had aggravated a long-standing problem which had been allowed to fester because of the inadequacy of international regulation. The logical outcome of each of these powerful nations acting as it did, in the name of the security of its citizens, was actually to increase insecurity for both and for the global economy as a whole.[111]

Although H5N1 and H1N1 differed significantly from SARS, the response to both diseases was shaped by the legacy of the new century's first pandemic. SARS came at a time when the optimism that accompanied the fall of the Soviet Bloc was evaporating amidst high-profile terrorist attacks and concerns about the future of global capitalism. A defensive mind set emerged in the West and many public health officials moved with the times, perhaps in the belief that the future of their discipline depended on its engagement with issues of national and global security. Forecasts of huge death tolls encouraged a 'not if but when' attitude to pandemic catastrophe and were used to garner more resources for a hitherto beleaguered branch of medicine. Although the public was often sceptical of these exaggerated claims, they played into the hands of those who wished to curtail the movement of trade and persons across borders. Quarantines and trade embargoes could now be presented as responsible acts according to the recently established conventions of 'germ governance'. The legitimacy of some of these measures was questionable but the WTO's rules permitted a diverse array of practices to continue. The SPS Agreement allowed countries to maintain a higher level of protection than could be achieved by measures based on international standards, subject to a proper scientific assessment of the risk of infection.[112] This provision provided a loophole which was ruthlessly and frequently exploited by countries seeking to restrict foreign competition. Yet faith placed in measures to prevent the spread of disease was sometimes misplaced. In the opinion of some critics, the prevalent notion of 'biosecurity' was a misnomer because it engendered a false sense of security, encouraging the public to believe that diseases such as influenza could be effectively contained when they could not. It seemed to be little more than a smokescreen to legitimize the intensive rearing of – and long-distance trade in – animals and meat products.[113]

CONCLUSION

Sanitary pasts, sanitary futures

This book has charted the passage of disease along the world's principal arteries of trade; from the caravans of medieval Asia to the myriad pathways of the global economy. It has shown that commerce has been a major factor in the redistribution of diseases, allowing pathogens and their vectors to circulate more widely than before, often with catastrophic results. Each step-change in commercial activity has had profound consequences for the health of humanity, with trade playing a major part in the Black Death, the passage of new infectious diseases to the Americas, and the great exchange of pathogens that accompanied global integration in the nineteenth century. As well as providing a vehicle for the passage of germs, long-distance trade often transformed the places it connected. Large-scale production for distant markets and rapid economic development altered the disease ecologies of port cities, manufacturing towns and agricultural areas, often in ways which permitted newly arrived infections to thrive. Through these means, the world was, in a sense, unified by disease.[1] But measures were gradually devised to block the transmission of trade-related diseases and to remove the environmental conditions in which they flourished. At two critical points, the most developed countries diverged from poorer ones, opening up a significant epidemiological gap. The first great divergence began in the late seventeenth century, as much of Europe rid itself of plague. The second occurred towards the end of the nineteenth century, during which time cholera, yellow fever and rinderpest were banished from most of the industrialized nations.

These divergences presented an enormous challenge to those who wished to maintain the flow of commerce, for as some countries began to enjoy freedom from pestilence, they looked nervously at founts of infection in 'less civilized' lands. This dread of infection from afar often produced responses which were unnecessarily damaging to trade. Indeed, the negative impact of quarantines and sanitary embargoes was often aggravated by the increasing propensity to use such measures as instruments of foreign policy. During the eighteenth century, quarantine became, in effect, a weapon of war and, on occasions, was the cause of conflict itself. The disruption of commerce by the misuse of sanitary measures was considerable, drawing forth the ire of merchants and their political allies. By the end of the eighteenth century, this opposition was tinged with a libertarian and humanitarian gloss, and given additional force by medical practitioners bent on the reform of what they saw as antiquated practices. These reformers enjoyed modest success but there were tight constraints on any country or port that wished to reduce its sanitary barriers. Such actions were watched keenly by fearful citizens and trading partners and any place which moved too far in removing quarantine was likely to face retaliation. It was this fear of a backlash and the difficulty of reforming quarantine unilaterally that led to the first international sanitary conferences of the nineteenth century. Like the international boards of health which were established in the Middle East from the 1830s, these conferences aimed to create a barrier to infection from the East (the presumed source of plague and cholera) while permitting quarantines in Europe to be liberalized. But progress was hindered by the resurgence of epidemic disease in the 1860s and by growing imperial rivalry. It was not until the 1900s that something like a consensus on how to manage disease in a global economy was reached and, even then, it excluded the numerous infections related to agriculture.

We are now in the midst of another great convergence, in which commodities, finance and people move at unprecedented speed. The same is true of disease. However, it is not the classic diseases of trade that stalk us but a host of infections that have crossed or threaten to cross the barriers between species. The first pandemic of the twenty-first century – SARS – originated in wild animals that were legally and illegally traded, while BSE raised concerns about an industry which many presumed to be safe. The last decade has also seen pandemic scares relating to influenza and the possible emergence of a new strain of the virus which combines the lethality of H5N1 with the human transmissibility of H1N1. While such an event may not be inevitable, it is certainly not unlikely and even the most advanced nations would struggle to deal with an emergency on the scale of that which would probably ensue. If the evidence presented in the previous chapter is anything to go by, the effects of a severe pandemic on commerce, industry and agriculture would be devastating. The cost of suppressing SARS and the business lost by travel restrictions and capital flight amounted to around US $50–100 billion but a truly virulent pandemic would cost far more. The World Bank has estimated that a pandemic of avian influenza, for example, would bring losses in the region of US $3 trillion and one can only imagine the bill left by a catastrophe on the scale of 1918–19.[2]

The potential of disease to destroy a volatile global economy has not gone unnoticed, for organizations such as APEC began to examine the issue of trade-related diseases and pandemics as early as the 1990s. In recent years, a great deal of thought has also gone into devising 'early warning systems' and into ensuring that governments and private organizations make contingency plans. We may or may not feel reassured by these measures but the recent history of responses to influenza suggests that governments will probably aggravate the economic damage that inevitably results from a pandemic. Indeed, their reaction to recent outbreaks of avian and swine influenza was in some ways reminiscent of the response to plague at the end of the nineteenth century. In both cases, the authorities in affected countries sought to deal with the threat by taking drastic measures in a desperate attempt to stamp out the disease, while their neighbours and trading partners erected sanitary barriers which some regarded as unjustified. The governments of the afflicted countries wished to reassure their trading partners and to put an end to damaging speculation and capital flight. But states also took advantage of the plight of other nations and imposed trade embargoes to gain economic and political advantage.

The tendency of governments to abuse sanitary precautions is perhaps even more evident now than it was in the 1890s. During the H1N1 pandemic of 2009, the vast majority of sanitary embargoes exceeded guidelines issued by international bodies such as the WHO and OIE and many were regarded as protectionist in intent. It does not take much effort to imagine what would happen if there were a pandemic of greater duration or severity. Quite apart from emergencies such as these, the abuse of sanitary measures causes hardship for many producers and consumers on a regular basis. Those hit hardest are usually located in developing countries, which tend to rely more heavily on exports of agricultural commodities. Poor countries are at a permanent disadvantage in this respect because they are unable to afford a sanitary infrastructure equal to those of richer nations and because they have fewer experts in subjects such as risk assessment. It is probably not insignificant that 61 per cent of the SPS-related concerns brought before the WTO in its first fifteen years were raised by developing country members.[3] These problems persist because the sanitary regulations governing trade are still too flexible; because scientific risk assessment is expensive; because dispute resolution is protracted; and because penalties for those in breach of international regulations are light.[4]

It is important that these issues be resolved, not simply for the health of the economy but for the well-being of humans and other animals across the planet. The present system is not only commercially disruptive but it affords scant protection for our health. In particular, there is a danger that heavy emphasis on surveillance and containment may engender a false sense of security. Early warning systems such as those recently devised for zoonoses by the FAO, OIE and WHO are designed to enable states to control outbreaks before they get out of hand. Biosecurity measures – tightened in the event of an imminent threat – are equally indispensable. But these precautions are not infallible and the pandemics of the present century show that persons carrying disease were able to cross borders and infect others despite quarantines and sophisticated systems of surveillance. Meat infected with avian influenza has also been detected in countries like Japan which are normally renowned for their biosecurity.[5] As Tseng Yen-Fen and Wu Chia-Ling recently observed in their study of the SARS pandemic, 'Since the microbial world is unobservable to the human eye, there is probably no such thing as true security.'[6]

The illusion of security is not the only problem, for the preoccupation with surveillance and containment tends to mask the conditions in which pathogens emerge and propagate. With respect to the zoonoses, the most important issues are those relating to large-scale, intensive animal production. In its recent report on factory farming in the USA, the Pew Commission, in conjunction with the Johns Hopkins Bloomberg School of Public Health, warned of the 'heightened risks of pathogens ... [passing] from animals to humans; the emergence of microbes resistant to antibiotics and antimicrobials, due in large part to widespread use of antimicrobials for nontherapuetic purposes [i.e. to stimulate growth]; food-borne disease; worker-health concerns; and dispersed impacts on the community at large'.[7] Perhaps the gravest of these problems is that factory farming provides what some commentators have referred to as an 'evolutionary fast-track' for the mutation of pathogens into deadlier forms.[8] As

Dr Robert Webster, a renowned expert on influenza recently put it: 'Previously we had backyard poultry. . . . Now we put millions of chickens into a chicken factory next door to a pig factory, and this [swine influenza] virus has the opportunity of getting into one of those chicken factories and mak[ing] billions of mutations simultaneously.'[9]

Some scientists believe that a more coordinated approach is vital if these problems are to be solved, for the issues raised by factory farming cut across a number of global bodies, most obviously the FAO, OIE, WTO and the WHO. A recent report from the International Livestock Research Institute has suggested the formation of joint task forces to deal with specific problems such as outbreaks of avian or swine influenza. In the longer term, it proposes stronger cross-sectoral collaboration including joint animal and human health units, integrated knowledge management and information-sharing, and joint training programmes.[10] The WHO, OIE and FAO have moved in this direction already, with an initiative on early warning systems in 2006 and a 2010 statement on further cooperation.[11] In its own right, the WHO is now placing strong emphasis upon what it terms the 'human–animal interface' as well as other trade-related health issues. However, at present it is difficult to determine what these initiatives will amount to in practice.

The success of cross-sectoral collaboration – and of what some advocates refer to as a one-health or pan-species approach to public health[12] – will depend very largely on how far producers in individual countries are prepared to cooperate with it and how seriously governments are prepared to scrutinize agricultural production and distribution. But even the most diligent authorities will struggle to keep pace with the challenges which confront us, for these are essentially structural in nature. As in Europe and North America a century ago, the shift of population from rural to urban areas has resulted in the consumption of greater quantities of meat.[13] Persons whose incomes are no longer at subsistence level have the wherewithal and usually the desire to afford more meat, while their hectic life style compels them to eat in different ways, shifting from a 'slow' to a 'fast' food culture which often entails the wasteful use of resources. These changing patterns of consumption have increased the demand for cheap meat and meat products which has, in turn, led to the intensification of production.[14] Many Asian countries have therefore shifted to intensive poultry rearing over the last few decades.[15] In 2010, China alone had an estimated 14,000 'confined animal feeding operations' (factory farms) which were said to have stocking densities conducive to the emergence of disease. These trends are likely to have far-reaching consequences both for the environment and for health, especially as the bulk of animal protein will probably come not from ruminants but from monogastric animals such as pigs and poultry, which are more amenable to large-scale, intensive farming.[16] The risk of diseases like influenza mutating and spreading within and beyond factory farms is therefore likely to increase. As some critics argue, it may be time to ban the long-distance trade in animals,[17] thereby reducing the chances of infection across national boundaries. But this ignores the fact that trade is often illegal and that disease can be spread in meat as well as in livestock. History shows us that disease, especially trade-related disease, is best tackled through a range of measures, by attention to the propagation of disease as well as its transmission. Unless we get the balance right, it is unlikely that we will enjoy either the security we crave or the commercial freedom essential to our prosperity.

This is easier said than done, for the matter of where sanitary responsibility lies is always controversial. A focus on cross-border infection allows governments and producing interests to locate responsibility elsewhere, while providing an ostensibly rational and compelling case for the exclusion of certain products. These protectionist impulses often exploit popular anxieties about disease and about the nature of trade itself, not least its propensity to destroy domestic industries and introduce new cultural influences. In an era of globalization, these fears are compounded by a general sense of unease and insecurity. Richer nations fear cheaper competition and everyone fears the prospect of contagion, whether in the form of a virulent pandemic or some stealthy infection such as BSE/vCJD. Organizations like the WTO and WHO have attempted to impose order on this unstable situation but assessments of the risk of infection are fraught with controversy and will remain so. All sides appeal to 'science' in an attempt to vindicate their arguments but there can never be a purely technical solution to the sanitary regulation of trade. To suggest otherwise, is at best naïve and at worst a dangerous fiction. Our greatest hope of maintaining freedom of trade and of providing sanitary protection is to ensure that no interest group, no country or trading bloc, is permitted to dominate.

Notes

Preface and acknowledgements

1. William Budd, *The Siberian Cattle Plague, or, the Typhoid Fever of the Ox*, Bristol: Kerslake & Co., 1865, p. 1.
2. Andrew Price-Smith, *Contagion and Chaos: Disease, Ecology, and National Security in the Era of Globalization*, Cambridge, Mass.: MIT Press, 2009.
3. Classically, F. Prinzing, *Epidemics Resulting from Wars*, Oxford: Clarendon Press, 1916. For a more recent example, see Matthew Smallman-Raynor and Andrew D. Cliff, *War Epidemics: An Historical Geography of Infectious Diseases in Military Conflict and Civil Strife, 1850-2000*, Oxford: Oxford University Press, 2000.
4. See B.K Gills and W.R. Thompson (eds), *Globalization and Global History*, London: Routledge, 2006; B. Mazlisch, 'Introduction', in B. Mazlisch and R. Buultjens (eds), *Conceptualizing Global History*, Boulder, Col.: Westview Press, 1993; Regina Grafe, 'Turning Maritime History into Global History: Some Conclusions from the Impact of Globalization in Early Modern Spain', in M. Fusaro and A. Polomia (eds), *Research in Maritime History. No. 43: Maritime History as Global History*, St John's: International Maritime History Association, 2010, 249–66.
5. See 'Globalization' in Frederick Cooper, *Colonialism in Question: Theory, Knowledge, History*, Berkeley: University of California Press, 2005.
6. See for example, Adrian Wilson, 'On the History of Disease-Concepts: The Case of Pleurisy', *History of Science*, 38 (2000), 271–319; Jon Arrizabalaga, John Henderson and Roger French, *The Great Pox: The French Disease in Renaissance Europe*, New Haven and London: Yale University Press, 1997.
7. For a more extensive discussion, see Mark Harrison, *Disease and the Modern World: 1500 to the Present Day*, Cambridge: Polity, 2004, pp. 12–13.
8. A classic study which is somewhat in this style, but very useful nevertheless, is George Rosen, *A History of Public Health*, Baltimore: Johns Hopkins University Press, 1993 [1958].
9. Christopher Hamlin, *Public Health and Social Justice in the Age of Chadwick: Britain, 1800-1854*, Cambridge: Cambridge University Press, 1998.
10. See especially: Michel Foucault, 'The Politics of Health in the Eighteenth Century', in C. Gordon (ed.), *Michel Foucault, Power/Knowledge: Selected Interviews and Other Writings 1972-1977*, Brighton: Harvester Press, 1988, pp. 166–82; Dorothy Porter, *Health, Civilization and the State: A History of Public Health from Ancient to Modern Times*, London: Routledge, 1999.

Chapter 1: Merchants of death

1. William Rosen, *Justinian's Flea: Plague, Empire and the Birth of Europe*, London: Viking, 2007; Lester K. Little (ed.), *Plague and the End of Antiquity: The Pandemic of 541–750*, Cambridge: Cambridge University Press, 2007.
2. See for example: John Hatcher, *Population and the English Economy 1348–1530*, London: Macmillan, 1987; G. Huppert, *After the Black Death*, Bloomington: Indiana University Press, 1986; David Herlihy, *The Black Death and the Transformation of the West*, Cambridge, Mass.: Harvard University Press, 1997; Norman Cantor, *In the Wake of Plague: The Black Death and the World it Made*, London: Simon & Schuster, 2001; Ronald Findlay and Kevin H. O'Rourke, *Power and Plenty: Trade, War, and the World Economy in the Second Millennium*, Princeton, NJ: Princeton University Press, 2007, pp. 111–20.
3. William H. McNeill, *Plagues and Peoples*, New York: Monticello, 1976, pp. 162–4.
4. See Ole J. Benedictow, *The Black Death 1346–1353: The Complete History*, Woodbridge: Boydell Press, 2004, p. 49; John Norris, 'East or West? The Geographic Origin of the Black Death', *Bulletin of the History of Medicine*, 51 (1977), 1–24.
5. George D. Sussman, 'Was the Black Death in India and China?', *Bulletin of the History of Medicine*, 85 (2011), 319–55. See also Li Bozhong, 'Was there a "fourteenth-century turning point"? Population, Land, Technology and Farm Management', in P. J. Smith and R von Glahn (eds), *The Song–Yuan Transition in Chinese History*, Cambridge, Mass.: Harvard University Press, 2003, pp. 134–75.
6. Denis Twitchett, 'Population and Pestilence in T'ang China', in *Studia Sino-Mongolica*, Wiesbaden: Franz Steiner Verlag, 1979, pp. 35–68.
7. Giovanna Morelli et al., 'Yersinia pestis Genome Sequencing Identifies Patterns of Global Phylogenetic Diversity', *Nature Genetics*, 42 (2010), 1140–43.
8. David Morgan, *The Mongols*, Oxford: Basil Blackwell, 1986, pp. 133–4.
9. Peter Spufford, *Power and Profit: The Merchant in Medieval Europe*, London: Thames & Hudson, 2002, pp. 12–29.
10. Michael W. Dols, *The Black Death in the Middle East*, Princeton, NJ: Princeton University Press, 1977, p. 53; Rosemary Horrox, 'Introduction', *The Black Death*, Manchester: Manchester University Press, 1994, p. 9.
11. Benedictow, *The Black Death*, p. 70.
12. John Kelly, *The Great Mortality: An Intimate History of the Black Death*, London and New York: Fourth Estate, 2005, p. 8; Joseph P. Byrne, *The Black Death*, Westport, Conn.: Greenwood Press, 2004, p. 7.
13. Horrox, 'Introduction', *The Black Death*, pp. 8–9.
14. Dols, *The Black Death*, pp. 36–40; Benedictow, *The Black Death*, pp. 51–2.
15. McNeill, *Plagues and Peoples*, p. 165.
16. Benedict Gummer, *The Scourging Angel: The Black Death in the British Isles*, London: The Bodley Head, 2009, pp. 60–1.
17. Vivian Nutton, 'Introduction', in V. Nutton (ed.), *Pestilential Complexities: Understanding Medieval Plague, Medical History*, Supplement No. 27, London: Wellcome Centre for the History of Medicine at UCL, 2008, p. 8.
18. J.N.-Biraben, *Les Hommes et la peste en France et dans les pays européens et méditerranées*, vol. 1, Paris: Mouton, 1975, pp. 71–92; *idem*, 'Les Routes maritimes des grandes épidémies au moyen âge', in C. Buchet (ed.), *L'Homme, la santé et la mer*, Paris: Champion, 1997, pp. 23–37.
19. Findlay and O'Rourke, *Power and Plenty*, pp. 125–6.
20. Stuart J. Borsch, *The Black Death in Egypt and England: A Comparative Study*, Austin: University of Texas Press, 2005, pp. 7–9.
21. E.g. Prosper Alpini, *De medicina Aegyptiorum*, Paris: G. Pele & I. Duval, 1646; Jean Martel, *Dissertation sur l'origine des maladies epidemiques et principalement sur l'origine de la peste*, Montpellier: Imprimeur Ordinaire du Roy, 1721, pp. 44–5, 55–8, 59–61, 87, 91–2.

22. Angela Ki Che Leung, 'Diseases of the Premodern Period in China', in K.F. Kiple (ed.), *The Cambridge World History of Human Disease*, Cambridge: Cambridge University Press, 1993, pp. 354–62; Jun Lian, *Zhungguo Gudai Yizheng Shilue*, Huhehaste: Nei Monggu Renmin, 1995, pp. 154–5.
23. Frank G. Clemow, *The Geography of Disease*, Cambridge: Cambridge University Press, 1903, p. 316.
24. Sussman, 'Was the Black Death in India and China?', pp. 340–1.
25. Clemow, *The Geography of Disease*, p. 319.
26. 'A Voyage Round the World by Dr. John Francis Gernelli Careri, Containing the Most Remarkable Things he saw in Indostan', in J.P. Guha (ed.), *India in the Seventeenth Century*, New Delhi: Associated Publishing House, 1976, vol. 2, p. 203.
27. T.S. Weir, 'Report from Brigade-Surgeon-Lieutenant-Colonel T.S. Weir, Executive Health Officer, Bombay', in P.C.H. Snow (ed.), *Report on the Outbreak of Bubonic Plague in Bombay, 1896–97*, Bombay: Times of India Steam Press, 1897, pp. 66–7.
28. John MacPherson, *Annals of Cholera: From the Earliest Periods to the Year 1817*, London: Ranken & Co., 1872, pp. 100, 106; Mark Harrison, *Climates and Constitutions: Health, Race, Environment and British Imperialism in India 1600–1850*, Oxford: Delhi University Press, 1999, p. 200.
29. Ashin Das Gupta, 'Indian Merchants and the Trade in the Indian Ocean, c.1500–1750', in T. Raychaudhuri (ed.), *The Cambridge Economic History of India, Volume I: c.1200–c.1750*, Cambridge: Cambridge University Press, 1982, pp. 407–33; idem, 'The Merchants of Surat, c.1700–50', in E. Leach and S.N. Mukherjee (eds), *Elites in South Asia*, Cambridge: Cambridge University Press, 1970, pp. 201–22.
30. Sanjay Subrahmanyam, *The Political Economy of Commerce: Southern India 1500–1650*, Cambridge: Cambridge University Press, 1990, p. 319.
31. Clemow, *The Geography of Disease*, p. 320; August Hirsch, *Handbook of Geographical and Historical Pathology*, vol. 1, London: The Sydenham Society, 1883, p. 507.
32. *Tuzuk-i-Jahangiri*, ed. Syud Ahmud, Aligarh: Private Press, 1864, pp. 209–10.
33. Ibid., pp. 219–20.
34. Biraben, *Les hommes e la peste*, p. 90.
35. Samuel K. Cohn, Jr., *The Black Death Transformed: Disease and Culture in Early Renaissance Europe*, London: Arnold, 2002; idem, 'Epidemiology of the Black Death and Successive Waves of Plague', in Nutton (ed.), *Pestilential Complexities*, pp. 74–100; Graham Twigg, *The Black Death: A Biological Reappraisal*, New York: Schocken, 1984.
36. Ann G. Carmichael, 'Universal and Particular: The Language of Plague, 1348–1500', and Lars Walløe, 'Medieval and Modern Bubonic Plague: Some Clinical Continuities', in Nutton (ed.), *Pestilential Complexities*, pp. 17–52 and 59–73 respectively: Stephane Barry and Norbert Gualde, 'La peste noire dans l'Occident chrétien et musulman, 1347–1353', *Canadian Bulletin of Medical History*, 25(2003), pp. 461–98.
37. E.g. Ludwig Friedrich Jacobi, *De Peste*, Erfurt: H. Grochius, 1708, *passim*; Rudolphus Guillelmus Crausius [Rudolf Wilhem Krause], *Excerpta quaedam ex observation in nupera peste Hambugensi*, Jena: I.F. Beerwinckel, 1714, p. 5; Martel, *Dissertation*, pp. 15–17.
38. Michel Drancourt et al., 'Detection of 400-Year Old *Yersinia pestis* DNA in Human Dental Pulp', *Proceedings of the National Academy of Sciences of the USA*, 95 (1998), 12637–40.
39. Robert S. Gottfried, *The Black Death: Natural and Human Disaster in Medieval Europe*, London: Robert Hale, 1983, pp. 52–3, 73–4; Barbara W. Tuchman, *A Distant Mirror: The Calamitous Fourteenth Century*, London: Macmillan, 1978, pp. 109–14.
40. Paul Slack, *The Impact of Plague in Tudor and Stuart England*, Oxford: Clarendon Press, 1985, p. 189.
41. McNeill, *Plagues and Peoples*, p. 170.
42. 'Ordinances against the Spread of Plague, Pistoia, 1348', in Horrox (ed.), *The Black Death*, pp. 195–6.

43. L. Fabian Hirst, *The Conquest of Plague: A Study of the Evolution of Epidemiology*, Oxford: Clarendon Press, 1953, p. 378.
44. Vivian Nutton, 'Medicine in Medieval Western Europe', in L. I. Conrad, M. Neve, V. Nutton, R. Porter and A. Wear, *The Western Medical Traditions 800 BC to AD 1800*, Cambridge: Cambridge University Press, 1995, pp. 196–7; McNeill, *Plagues and Peoples*, p. 170.
45. See William G. Naphy, *Plagues, Poisons and Potions: Plague-Spreading Conspiracies in the Western Alps c.1530–1640*, Manchester: Manchester University Press, 2002.
46. F.P. Wilson, *The Plague in Shakespeare's London*, Oxford: Clarendon Press, 1927, pp. 85–6, 107–10, 120–3, 151–2, 163–6.
47. Daniel Defoe, *A Journal of the Plague Year*, London: Penguin, 2003 [1722], p. 187.
48. Adrien Proust, *Essai sur l'hygiène internationale ses applications contre la peste, la fièvre jaune et le cholera asiatique*, Paris: G. Masson, 1873, p. 158.
49. Thoman Brinton, 'The Sins of the English', in Rosemary Horrox (ed.), *The Black Death*, Manchester: Manchester University Press, 1994, p. 139.
50. Michael Neill, *Issues of Death: Mortality and Identity in English Renaissance Tragedy*, Oxford: Clarendon Press, 1997, p. 22.
51. William Bullein, 'A Dialogue both pleasant and pietyful' (1564), in Rebecca Totaro (ed.), *The Plague in Print: Essential Elizabethan Sources, 1558–1603*, Pittsburgh: Duquesne University Press, 2010, p. 164.
52. Vivian Nutton, 'The Seeds of Disease: An Explanation of Contagion and Infections from the Greeks to the Renaissance', *Medical History*, 27 (1983), 1–34; idem, 'Did the Greeks Have a Word For It?, in L.I. Conrad and D. Wujastyk (eds), *Contagion: Perspectives from Pre-Modern Societies*, Aldershot: Ashgate, 2000, pp. 137–62.
53. Ann G. Carmichael, *Plague and the Poor in Renaissance Florence*, Cambridge and New York: Cambridge University Press, 1986.
54. See Margaret Pelling, 'The Meaning of Contagion: Reproduction, Medicine and Metaphor', in A. Bashford and C. Hooker (eds), *Contagion: Historical and Cultural Studies*, London: Routledge, 2001, pp. 15–38.
55. Anon., *Traité de la peste*, Paris: Guillaume Caelier, 1722, pp. 21–3; Johannes Kanold, *Einiger Medicorus Schreiben von der in Preussen an. 1708, in Dantzig an. 1709, in Rosenberg an. 1708, und in Fraustadt an. 1709 grassireten Pest, wie auch von der wahren Beschaffenheir des Brechens, des Schweisses under der Pest-Schwären, sonderlich der Beulen*, Breslau: Fellgiebel, 1711, p. 101.
56. Paul Slack, 'Responses to Plague in Early Modern Europe: The Implications of Public Health', *Social Research*, 55 (1988), 433–53; pp. 436–7.
57. Samuel K. Cohn, Jr., *Cultures of Plague: Medical Thinking at the End of the Renaissance*, Oxford: Oxford University Press, 2010, pp. 296–8.
58. For the persistence of astrological notions, see Christiano Laitner, *De febrius et morbis acutis*, Venice: H. Albricium, 1721, pp. 140–2; Richard Mead, *Of the Power of the Sun and Moon on Humane Bodies; and of the Diseases that Rise from Thence*, London: Richard Wellington, 1712. Yet these 'astrological' ideas were quite different from those of the early Renaissance, being more concerned with the physical effects of particles or gravity upon the body and the atmosphere: see Mark Harrison, 'From Medical Astrology to Medical Astronomy: Sol-Lunar and Planetary Theories of Disease in British Medicine, c.1700–1850', *British Journal for the History of Science*, 33 (2000), 25–48.
59. Richard Mead, *A Short Discourse concerning Pestilential Contagion, and the Methods used to Prevent it*, London: S. Buckley, 1720, pp. 3–5.
60. E.g. Guy de la Brosse, *Traité de la peste*. Paris; L. & C. Perier, 1623, p. 72.
61. Dols, *The Black Death*, pp. 285–98; Lawrence I. Conrad, 'Epidemic Disease in Formal and Popular Thought in Early Islamic Society', in T. Ranger and P. Slack (eds), *Epidemics and Ideas: Essays on the Historical Perception of Pestilence*, Cambridge: Cambridge University Press, 1992, 77–100.

62. Lawrence I. Conrad, 'A Ninth-Century Muslim's Scholar's Discussion of Contagion', ibid., 163–78; Peter E. Pormann and Emilie Savage-Smith, *Medieval Islamic Medicine*, Edinburgh: Edinburgh University Press, 2007, pp. 58–9.
63. *The Shah Jahan Nama of 'Inayat Khan*, trans. A.R. Fuller, New Delhi: Oxford University Press, 1990, pp. 305, 535.
64. Nükhet Varlik, 'Disease and Empire: A History of Plague Epidemics in the Early Ottoman Empire (1453–1600)', University of Chicago PhD thesis, 2008, chap. 6; Lian, *Zhongguuo Gudai Yizheng Shilue*, pp. 154–6.
65. Angela K.C. Leung, 'The Evolution of the Idea of *Chuanran* Contagion in Imperial China', in A.K.C. Leing and C. Furth (eds), *Health and Hygiene in Chinese East Asia*, Durham, NC and London: Duke University Press, 2010, 25–50.
66. See the documents collected in *L' Ordre public pour la ville de Lyon, pendant la maladie contagieuse*, Lyons: A. Valancol, 1670; also Slack, 'Responses to Plague', p. 438; Brian Pullan, 'Plague Perceptions and the Poor in Early Modern Italy', in T. Ranger and P. Slack (eds), *Epidemics and Ideas*, Cambridge: Cambridge University Press, 1992, 101–24.
67. M.W. Flinn, 'Plague in Europe and the Mediterranean Countries', *Journal of European Economic History*, 8 (1979), 31–48.
68. Quentin Skinner, *The Foundations of Modern Political Thought. Volume I: The Renaissance*, Cambridge: Cambridge University Press, 1978, pp. 42–6, 49–65.
69. John Henderson, *Piety and Charity in Late Medieval Florence*, Oxford: Clarendon Press, 1994, pp. 16–20, 354–9; idem, *The Renaissance Hospital: Healing the Body and Healing the Soul*, London and New Haven: Yale University Press, 2006, pp. 28–31.
70. Pullan, 'Plague Perceptions and the Poor,' p. 119.
71. Cohn, *Cultures of Plague*.
72. Ibid., p. 280.
73. Ibid., p. 206.
74. Quentin Skinner, *Visions of Politics. Volume II: Renaissance Virtues*, Cambridge: Cambridge University Press, 2002.
75. Randal P. Garza, *Understanding Plague: The Medical and Imaginative Texts of Medieval Spain*, New York: Peter Lang, 2008, p. 40.
76. F.P. Wilson, *The Plague in Shakespeare's London*, pp. 85–6.
77. Ibid., pp. 108–9.
78. Margaret Pelling, 'Illness among the Poor', in M. Pelling, *The Common Lot: Sickness, Medical Occupations, and the Urban Poor in Early Modern England*, London: Longman, 1998, pp. 64–5.
79. Carmichael, *Plague and the Poor*, pp. 121–6.
80. Daniel Panzac, *Quarantaines et lazarets: l'empire et la peste*, Aix-en-Provence: Édisud, 1986, pp. 32–4.
81. William Naphy and Andrew Spicer, *The Black Death: A History of Plagues 1345–1730*, Stroud: Tempus, 2001, pp. 86–8, 102; Michael Limberger, 'City, government and public services in Antwerp, 1500–1800', https://lirias.hubrussel.be, accessed 28/11/10.
82. Thursday meeting book, Kingston-upon-Hull Corporation, 8 September 1668, Western MS.3109, Wellcome Library for the History and Understanding of Medicine, London [WL]; Booker, *Maritime Quarantine*, chap. 2.
83. Laurence Brockliss and Colin Jones, *The Medical World of Early Modern France*, Oxford: Clarendon Press, 1997, p. 351.
84. Anon., 'Quarantaines', *Dictionnaire encyclopédie des sciences medicalés*, Paris: P. Asselin & G. Masson, 1874, p. 24.
85. Alexandra Parma Cook and Noble David Cook, *The Plague Files: Crisis Management in Sixteenth-Century Seville*, Baton Rouge: Louisiana State University Press, 2009, pp. 246–51.
86. Naphy and Spicer, *The Black Death*, p. 93.
87. Daniel Panzac, *Populations et santé dans l'empire Ottoman (XVIIIe–XXe siècles)*, Istanbul: Isis, 1996, pp. 20–1.

88. Carlo M. Cipolla, *Fighting the Plague in Seventeenth-Century Italy*, Madison: University of Wisconsin Press, 1981, pp. 19-50.
89. Alfred W. Crosby, *The Columbian Exchange: The Biological and Cultural Consequences of 1492*, Westport, Conn.: Greenwood Press, 1974. See also, *idem*, *Ecological Imperialism: The Biological Expansion of Europe, 900-1900*, Cambridge: Cambridge University Press, 1986.
90. See David E. Stannard, *American Holocaust: Columbus and the Conquest of the New World*, New York: Oxford University Press, 1992; Ronald Wright, *Stolen Continents: The Americas through Indian Eyes since 1492*, Boston: Houghton Mifflin, 1992; Francis J. Brooks, 'Revising the Conquest of Mexico', in R.I. Rotberg (ed.), *Health and Disease in Human History: A Journal of Interdisciplinary History Reader*, Cambridge, Mass.: MIT Press, 2000, 15-28; Robert McCaa, 'Spanish and Nahuatl Views on Smallpox and Demographic Collapse in Mexico' in Rotberg (ed.), *Health and Disease*, pp. 167-202.
91. See Claude Quétel, *History of Syphilis*, Baltimore: Johns Hopkins University Press, 1992; R.S. Morton, *Venereal Diseases*, London: Penguin, 1974.
92. David Noble Cook, *Born to Die: Disease and New World Conquest, 1492-1650*, New York: Cambridge University Press, 1998, pp. 83-5. The importation of smallpox, however, has been the subject of some controversy. See George W. Lovell, 'Disease and Depopulation in Early Colonial Guatemala', in D.N. Cook and W.G. Lovell (eds), *The Secret Judgments of God: Native Peoples and Old World Disease in Colonial Spanish America*, Norman: University of Oklahoma Press, 1992, 51-85.
93. Cook, *Born to Die*, chap. 2; T.M. Whitmore, *Disease and Death in Early Colonial Mexico: Simulating Amerindian Depopulation*, Boulder, Col.: Westview Press, 1991.
94. In Portuguese Brazil, for example, population density was lower than in most parts of Spanish America and this rendered the native peoples of the interior less vulnerable to disease. See A.J.R. Russell-Wood, *The Portuguese Empire, 1415-1808: A World on the Move*, Baltimore: Johns Hopkins University Press, 1992, p. 121.
95. See, for example: S.B. Schwartz (ed.), *Tropical Babylons: Sugar and the Making of the Atlantic World, 1450-1680*, Chapel Hill and London: University of North Carolina Press, 2004; Horst Pietschmann (ed.), *Atlantic History: History of the Atlantic System 1580-1830*, Göttingen: Vandenhoeck & Ruprecht, 2002; Fernand Braudel, *Civilization and Capitalism, 15th-18th Century: Volume III, The Perspective of the World*, London: Collins, 1988, chaps 2-3; A. Inikori, 'Africa and the Globalisation Process: West Africa 1450-1850', *Journal of Global History*, 2 (2007), 63-86.
96. Immanuel Wallerstein, *The Modern World System: Capitalist Agriculture and the Origins of the European World-Economy in the Sixteenth Century*, New York: Academic Press, 1974. See also S.K. Sanderson, *Civilizations and World Systems: Studying World-Historical Change*, Walnut Creek, Calif.: AltiMira Press, 1995.
97. See Johannes M. Postma, *The Dutch in the Atlantic Slave Trade*, Cambridge: Cambridge University Press, 1990; Robin Law, *The Slave Coast of West Africa*, Oxford: Clarendon Press, 1991; Hugh Thomas, *The Slave Trade: The History of the Atlantic Slave Trade 1440-1870*, London: Picador, 1997; Philip D. Curtin, *The Atlantic Slave Trade: A Census*, New York: Norton, 1981.
98. D. Alden and J.C. Miller, 'Out of Africa: The Slave Trade and the Transmission of Smallpox to Brazil, 1560-1831', in R. I. Rotberg (ed.), *Health and Disease in Human History: A Journal of Interdiciplinary Studies Reader*, Cambridge, Mass.: MIT Press, 2000, 203-30.
99. See, for example, Miguel Ángel Cuerya Mateos, *Puebla de los Ángeles en Tiempos de Una Peste Colonial: Una Mirada en Torno al Matlazahuatl de 1737*, Puebla: El Colegio de Michoacan, 1999.
100. Donald B. Cooper, *Epidemic Disease in Mexico City 1761-1813: An Administrative, Social, and Medical Study*, Austin: University of Texas Press, 1965, pp. 95-6.
101. Larry Stewart, 'The Edge of Utility: Slaves and Smallpox in the Early Eighteenth Century', *Medical History*, 29 (1985), 54-70; H.S. Klein and S.L. Engermann, 'A Note

on Mortality in the French Slave Trade in the Eighteenth Century', in H.A. Gemery and J.S. Hogendorn (eds), *The Uncommon Market: Essays on the Economic History of the Atlantic Slave Trade*, New York: Academic Press, 1979, p. 271.
102. Alden and Miller, 'Out of Africa', pp. 218–19.
103. G.M. Findlay, 'The First Recognized Epidemic of Yellow Fever', *Transactions of the Royal Society of Tropical Medicine and Hygiene*, 35 (1941), 143–54.
104. Charles de Rochefort, *Histoire naturelle et morale des Antilles de l'Amerique*, 2nd edn, Rotterdam: Arnout Leers, 1665, p. 2.
105. John Darwin, *After Tamerlane: The Global History of Empire since 1405*, London: Allen Lane, 2007, p. 107; Philip D. Curtin, *The Rise and Fall of the Plantation Complex*, Cambridge: Cambridge University Press, 1998, pp. 73–85.
106. Kenneth F. Kiple, *The Caribbean Slave: A Biological History*, Cambridge: Cambridge University Press, 1984; *idem* (ed.), *The African Exchange: Towards a Biological History of Black People*, Durham, NC: Duke University Press, 1987.
107. K. David Patterson, 'Yellow Fever Epidemics and Mortality in the United States, 1693–1905', *Social Science and Medicine*, 34 (1992), 855–6.
108. William Hillary, *Observations on the Changes of the Air, and the Concomitant Epidemical Diseases in the Island of Barbadoes. To which is added, a Treatise on the Putrid Bilious Fever, commonly called the Yellow Fever; and such other Diseases as are indigenous or endemial in the West India Islands or in the Torrid Zone*, London: C. Hitch & L. Hawes, 1759.
109. Gerald N. Grob, *The Deadly Truth: A History of Disease in America*, Cambridge, Mass.: Harvard University Press, 2002, p. 76; Patterson, 'Yellow Fever Epidemics', p. 857.
110. J.R. McNeill, *Mosquito Empires: Ecology and War in the Greater Caribbean, 1620–1914*, Cambridge: Cambridge University Press, 2010, pp. 47–9; James D. Goodyear, 'The Sugar Connection: A New Perspective on the History of Yellow Fever in West Africa', *Bulletin of the History of Medicine*, 52 (1978), 5–21; Richard B. Sheridan, *Doctors and Slaves: A Medical and Demographic History of Slavery in the British West Indies, 1680–1834*, Cambridge and New York: Cambridge University Press, 1985.
111. Henry Warren, *A Treatise concerning the Malignant Fever in Barbadoes, and the Neighbouring Islands: with an Account of the Seasons there, from the Year 1734 to 1738. In a Letter to Dr. Mead*, London: Fletcher Gyles, 1740, pp. 3–4.
112. Ibid., pp. 7–8.
113. Ibid., pp. 13–14.
114. Ibid., pp. 21–7.
115. Arnold Zuckerman, 'Plague and Contagionism in Eighteenth-Century England: The Role of Richard Mead', *Bulletin of the History of Medicine*, 78 (2004), 273–308.
116. Mead, *Short Discourse*, pp. 4–5.
117. The first use of the term is generally attributed to Pouppé Desportes, who practised physic on St Domingue from 1732 to 1748. See Benjamin Moseley, *A Treatise on Tropical Diseases*, London: T. Cadell, 1789, p. 387.
118. The earliest recorded incidence being in 1692; Slack, *Impact of Plague*, p. 342.
119. Findlay, 'The First Recognized Epidemic', p. 146.
120. 'An Act to oblige Ships and other Vessels coming from Places infected with Epidemical Distempers, to perform Quarantain', Georgia Council Chamber, 24 April 1760, http://infoweb.newsbank.com, accessed 31/08/10.
121. Proclamation by the Lieutenant-Governor of the Commonwealth of Virginia, http://infoweb.newsbank.com, accessed 31/08/10.
122. See for example: John Booker, *Maritime Quarantine: The British Experience, c.1650–1900*, Aldershot: Ashgate, 2007, pp. 256–61; R.C. Williams, *On Guard against Disease from Without: The United States Public Health Service, 1798–1950*, Washington, DC: Commissioned Officers Association of the United States Public Health Service, 1951, chap. 2.

Chapter 2: War by other means

1. See Jeremy Black, *European Warfare 1660-1818*, London and New Haven: Yale University Press, 1994; Robert I. Frost, *The Northern Wars 1558-1721*, Harlow: Pearson, 2000.
2. On the evolution of the Dutch economy, see J. Vries and A. Worde, *The First Modern Economy: Success, Failure and the Perseverance of the Dutch Economy, 1500-1800*, Cambridge: Cambridge University Press, 1997.
3. Mr Du Bacquoy to Mr Williamson, The Hague, 16 December 1663, SP 84/168, The National Archives, London [TNA].
4. Peter Christensen, '"In These Perilous Times": Plague and Plague Policies in Early Modern Denmark', *Medical History*, 47 (2003), p. 442.
5. Sir George Downing, to the Rt Hon. Sir Henry Bennett, Secretary of State, The Hague, 4 March 1663, TNA.
6. Mr Van Gogh, The Hague, to [illegible], 29 August 1664, SP 84/171, TNA.
7. Downing to Bennett, 8 April 1664, SP 84/170, TNA.
8. A. Lloyd Moote and Dorothy C. Moote, *The Great Plague: The Story of London's Most Deadly Year*, Baltimore and London: Johns Hopkins University Press, 2004, p. 51.
9. Privy Council to Mayor and Aldermen of Chester, ZM/L/4, 17 June 1664, Chester and Cheshire Archives [CCA].
10. Jeremy Black, *The British Seaborne Empire*, New Haven and London: Yale University Press, 2004, p. 92.
11. Stephen Porter, *The Great Plague*, Stroud: Sutton, 1999, p. 35.
12. J.R. Jones, *The Anglo-Dutch Wars of the Seventeenth Century*, New York: Longman, 1996.
13. M. Borcel to States General, Paris, 23 August 1664 [author's translation], SP 84/171, TNA.
14. Extrait des Régistres de Parlement, confirming a ban on commerce with the provinces of Holland and Zealand; Paris, 19 November 1664, BOD Vet. E3d.83 (53), Special Collections Reserve, Bodleian Library, University of Oxford.
15. Southwell to Joseph Williamson, 2 February 1667, Lisbon, SP 89/8, TNA.
16. Karl-Erik Frandsen, *The Last Plague in the Baltic Region 1709-1713*, Copenhagen: Museum Tusculanum Press, 2010.
17. Booker, *Maritime Quarantine*, p. 36.
18. Mr Wheeler to Sir Robert Sutton, Aix, 8 August 1720, SP 78/168, TNA; Jean-Baptiste Bertrand, *A Historical Relation of the Plague at Marseilles, in the Year 1720*; reprint of 1805 edn, trans. A. Plumptre, London: Mawman, 1973.
19. Sutton to Secretary of State, Paris, 10 August 1720, SP 78/168, TNA; Naphy and Spicer, *The Black Death*, p. 134.
20. Martin Arnoul, *Histoire de la derniere peste de Marseilles, Aix, Arles, et Toulon. Avec plusiers avantures arrives pendant la contagion*, Paris: Paulus du Mesnil, 1732, pp. 7-8.
21. Jean Astruc, *Dissertation sur l'origine des maladies epidemiques et principalement sur l'origine de la peste*, Montpellier: Jean Martel, 1721; Le Sr. Manget, *Traité de la peste*, Geneva: Philippe Planche, 1721.
22. Sutton to Secretary of State, 7 August 1720, SP 78/168, TNA; Pichaty de Croissante, Attorney General of Marseilles, 'Some Account of the Plague at Marseilles in the Year 1720', *Gentleman's Magazine*, 24 (1754), 32-6; Arnoul, *Histoire*, pp. 2, 6-7.
23. In the years 1713-22, Marseilles received 419 vessels from Smyrna. See Daniel Panzac, *Commerce et navigation dans l'empire Ottoman au XVIIIe siècle*, Istanbul: Isis, 1996, p. 29.
24. Wheeler to Sutton, 8 May 1720, SP 78/168, TNA.
25. Panzac, 'Quarantaines', pp. 26-30.
26. Martel, *Dissertation*, pp. 65-6.
27. Daniel Gordon, 'The City and the Plague in the Age of Enlightenment', *Yale French Studies*, 92 (1997), 81-2.

28. Junko T. Takeda, *Between Crown and Commerce: Marseille and the Early Modern Mediterranean*, Baltimore: Johns Hopkins University Press, 2011, pp. 97–101.
29. Ibid., chap. 8.
30. Ibid., pp. 104–5.
31. Biraben, *Les Hommes et la peste*, pp. 245–51.
32. Mr J. Pulteney to James Craggs, Paris, 27 August and 17 September 1720, SP 78/166, TNA.
33. Mr Wheake to Sir Robert Sutton, Aix, 8 August 1720, SP 78/168, TNA.
34. Sutton to Craggs, 26 October 1720; 'Arrest du Conseil d' Estat du Roy, Au Sujet de la Maladie Contagieuse de la Ville de Marseilles', 14 September 1720, SP 78/169, TNA.
35. 'Memoire servant à justifier qu'il n'y a aucun lieu à interdire le commerce avec la ville de Genève pour cause de la peste qui aflige Marseilles et une partie de la Provence', Geneva, 11 September 1720, SP 78/169, TNA. See also Manget, *Traité de la peste*.
36. Zuckerman, 'Plague and Contagionism in Eighteenth-Century England', 273–308.
37. Slack, *Impact of Plague*, pp. 330–2.
38. Sir Luke Schaub to Lord Carteret, Paris, 4 September 1727, SP 78/170/1, TNA.
39. M. Laurenz to Mr Stanyan, 10 September, 1721; 'Instruction sur les Précautions qui doivent ester observées dans les Provinces où il y a des Lieux attaquez de la Maladie contagieuse, et dans les Provinces voisines', Paris 1721, SP 78/170/2, TNA.
40. Report of Intendants de la Santé de Marseilles, 27 August 1721; Laurenz to Mr Temple Morgan, 3 September 1721; Sutton to Carteret, 4 September 1721, SP 78/170/1, TNA.
41. Sutton to Carteret, 10 September 1721, SP 78/170/1, TNA.
42. Booker, *Maritime Quarantine*, p. 127.
43. Paul Langford, *A Polite and Commercial People: England 1727–1783*, Oxford: Clarendon Press, 1989, pp. 176–7.
44. Henry Worsley, Lisbon, to Lord Carteret, Secretary of State, 12 May 1721, SP 89/29, TNA.
45. Worsley to James Craggs, 8 March 1721, SP 89/29, TNA.
46. Worsley to Carteret, 7 November 1721, SP 89/29, TNA.
47. Mr J. Burnett to Carteret, Lisbon, 6 September 1723, SP 89/30, TNA.
48. De Mendonia to Sen. Claudio Gorgel de Amaral, 8 November 1726, SP 89/33, TNA.
49. Marquis Don Juan Baptista de Orendayn to Don Jorge de Macazaga, 1 November 1726, SP 89/33; Mr Robinson to the Hon. M. Delafage, Paris, 4 July 1728, SP 78/196; Newcastle to Lord Stanhope, 1 July 1728, SP 78/196, TNA.
50. Tyrawley to Newcastle, 14 August 1728, SP/35, TNA.
51. Quoted ibid.
52. Charles Compton to Lord Newcastle, Lisbon, 28 August 1728, SP 89/35, TNA.
53. Tyrawley to Newcastle, Lisbon, 14 September 1728, SP 89/35, TNA.
54. Compton to Newcastle, 7 November 1728, SP 89/35, TNA.
55. Newcastle to Mr Poyntz, 17 February 1729, SP 78/193, TNA.
56. Braudel, *Civilization and Capitalism*, Vol. III, p. 54.
57. General Sabine, Commanding Gibraltar garrison, to Sir John Norris, British Ambassador, Lisbon 19 June 1735, SP 89/38, TNA.
58. Lord Waldegrave to Newcastle, 11 August 1740, SP 78/223, TNA. The British also regularly imposed quarantine against vessels from West Barbary along the Mediterranean coast to Fez. See W. Sharpe, Office of the Privy Council, to Mr Weston, Secretary to the Viceroy in Ireland, 18 June 1747, D 3135/C804; Order in Council, 28 June 1749, D 3155/C1018, and W. Sharpe to Lord George Sackville, 4 September 1751, C 3155/C1271, Derbyshire County Record Office [DCRO].
59. George Rosen, 'Cameralism and the Concept of Medical Police', *Bulletin of the History of Medicine*, 27 (1953), 21–42; *idem*, 'The Fate of the Concept of Medical Police, 1780–1890', *Centaurus*, 5 (1957), 97–113; Foucault, 'The Politics of Health in the Eighteenth Century', in E. Gordon (ed.), *Michel Foucault, Power/Knowledge*; Dorothy Porter, *Health, Civilization and the State: A History of Public Health from Ancient to Modern Times*, London: Routledge, 1999, chap. 3.

60. Christopher Lawrence, 'Disciplining Disease: Scurvy, the Navy, and Imperial Expansion, 1750-1825', in D.P. Miller and P.H. Reill (eds), *Visions of Empire: Voyages, Botany, and Representations of Nature*, Cambridge: Cambridge University Press, 1996, pp. 80-106.
61. Johann Peter Frank, *A System of Complete Medical Police*, ed. E. Lesky, Baltimore: J.H.V. Press, 1976, trans. E. Vlim from 3rd edn, Vienna, 1786, p. 446.
62. Gunther E. Rothenberg, 'The Austrian Sanitary Cordon and the Control of Bubonic Plague: 1710-1871', *Journal of the History of Medicine and Allied Sciences*, 28 (1973), 15-23; Panzac, *Quarantaines*, pp. 65-78.
63. For the debate over the effectiveness of such measures see: Slack, 'Reponses to Plague', pp. 442, 449; idem, 'The Disappearance of Plague: An Alternative View', *English Historical Review*, 34 (1981), 469-76; Andrew Appleby, 'The Disappearance of Plague: A Continuing Puzzle', *English Historical Review*, 33 (1980), 161-73.
64. Jean-Jacques Manget, *Traité de la Peste*, Geneva: Philippe Planche, 1721; Anon., *Dissertations sur l'origine des maladies epide'miques*, pp. 110-11; Paskal Joseph Ferro, *Untersuchung der Pestanstekung, nebst zwei Aufsätzen von der Glaubwürdigkeit der meisten Pestberichte aus der Moldau und Wallachia, unter der Schädlichkeit der bisherigen Contumanzen von D. Lange and Fronius*, Vienna: Joseph Edlen, 1787, Foreword, p. iii.
65. Ferro, *Untersuchung der Pestanstekung*; Martin Lange, *Rudimenta doctrinae de peste*, Vienna: Rudolph Graeffer, 1784.
66. Anon., *Della peste ossia della cura per preservarsene, e guarire da questo fatalismo morbo*, Venice: Leonardo & Giammaria, 1784, pp. 195-6.
67. Herbert H. Kaplan, *The First Partition of Poland*, New York and London: Columbia University Press, 1969, pp. 129-30.
68. Nancy E. Gallagher, *Medicine and Power in Tunisia, 1780-1900*, New York and Cambridge: Cambridge University Press 1983, p. 24.
69. D. McKay and H.M. Scott, *The Rise of the Great Powers 1645-1815*, London: Longman, 1983.
70. John T. Alexander, *Bubonic Plague in Early Modern Russia: Public Health and Urban Disaster*, Oxford: Oxford University Press, 2003, pp. 249-51.
71. Ibid., pp. 284-96.
72. Daniel Samoilowitz (Samoilovich), *Mémoire sur la peste, qui, en 1771, ravage l'Empire Russe, sur-tout Moscou, la Capitale*, Paris: Leclerc, 1783, pp. 10-11.
73. Ibid., p. xv.
74. Ibid., pp. 229, 243, 254, 265, 273-6.
75. Ibid., p. 205; Alexander, *Bubonic Plague*, pp. 113-14, 286-8, 289-90.
76. J.P. Papon, *De la Peste, ou les époques mémorables de ce fléau, et les moyen de s'en préserver*, 2 vols, Paris: Lavillette, 1799, vol. 2, p. 142.
77. Brockliss and Jones, *The Medical World of Early Modern France*, p. 354.
78. Panzac, *Quarantaines*, p. 33.
79. Balthasar de Aperregui, 'Orderes relatives a sanidad y lazarettos en el Puerto de Barcelona, con motivo de la peste, en el año de 1714, y siguientes', Barcelona, 1752, Western MS.963, WL; William Brownrigg, *Considerations on the Means of Preventing the Communication of Pestilential Contagion and of Eradicating it in Infected Places*, London: Lockyer Davis, 1771, p. 4.
80. Brownrigg, *Considerations*.
81. Ibid., pp. 5-6.
82. Panzac, *Population et santé*, pp. 20-1.
83. John Howard, *An Account of the Principal Lazarettos in Europe*, Warrington: William Eyres, 1789, pp. 25-7. The city enjoyed only twenty-three plague-free years between 1713 and 1792: see Panzac, *Population et santé*, pp. 42-4.
84. E.g. Petition from Batholemew Midy, undated, PC 1/3/101; petition from Jonas Alling and others, 20 December 1713, PC 1/14/115, TNA. For a detailed study of mercantile opposition to quarantine in this period, see Booker, *Maritime Quarantine*, chap. 3.

85. Petition of Henry Morris, 24 December 1713; see also petition of Mr Wallace, 20 December 1713, PC 1/14/115, TNA.
86. Lord Leven to the Lords of the Privy Council, 28 August 1711; Adam Brown to Lords of the Privy Council, 26 September 1711, SP 54/4, TNA.
87. Petition from the Council Chamber, Whitehall, 11 May 1722, PC 1/3/106, TNA.
88. E.g. Petition from John March, Deputy Governor of the Levant Company, n.d., c.1763, PHA/35, West Sussex Record Office [WSRO].
89. Report on quarantine by the Privy Council, 1780, PC 1/13/101, TNA.
90. Langford, *A Polite and Commercial People*, pp. 166–7.
91. Untainted Englishman, *The Nature of a Quarantine, as it is performed in Italy: To Guard against . . . the Plague: with Important Remarks on the Necessity of Laying Open the Trade to the East Indies*, London: J. Williams, 1767.
92. A Gentleman, *A Journal Kept on a Journey from Bassora to Bagdad; over the Little Desert, to Cyprus, Rhodes, Zante, Corfu; and Otranto, in Italy; in the Year 1779*, Horsham: Arthur Lee, 1784, p. 133.
93. W. Sharpe, Office of the Privy Council, forwarding copy of letter from Venetian ambassador Colombo to William Pitt, enclosing orders issued in Venice to impose quarantine against plague, 'Terminazione Sopra Proveditori, e Proveditori alla Sanita', T 402, TNA.
94. F. Vernon, Office of the Privy Council, to Richard Rigby, Principal Secretary to the Lord Lieutenant of Ireland, 9 June 1758, D 3155/C 2146, DCRO.
95. John Dick, British consul, Leghorn, 26 April 1766, forwarding memorial from British merchants and translation of a Memorial from the Court of Tuscany, PC 1/8/29, TNA.
96. Memorial of Lord Dartmouth and other Privy Councillors, 4 July 1766, PC 1/8/29, TNA.
97. Dartmouth and other Privy Councillors, PC 1/8/29, TNA.
98. Adorno to Sir Horace Mann, 3 May 1766, PC 1/8/29, TNA.
99. Petition from merchants of Liverpool to Privy Council, 6 January 1767, PC 1/8/55, TNA.
100. Privy Council, report on quarantine and copy of Quarantine Act, 1780, PC 1/13/101, TNA.
101. Papon, *De la Peste*, pp. 216–19.
102. Daniel Panzac, *La Peste dans l'empire Ottoman 1700–1850*, Leuven: Peeters, 1985, pp. 58–62; idem, *Commerce et navigation*, p. 129ff.
103. Panzac, *Quarantaines*, pp. 34–5.
104. Customhouse letter, 10 November 1770 and legal opinion, 13 November 1770, T 1/475/130–135, TNA.
105. E.g. Petition from London merchants, 9 February 1787, PC 1/3/4, TNA.
106. David Cantor (ed.), *Reinventing Hippocrates*, Aldershot: Ashgate, 2002.
107. Jean Baptiste Sénac, *Traité des causes des accidens, et de la cure de la peste*, Paris: P.-J. Mariette, 1744.
108. John Huxham, *Essay on Fevers*, London: S. Austen, 1750, p. 144.
109. Dr Timoni, 'An Account of the Plague at Constantinople', in S. Miles (ed.), *Medical Essays and Observations relating to the Practice of Physic and Surgery*, London: S. Birt, 1745, pp. 69, 72.
110. Dale Ingram, *An Historical Account of the Several Plagues that have appeared in the World since the Year 1346 with An Enquiry into the Present prevailing Opinion, that the Plague is a Contagious Disease, capable of being transported in Merchandize, from one Country to Another*, London: R. Baldwin, 1755, p. 68. For more on the mythology surrounding contagion stories during this epidemic, see Patrick Wallis, 'A Dreadful Heritage: Interpreting Epidemic Disease at Eyam, 1666–2000', *History Workshop Journal*, 61 (2006), 31–56.
111. Ingram, *Historical Account*, pp. 73–4.
112. Ibid., pp. 119, 139.

113. Ibid., pp. ii–iii.
114. E.g. Richard Manningham, *A Discourse concerning the Plague and Pestilential Fevers*, London: Robinson, 1758.
115. 'Extracts of Several Letters of Mordach Mackenzie, M.D. concerning the Plague at Constantinople', *Philosophical Transactions of the Royal Society*, 47 (1752), 385.
116. 'A Further Account of the late Plague at Constantinople, in a Letter of Dr. Mackenzie from thence', *Philosophical Transactions of the Royal Society*, 47 (1752), 515.
117. Anon., 'Some Account of the Life and Writings of the late Dr. Richard Mead', *Gentleman's Magazine*, 24 (1754), 512.
118. See Mark Harrison, *Medicine in an Age of Commerce and Empire; Britain and its Tropical Colonies, 1660-1830*. Oxford: Oxford University Press, 2010, Parts I and III.

Chapter 3: The evils of quarantine

1. Carla G. Pestana, *The English Atlantic in an Age of Revolution 1640-1661*, Cambridge, Mass.: Harvard University Press, 2004.
2. The classic statement of the relationship between political ideology and sanitary policy is Erwin H. Ackerknecht's essay, 'Anticontagionism between 1821 and 1867', *Bulletin of the History of Medicine*, 22 (1948), 561-93.
3. E.g. Corneille le Brun, *Voyages de Corneille le Brun au Levant, c'est-à-dire, dans les principaux endroits de l'Asie Mineure, dans les isles de Chio, Rhodes, Chypres, etc.*, Paris: P. Gosse & J. Neautme, 1732, p. 554.
4. Howard, *An Account of the Principal Lazarettos in Europe*.
5. Ibid., pp. 2, 26.
6. Ibid., pp. 26-7, 31.
7. Patrick Russell, *A Treatise of the Plague*, London: G.G.J. & J. Robinson, 1791.
8. Harrison, *Medicine in an Age of Commerce and Empire*, pp. 57-8; Alexander Russell, *The Natural History of Aleppo, and Parts Adjacent, containing a Description of the City, and the Principal Natural Productions in its Neighbourhood, together with an Account of the Climate, Inhabitants, and Diseases; particularly of the PLAGUE, with the Methods used by Europeans for their Prevention*, London: A. Millar, 1756.
9. Booker, *Maritime Quarantine*, pp. 225, 234-54.
10. Lisabeth Haakonssen, *Medicine and Morals in the Enlightenment: John Gregory, Thomas Percival and Benjamin Rush*, Amsterdam and Atlanta: Rodopi Press, 1997, pp. 214-16.
11. Rush to Cullen, 16 September 1784, in L.H. Butterfield (ed.), *Letters of Benjamin Rush*, Princeton, NJ: Princeton University Press, 1951, vol. 1, p. 310.
12. Rush to Howard, 14 October 1789, ibid., p. 527.
13. J.H. Powell, *Bring Out Your Dead: The Great Plague of Yellow Fever in Philadelphia in 1793*, Philadelphia: University of Pennsylvania Press, 1949, pp. 281-2.
14. Benjamin Rush, *An Account of the Bilious Remitting Yellow Fever*, Philadelphia: Thomas Dobson, 1794, p. 313.
15. Patterson, 'Yellow Fever Epidemics', p. 857.
16. Simon Finger, 'An Indissoluble Union: How the American War for Independence Transformed Philadelphia's Medical Community and Created a Public Health Establishment', *Pennsylvania History*, 77 (2010), 37-72.
17. Martin S. Pernick, 'Politics, Parties and Pestilence: Epidemic Yellow Fever in Philadelphia and the Rise of the First Party System', in J. Worth Estes and B.G. Smith (eds), *A Melancholy Scene of Devastation: The Public Response to the 1793 Philadelphia Yellow Fever Epidemic*, Philadelphia: College of Physicians of Philadelphia, 1997, 119-46.
18. Sean P. Taylor, '"We Live in the Midst of Death": Yellow Fever, Moral Economy and Public Health in Philadelphia, 1793-1805', Northern Illinois University PhD thesis, 2001, pp. 134-7; p. 157.

19. William Coleman, *Yellow Fever in the North: The Methods of Early Epidemiology*, Madison: University of Wisconsin Press, 1987; Martin S. Pernick, 'Politics, Parties and Pestilence: Epidemic Yellow Fever in Philadelphia and the Rise of the First Party System', in J. Walzer Leavitt and R.L. Numbers (eds) *Sickness and Health in America: Readings in the History of Medicine and Public Health*, Madison: University of Wisconsin Press, 1985, pp. 356–71.
20. Taylor, '"We Live in the Midst of Death"', p. 161.
21. *An Act for the Establishing an Health Office, for Securing the City and Port of Philadelphia, from the Introduction of Pestilential and Contagious Diseases*, Philadelphia: True American, 1799, pp. 6–7.
22. John Duffy, *A History of Public Health in New York City 1625–1866*, New York: Russell Sage Foundation, 1968, pp. 97–150.
23. A Philadelphian, *Occasional Essays on the Yellow Fever*, Philadelphia: John Ormorod, 1800, p. 9.
24. Rush to John Coakley Lettsom, 13 May 1804, in Butterfield, *Letters*, vol. ii, p. 880.
25. Rush to Thomas Jefferson, 6 October 1800, ibid., p. 826; Rush to John Adams, 14 August 1805, ibid., p. 901; Rush to Adams, 21 September 1805, ibid., p. 206.
26. Rush to Madison, 23 June 1801, ibid., p. 835
27. Rush to Jefferson, 5 August 1803, ibid., p. 872.
28. Academy of Medicine of Philadelphia, *Proofs of the Origin of Yellow Fever, in Philadelphia & Kensington, in the Year 1797, from Domestic Exhalation; and from the Foul Air of the Snow Navigation, from Marseilles: and from that of the Ship Huldah, from Hamburgh, in two Letters addressed to the Governor of the Commonwealth of Philadelphia*, Philadelphia: T. & S.F. Bradford, 1798.
29. Charles Caldwell, *Anniversary Oration on the Subject of Quarantines, delivered to the Philadelphia Medical Society, on the 21st of January, 1807*, Philadelphia: Fry & Kammerer, 1807, pp. 7–11, 21–2.
30. William Travis Howard, Jr., *Public Health Administration and the Natural History of Disease in Baltimore, Maryland 1797–1920*, Washington, DC: Carnegie Institution of Washington, 1924, p. 60.
31. Caldwell, *Anniversary Oration*, pp. 26–8.
32. Marcus Ackroyd, Laurence Brockliss, Michael Moss, Kate Retford and John Stevenson, *Advancing with the Army: Medicine, the Professions, and Social Mobility in the British Isles 1790–1850*, Oxford: Oxford University Press, 2006, pp. 198–202, 318.
33. James M'Gregor [McGrigor], *Medical Sketches of the Expedition to Egypt*, London: John Murray, 1804, p. 68; Harrison, *Medicine in the Age of Commerce and Empire*, pp. 259–60.
34. See Gilbert Blane, *Observations on the Diseases Incident to Seamen*, London: Joseph Cooper, 1785, pp. 128, 187, 252.
35. Colin Chisholm, *An Essay on the Malignant Pestilential Fever introduced into the West Indian Islands from Boullam, on the Coast of Guinea, as it appeared in 1793 and 1794*, London: C. Dilly, 1795; idem, *An Essay on the Malignant Pestilential Fever, introduced into the West Indian Islands from Boullam, on the Coast of Guinea, as it appeared in 1793, 1794, 1795, and 1796. Interspersed with Observations and Facts, tending to Prove that the Epidemic existing at Philadelphia, New-York, &c. was the same Fever introduced by Infection imported from the West Indian Islands: and Illustrated by Evidences found on the State of those Islands, and Information of the most eminent Practitioners residing on them*, London: J. Mawman, 1801.
36. E.g. John Clark, *Observations on the Diseases in Long Voyages to Hot Countries, and Particularly to those which prevail in the East Indies*, London: D. Wilson & G. Nicol, 1773; Charles Curtis, *An Account of the Diseases of India*, Edinburgh: W. Laing, 1807. For a discussion of this literature, see W.F. Bynum, 'Cullen and the Study of Fevers in Britain, 1760–1820', in W.F. Bynum and V. Nutton (eds), *Theories of Fever from Antiquity to the Enlightenment, Medical History* Supplement, No. 1, London: Wellcome Institute for the History of Medicine, 1981, 135–48.

37. Mark Harrison, *Climates and Constitutions: Health, Race, Environment and British Imperialism in India 1600-1850*, New Delhi: Oxford University Press, 1999.
38. John P. Wade, *A Paper on the Prevention and Treatment of the Disorders of Seamen and Soldiers in Bengal*, London: J. Murray, 1793, pp. 5, 9.
39. John R. McNeill, 'The Ecological Basis of Warfare in the Caribbean, 1700-1804', in M. Utlee (ed.), *Adapting to Conditions: War and Society in the Eighteenth Century*, Tuscaloosa: University of Alabama Press, 1986, pp. 26-42; David Geggus, *Slavery, War, and Revolution: The British Occupation of Saint Dominique, 1793-1798*, Oxford: Clarendon Press, 1982, pp. 347-72; Roger N. Buckley, *The British Army in the West Indies: Society and the Military in the Revolutionary Age*, Gainesville: University Press of Florida, 1998, pp. 272-324.
40. P. Assalini, *Observations on the Disease called The Plague, on the Dysentery, The Opthalmy of Egypt, and on the Means of Prevention, with some Remarks on the Yellow Fever of Cadiz*, trans. A. Neale, New York: T.J. Swords, 1806.
41. E.g. Hector M'Lean, *An Enquiry into the Nature, and Causes of the Great Mortality among the Troops at St. Domingo*, London: T. Cadell, 1797; James Clark, *A Treatise on the Yellow Fever, as it appeared in the Island of Dominica, in the Years 1793-4-5*, London: J. Murray & S. Highley, 1797; J. Mabit, *Essai sur les maladies de l'armée de St.-Domingue en l'an XI, et principalement sur la fièvre jaune*, Paris: Ecole de Médecine, 1804; Victor Bally, *Du typhus d'Amérique ou fièvre jaune*, Paris: Smith, 1814.
42. E.g. George Davidson, 'Practical and Diagnostic Observations on Yellow Fever, as it occurs in Martinique', *Medical Repository*, 2 (1805), 244-52; S. Ffirth, 'Practical Remarks on the Similarity of American and Asiatic Fevers', *Medical Repository*, 4 (1807), 21-7.
43. Richard J. Kahn and Patricia G. Kahn, 'The *Medical Repository* - The First US Medical Journal (1797-1824)', *New England Medical Journal*, 337 (1997), 1926-30.
44. Chisholm, *An Essay* (1801 edn), pp. xxi-xxiii, 95-6, 99-108; idem, *A Letter to John Haygarth, M.D. F.R.S., London and Edinburgh, &c., from Colin Chisholm, M.D. F.R.S., &c., Author of An Essay on the Pestilential Fever: Exhibiting farther Evidence of the Infectious Nature of this Fatal Distemper in Grenada, during 1793, 4, 5, and 6; and in the United States of America, from 1793 to 1805: in order to correct the pernicious Doctrine promulgated by Dr Edward Miller, and other American Physicians, relative to this destructive Pestilence*, London: Joseph Newman, 1809; David Hosack to Dr Chisholm, New York, 12 August 1809, Letters and Papers of David Hosack, 1795-1835, American Philosophical Society [APS], Philadelphia. I am indebted to Katherine Arner for this reference and for allowing me to see her paper 'Making Yellow Fever American: The United States, the British Empire and the Geopolitics of the Disease in the Atlantic World, 1793-1825'.
45. Jonathan Cowdrey, 'A Description of the City of Tripoli, in Barbary; with Observations on the Local Origin and Contagiousness of the Plague', *Medical Repository*, 3 (1806), 154-9.
46. John Taylor, *Travels from England to India, in the Year 1789*, 2 vols, London: S. Low, 1799, vol. 1, pp. 114-15.
47. See A.G. Hopkins (ed.), *Globalization in World History*, London: Pimlico, 2002; C.A. Bayly, *The Birth of the Modern World 1780-1914*, Oxford: Blackwell, 2004; Robbie Robertson, *The Three Waves of Globalization*, London: Zed Books, 2003; Rondo Cameron and Larry Neal, *A Concise Economic History of the World*, New York and Oxford: Oxford University Press, 2003.
48. See 'Globalization', in Cooper, *Colonialism in Question*.
49. *Second Report of the Select Committee Appointed to consider the means of improving and maintaining the Foreign Trade of the Country. Quarantine*, Parliamentary Papers [PP] 417, 1824.
50. See Michael Brown, 'From Foetid Air to Filth: The Cultural Transformation of British Epidemiological Thought, ca.1780-1848', *Bulletin of the History of Medicine*, 82 (2008), 515-44; Roger Cooter, 'Anticontagionism and History's Medical Record', in P. Wright

and A. Treacher (eds), *The Problem of Medical Knowledge: Examining the Social Construction of Medicine*, Edinburgh: Edinburgh University Press, 1982, pp. 87–108; Charles F. Mullet, 'Politics, Economics and Medicine: Charles Maclean and Anticontagion in England', *Osiris*, 10 (1952), 224–51; Charles Maclean, *Results of an Investigation respecting Epidemic and Pestilential Diseases: Including Researches in the Levant concerning the Plague*, 2 vols, London: Thomas & George Underwood, 1817.

51. R.C. Wellesley to Henry Dundas, 21 March 1799, in E. Ingram (ed.), *Two Views of British India: The Private Correspondence of Mr Dundas and Lord Wellesley: 1798–1801*, London: Adams & Dart, 1969, pp. 235–6; Charles Maclean, *To the Inhabitants of British India*, Calcutta: s.n., 1798; Harrison, *Medicine in the Age of Commerce and Empire*, Part II, chap. 4.
52. Charles Maclean, *The Affairs of Asia considered in their Effects on the Liberties of Britain*, London: s.n., 1806.
53. E.g. Charles Maclean, *Specimens of Systematic Misrule*, London: H. Hay, 1820.
54. Charles Maclean, *A View of the Consequences of Laying Open the Trade to India, to Private Ships*, London: Black, Perry & Co., 1813.
55. For more on this, see Catherine Kelly, '"Not from the College, but through the Public and the Legislature": Charles Maclean and the Relocation of Medical Debate in the Early Nineteenth Century', *Bulletin of the History of Medicine*, 82 (2008), 545–69.
56. Booker, *Maritime Quarantine*, pp. 380–1.
57. *Report from the Select Committee appointed to consider the validity of the Doctrine of Contagion in the Plague*, 14 June 1819, PP 449, 1819, p.ii.
58. Booker, *Maritime Quarantine*, p. 383.
59. Ibid., pp. 383–4.
60. Margaret Pelling, *Cholera, Fever and English Medicine 1825–1865*, Oxford: Oxford University Press, 1978, p. 28.
61. Booker, *Maritime Quarantine*, pp. 401–2.
62. Office of Sick and Wounded Seamen to Admiralty Board, 19 August 1797, ADM/F/27, National Maritime Museum [NMM].
63. In 1794, for instance, the Royal Navy was annoyed by the prolonged quarantine in Lisbon of a captured French vessel containing valuable merchandise from Saragossa. The seemingly arbitrary extension of the quarantine led to protracted negotiations with the Portuguese Secretary of State and other officials. See Thomas Mayne, Lisbon, 8 November 1794, to Sir Charles Hamilton, Commander, HMS *Rodney*, Portsmouth, Western MS. 7313, WL.
64. See for example, J. Tommasini, *Recherches pathologiques sur la fièvre de Livorne de 1804, sur la fièvre jaune d'amérique*, Paris: Arthus-Bertrand, 1812.
65. Maj.-Gen. Sir Charles Phillips, 'Letters and Instructions to the Officers during the Plague at Corfu, 1816', Western MS.3883, WL; Gallagher, *Medicine and Power*, p. 31.
66. A.B. Granville, *A Letter to the Right Honble. W. Huskisson, M.P., President of the Board of Trade, on the Quarantine Bill*, London: J. Davy, 1825, p. 3.
67. Ibid., p. 9.
68. Ibid., p. 14.
69. Howard, *Public Health Administration*, p. 61; 'Report of the Health Officer' (1827), in Baltimore City Health Department, *The First Thirty-Five Annual Reports*, Baltimore: The Commissioner of Health of Baltimore, Maryland, 1853.
70. 'Report of the Health Officer' (1829), ibid.
71. Ibid.
72. John Buckley & Sons, Lisbon, to Stephan Girard, Philadelphia, 12 August 1816 (No. 695 1816); Daniel Crommelin & Sons, Amsterdam, to Girard, 30 August 1816 (No. 737 1816); [Illegible] to Girard, Elsinore, 11 September 1816 (No. 791 1816); John R. Warder, Amsterdam, 30 1818 (No. 121 1818); John Buckley & Sons, Lisbon, undated (No. 820 1819), Stephan Girard Collection, APS.

73. E.g. Girard to Devèze, 17 December 1823 (LB 19 No. 76), 26 January 1825 (LB 20 No. 174); Devèze to Girard, 4 July 1821 (No. 492 1821); 22 September 1821 (No. 675 1821), Stephan Girard Collection, APS.
74. 'Report of the Consulting Physician' (1830), *The First Thirty-Five Annual Reports*.
75. Thomas O'Halloran, *Remarks on the Yellow Fever of the South and East Coasts of Spain*, London: Callow & Wilson, 1823; Jean Devèze, *Memoire sur la fièvre jaune afin de la proven non-contagieuse*, Paris: Ballard, 1821; Nicolas Chervin, Certified copies and original documents relating to yellow fever in Guadeloupe and the West Indies, Western MS. Amer. 113, WL.
76. Étienne Pariset, *Observations sur la fièvre jaune, faites a Cadiz*, Paris: Audot, 1820.
77. Ibid., p. 128.
78. Charles Maclean, *Evils of Quarantine Laws, and Non-Existence of Pestilential Contagion; deduced from the Phoenomena of the Plague of the Levant, the Yellow Fever of Spain, and the Cholera Morbus of Asia*, London: T.& G. Underwood, 1824, pp. 110–15.
79. See: Ackerknecht, 'Anticontagionism'; Ann F. La Berge, *Mission and Method: The Early Nineteenth-Century French Public Health Movement*, Cambridge: Cambridge University Press, 1992, pp. 90–4; E.A. Heaman, 'The Rise and Fall of Anticontagionism in France', *Canadian Bulletin of the History of Medicine*, 12 (1995), 3–25.
80. Review of Charles Maclean's *Evils of Quarantine Laws* (1824), *Medico-Chirurgical Review*, n.s., 2 (1825), 21.
81. Henry Halford, Gilbert Blane, et al., *Cholera Morbus: Its Causes, Prevention, and Cure; with Disquisitions on the Contagious or Non-Contagious Nature of this dreadful Malady, by Sir Henry Halford, Sir Gilbert Blane, and eminent Birmingham Physicians, and the Lancet, and Medical Gazette, together with ample Directions regarding it, by the College of Physicians and Board of Health*, Glasgow: W.R. M'Phan, 1831, pp. 3, 9–21.
82. Richard Evans, *Death in Hamburg: Society and Politics in the Cholera Years 1830–1910*, Oxford: Clarendon Press, 1987, p. 245.
83. James McCabe, *Observations on the Epidemic Cholera of Asia and Europe*, Cheltenham: G.A. Williams, 1832, pp. 5–6; William White, *The Evils of Quarantine Laws, and Non-existence of Pestilential Contagion*, London: Effingham Wilson, 1837. See also Pelling, *Cholera*, pp. 24–5; Michael Durey, *Return of the Plague: British Society and the Cholera of 1831–2*, London: Macmillan, 1979; Harrison, *Climates*, chap. 4; Charles E. Rosenberg, *The Cholera Years: The United States in 1832, 1849, and 1866*, Chicago and London: Chicago University Press, 1968.
84. Evans, *Death in Hamburg*, pp. 245–9.
85. Richard Evans, 'Epidemics and Revolutions: Cholera in Nineteenth-Century Europe', in. P. Slack and T. Ranger (eds), *Epidemics and Ideas*, Cambridge: Cambridge University Press, 1992, pp. 167–8; *idem*, *Death in Hamburg*.
86. Peter Baldwin, *Contagion and the State in Europe 1830–1930*, Cambridge: Cambridge University Press, 1999, pp. 97–8.
87. See, A.D. Vasse St. Ouen, French Consul at Larnaca, Cyprus, to A.R. Roussin, French Ambassador at Constantinople, 27 November 1834 to 26 April 1836, Western MS. 4911, WL.
88. La Verne Kuhnke, *Lives at Risk: Public Health in Nineteenth-Century Egypt*, Berkeley: University of California Press, 1990; Panzac, *La Peste*, pp. 446–9; Sheldon Watts, *Epidemics and History: Disease, Power and Imperialism*, New Haven: Yale University Press, 1997, pp. 35–9; Gallagher, *Medicine and Power*, p. 33.
89. J.E. Johnson, *An Address to the Public on the Advantages of a Steam Navigation to India*, London: D. Sidney & Co., 1824, pp. 18–21.
90. Ataf Lufti al-Sayyid Marsot, *Egypt in the Reign of Muammad Ali*, Cambridge: Cambridge University Press, 1994, pp. 162–72.
91. R. Owen, *The Middle East in the World Economy 1800–1814*, London: I.B. Tauris, 1981, pp. 92–3; P. Harnetty, *Imperialism and Free Trade: Lancashire in the Mid-Nineteenth Century*, New York: Columbia University Press, 1972.

92. A.-B. Clot-Bey, De la Peste observée en Égypte; recherches et considerations sur cette maladie, Paris: Fortin, Masson, 1840, pp. 407–8.
93. Panzac, La Peste, pp. 458–9.
94. Ibid., pp. 465–70.
95. Gallagher, Medicine and Power, pp. 41–2.
96. Hamdan ibn 'Uthman Khawajah, Ithaf al-munsifinwa-al-udaba fi al-ihtiras an al-waba, Algiers: al-Sharikah, 1968, p. 26.
97. De Ségur Dupeyron, Rapport adressé a son exc. le ministre du commerce, chargé de procéder a une enquête sur les divers régimes sanitaires de la Mediterranée, Paris: L'Imprimerie Royale, 1834.
98. C.K. Webster, Palmerston, Metternich and the European System 1830–1841, London: The British Academy, 1934, pp. 19–21.
99. Jospeh Ayre, A Letter addressed to the Right Honourable Lord John Russell, M.P., Secretary of State for the Home Department, on the evil Policy of those Measures of Quarantine, and Restrictive Police, which are employed for arresting the Progress of Asiatic Cholera, London: Longman, Orme, Brown, Green & Longman, 1837; John Bowring, Observations on the Oriental Plague, and on Quarantine as a Means of Arresting its Progress, Edinburgh: W. Tait, 1838; see also Arthur T. Holroyd, The Quarantine Laws, their Abuses and Inconsistencies. A Letter, addressed to the Rt. Hon. Sir John Cam Hobhouse, Bart., M.P., President of the Board of Control, London: Simpkin, Marshall & Co., 1839.
100. Earl of Aberdeen to Lord Cowley, British Ambassador to France, 27 June 1843, Correspondence respecting the Quarantine Laws since the Correspondence last presented to Parliament, London: T.R. Harrison, 1846, PP 1846 [718], XLV.
101. John Murray, The Plague and Quarantine. Remarks on some Epidemic and Endemic Diseases; (including the Plague of the Levant) and the Means of Disinfection: with a Description of the Preservative Phial. Also a Postscript on Dr. Bowring's Pamphlet, 2nd edn, London: John Murray, 1839.
102. Palmerston to Sir Frederick Lamb, British Ambassador to Austria, 11 June 1838, in Correspondence respecting the Quarantine Laws.
103. Prince Esterhazy, Austrian Ambassador to Britain, to Palmerston, 19 November 1936, Correspondence relative to the Contagion of Plague and the Quarantine Regulations of Foreign Countries, 1836–1943, London: T.R. Harrison, 1843, PP 1843 [475], LIV.
104. Harold Nicolson, The Congress of Vienna: A Study in Allied Unity 1812–1822, London: Constable & Co., 1946; Henry A. Kissinger, A World Restored: Metternich, Castlereagh and the Problems of Peace 1812–22, London: Weidenfeld & Nicolson, 1957; Charles Webster, The Congress of Vienna 1814–1815, London: Thames and Hudson, 1963; Tim Chapman, The Congress of Vienna: Origins, Processes and Results, London: Routledge, 1998; Adam Zamoyski, Rites of Peace: The Fall of Napoleon & the Congress of Vienna, London: Harper Press, 2007.
105. These efforts were grounded on a report on quarantine in the Mediterranean (1846) by the chairman of the French Academy of Medicine, Dr R.C. Prus. See George Weisz, The Medical Mandarins: The French Academy of Medicine in the Nineteenth and Early Twentieth Centuries, New York and Oxford: Oxford University Press, 1995, p. 77.
106. F.S.L. Lyons, Internationalism in Europe 1815–1914, Leiden: A.W. Sythoff, 1963, pp. 56–64.
107. F.R. Bridge and Roger Bullen, The Great Powers and the European States System 1815–1914, London: Longman, 1980, pp. 41–2.
108. Palmerston to Marquess of Clanricarde, 9 July 1839, Correspondence relative to the Affairs of the Levant, PP 1841 [304], VIII, Session 2.
109. Coleman Phillipson and Noel Buxton, The Question of the Bosphorus and Dardanelles, London: Stevens & Hayes, 1917, pp. 74–80.
110. Roger Bullen, Palmerston, Guizot and the Collapse of the Entente Cordiale, London: Athlone Press, 1974, p. 334.

111. Gavin Milroy, *Quarantine and the Plague: Being a Summary of the Report on these Subjects recently addressed to the Royal Academy of Medicine in France*, London: Samuel Highley, 1846, p. 5.
112. Donald Quataeret, 'Population', in *An Economic and Social History of the Ottoman Empire*, ed. H. Inalcik and D. Quataert, vol. 2, Cambridge: Cambridge University Press, 1994, pp. 787–9.
113. The nineteenth century saw an escalation in the volume of international trade, which had been rising steadily during the preceding century, yet commerce within the Ottoman Empire remained relatively more important than external trade until after 1900. See Donald Quataert, *The Ottoman Empire: 1700–1922*, Cambridge: Cambridge University Press, 2005, pp. 127–8.
114. The regulations were approved in May 1841 and were accompanied by detailed guidelines for all doctors in the Sanitary Service of the Ottoman Empire. See *Papers respecting Quarantine in the Mediterranean*, London: Harrison & Sons, 1860, pp. 81–7.
115. Metternich to Baron Langsdorff, French Chargé de Affaires at Vienna, 13 July 1838, PP 1843 [475] LIV.
116. The Council consisted of sixteen members, with an Ottoman official as president. It has not been possible to determine the nationality and profession of all members, but it would appear to have included at least five with medical qualifications, and foreign members comprised roughly half the Council; of the latter, two members appear to have been drawn from each of the major European powers represented – Britain, France and Austria-Hungary. See *Papers respecting Quarantine*, p. 94.
117. Convention between Great Britain, Austria, France, Prussia, Russia, and Turkey respecting the Straits of the Dardanelles and of the Bosphorus, PP 1842 [350] XLV.
118. Webster, *Palmerston*, p. 24.
119. Ibid., pp. 6–7.
120. Aberdeen to Lord Cowley, 27 June 1843, PP 1846 [318], XLV.
121. J. Macgregor to Viscount Canning, 2 March 1844, PP 1846 [718] XLV.
122. Metternich to Sir Robert Gordon, British Ambassador to Austria, 24 May 1844; Gordon to Aberdeen, 31 May 1844; Canning to M. Lefevre, 17 April 1845; Canning to Lefevre, 12 September 1845, PP 1846 [718] XLV.
123. Pym to the Earl of Dalhousie, 5 June 1845; Pym to Lefevre, 6 June 1845, PP 1846 [718] XLV.
124. Pym to Lefevre, 22 September 1845, PP 1846 [718] XLV.
125. Mr Magenis, Austrian Ambassador to Britain, to Aberdeen, 15 December 1845, PP 1846 [718] XLV.
126. See tables of vessels subjected to quarantine at Rhodes, *Papers respecting Quarantine*, pp. 66–70.
127. Magenis to Aberdeen, 15 November 1845, PP 1846 [718], XLV.
128. General Board of Health, *Report on Quarantine*, London: W. Clowes & Sons, 1849, pp. 78–9.
129. Clot-Bey, *De la Peste*, p. 383.
130. Speech by Bowring, 15 March 1842, Hansard, *Parl. Debates*, 3rd ser. col. 610.
131. See: 'Copy of the Tariff agreed upon by the Commissioners appointed under the Seventh Article of the Convention of Commerce and Navigation between Turkey and England', PP 1839 [549] XLVII; *Convention of Commerce and Navigation between Her Majesty, and the Sultan of the Ottoman Empire*, London: J. Harrison, 1839, PP 1839 [157] L; Correspondence respecting the Operation of the Commercial Treaty with Turkey, of August 16, 1838, PP [341], Session 2, VIII.
132. Collective Note of the Representatives of Austria, France, Great Britain, Prussia, and Russia at Constantinople, to the Porte, July 27, 1839, PP 1839 [205] L; Correspondence relative to the Affairs of the Levant, Part III, PP 1841 [337], Session 2, VIII.
133. E.g. Holroyd, *The Quarantine Laws*.
134. *Papers respecting Quarantine*, p. 26.

135. Campbell to Lord Palmerston, 27 March 1838, G/17/10, Asia, Pacific and African Collections [APAC], British Library [BL].
136. Panzac, *La Peste*, p. 470.
137. Lt-Col. P. Campbell, East India Company agent, Cairo, to Peter Amber, 14 July 1835; Campbell to James Melville, 14 July 1837; Alexander Waghorn, East India Company agent, Alexandria, to French Post Office, Alexandria, 18 July 1837, G/17/10, APAC, BL.
138. Hansard, *Parl. Debates*, 15 March 1842, 3rd ser. LXI, cols 608–18.
139. *Procès-verbaux de la conférence sanitaire internationale, ouverte à Paris le 27 juillet 1851*. Paris: Imperimerie Nationale, 1851, vol. 1, 5 August 1851, pp. 3–4.
140. Bullen, *Palmerston*, pp. 337–8.
141. A.J.P. Taylor, *The Struggle for Mastery in Europe 1848–1918*, Oxford: Clarendon Press, 1954, p. 46.
142. Howard-Jones, *Scientific Background*, pp. 15–16.
143. *Procès-verbaux de la conférence sanitaire internationale*, 24 October 1851, pp. 23–5; 4 October 1851, pp. 8–9; 18 September 1851, pp. 3–12.
144. Baldwin, *Contagion*, p. 198.
145. *Procès-verbaux de la conférence sanitaire internationale*, vol. 2, Annex to Proc.29, 11 November 1851.
146. Panzac, *La Peste*, p. 475.

Chapter 4: Quarantine and the empire of free trade

1. On the Navy's role in the suppression of the slave trade, see Christopher Lloyd, *The Navy and the Slave Trade*, London, Longmans, Green & Co., 1949; E.P. Leveen, *British Slave Trade Suppression Policies*, New York: Arno Press, 1977; Raymond Howell, *The Royal Navy and the Slave Trade*, London: Croom Helm, 1987.
2. Letter from 'A Naval Officer' to *The Times*, undated [1846], giving mortality statistics of fourteen vessels on the West Africa service, over an eighteen-month period in 1837–8. Out of a total of 1,000 men serving on these vessels, 336 died, almost all as a result of disease. Mortality rates for each vessel varied from 3 per cent to 71 per cent. Letters and newspaper cuttings concerning the *Eclair*, Sotherton-Estcourt Papers, F.529, Gloucestershire County Record Office [GCRO]. For mortality rates of British servicemen in West Africa, see also J.L.S. Coulter and C. Lloyd, *Medicine and the Navy 1200–1900. Vol. IV – 1815–1900*, Edinburgh and London: E. & S. Livingstone, 1963, pp. 155–64; Philip D. Curtin, '"The White Man's Grave": Image and Reality, 1750–1850', *Journal of British Studies*, 1 (1961), 94–110; idem, *Death by Migration: Europe's Encounter with the Tropical World in the Nineteenth Century*, Cambridge: Cambridge University Press, 1989.
3. The literature on this subject is extensive. See, for example: Black, *The British Seaborne Empire*, chaps 5–6; A.G. Hopkins, 'The History of Globalization – and the Globalization of History?', in A.G. Hopkins (ed.), *Globalization in World History*, London: Pimlico, 2002; P.J. Cain and A.G. Hopkins, *British Imperialism, 1688–2000*, London: Routledge, 2001; Patrick O'Brien, 'Europe in the World Economy', in H. Bull and A. Watson (eds), *The Expansion of International Society*, Oxford: Clarendon Press, 1984, 43–60.
4. John Darwin, *The Empire Project: The Rise and Fall of the British World System 1830–1970*, Oxford: Oxford University Press, 2009; Deepak Lal, *Reviving the Invisible Hand: The Case for Classical Liberalism in the Twenty-First Century*, Princeton, NJ: Princeton University Press, 2006.
5. Gary B. Magee and Andrew S. Thompson, *Empire and Globalisation: Networks of People, Goods and Capital in the British World, c.1850–1914*, Cambridge: Cambridge University Press, 2010; Jeffery G. Williamson, 'Globalization, Convergence, and History', *Journal of Economic History*, 56 (1996), 277–306; idem, 'The Evolution of Global Labour Markets since 1830: Background Evidence and Hypotheses', *Explorations in Economic History*, 32 (1995), 141–96; Alan M. Taylor and Jeffery G. Williamson, 'Convergence in the Age of Mass Migration', *European Review of Economic History*, 1 (1997), 27–63.

6. See Philip D. Curtin, *The Image of Africa: British Ideas and Action 1780-1850*, Madison: University of Wisconsin Press, 1964, pp. 289-317; James O. M'William, *Medical History of the Expedition to the Niger during the Years 1841-2, comprising an Account of the Fever which led to its abrupt Termination*, London: HMSO, 1843.
7. Capt. Walter Estcourt to Commodore Jones, 8 September, 1845, Boa Vista, F.533, GCRO.
8. *Statistical Reports on the Health of the Navy, for the Years 1837-43: Part III. North Coast of Spain Station, West Coast of Africa Station, Packet Service, Home Station, Ships employed variously*, London: HMSO, 1854, p. 62.
9. H.W. Macaulay, British Consul, Cape Verde Islands, to the Earl of Aberdeen, 24 December 1846, Estcourt Papers, F.534, GCRO.
10. Estcourt to Jones, 8 September 1845, F.533, GCRO.
11. Statement by Mr William Pym, Supt.-Genl. of Quarantine, after interviewing survivors on board the *Eclair* at Motherbank, F.529, GCRO.
12. Cutting from *The Times*, undated, F.529, GRCO.
13. Mr Sterrit, Surgeon of HMS *Ocean*, to Vice-Admiral Sir Edward King, 7 October 1845, F.546, GCRO.
14. Dr J.G. Stewart, R.N., to Sir Edward King, 9 October 1845, F.546, GCRO.
15. Dr Stewart to Sir William Burnett, Director General of the Naval Medical Service, 10 October 1845, F.546; Capt. F.E. Loch, R.N., Supt. of Quarantine, to Sir Edward King, HMS *Rhin*, Standgate Creek, 11 October 1845, F.546, GCRO.
16. Dr Stewart, Medical Report, 13 October 1845; Loch to King, 13 October 1845, F.546, GCRO.
17. King to Secretary to the Admiralty, 31 October HMS *Trafalgar*, Sheerness, F.546, GCRO.
18. Loch to King, 31 October 1845, F.546, GCRO.
19. Return of officers and men who volunteered to serve on board HMS *Eclair*, F.534, GCRO.
20. Editorial, *The Times*, undated cutting, F.529, GCRO.
21. 'The Pest Ship', cutting from an unnamed newspaper, F.529; *Gloucester Journal*, 'The Plague on board the Eclair,' 18 October 1845.
22. Cutting from *The Times*, undated, F.529, GCRO.
23. *Gloucester Journal*, 4 October 1845, p. 2; 11 October 1845, p. 2.
24. *Gloucestershire Chronicle*, 4 October 1845, p. 3.
25. *Lancet*, Editorial, 11 October 1845, pp. 402-3.
26. Cutting from the *Naval Intelligencer*, undated, F.529, GCRO.
27. Dr Richardson, Inspector of Haslar Hospital, to Rear Admiral Parker, 29 September 1845, F.546, GCRO.
28. Sir William Pym to Greville, 1 October 1845, F.546, GCRO.
29. Sir William Burnett, report to the Admiralty, 21 November 1845, F.546, GCRO.
30. Pym, *Observations on the Bulam Fever*.
31. Pym to Mr Grenville, Admiralty, 29 November 1845, F.546, GCRO.
32. Pym to Greville, 22 October 1845, GCRO.
33. Burnett to Admiralty, 21 November 1845, GCRO.
34. Burnett to Admiralty, 11 December 1845, GCRO.
35. Burnett to Admiralty, 21 November 1845, GCRO.
36. See Curtin, *Image of Africa*, chap. 14.
37. Mr Lefevre, Foreign Office, to Viscount Canning, enclosing letters from William Pym, 22 August 1845, *Correspondence on the Subject of the 'Eclair' and the Epidemic which broke out on the said Vessel*, PP 1846 [707] XXVI.
38. Pym to Lefevre, 16 June 1845, ibid.
39. Earl of Aberdeen to Lord Cowley, British Ambassador in Paris, 27 June 1843; M. Guizot to Lord Cowley, 9 September 1843; Mr MacGregor to Viscount Canning, 2 March 1844, ibid.

40. Pym to Lefevre, 22 September 1845, ibid.
41. 'The Pest Ship', cutting from an unnamed newspaper, F.529, GCRO.
42. See E.L. Rasor, *Reform in the Royal Navy: A Social History of the Lower Deck 1850 to 1880*, Hamden, Conn.: Archon Books, 1976, pp. 9–10.
43. The Estcourt family had acquired an estate at Shipton Moyne, near Tetbury, in the thirteenth century.
44. Obituary of Cmdr Walter Grimston Bucknall Estcourt, *Nautical Magazine and Naval Chronicle*, 14 and 12 December 1845, pp. 686–8.
45. Burnett, Report to the Admiralty, 21 November 1845, F.546, GCRO.
46. A.H. Addington, Foreign Office, to Capt. W. Hamilton, Admiralty, 29 May 1846, F.534, GCRO.
47. Capt. Buckle to Capt. Hamilton, 27 April 1846, F.534, GCRO.
48. Buckle insisted that 'nothing but the constant watchfulness of our Boats and Cruisers during the fine season prevented the slave trade from being successfully carried on this Quarter', in Buckle to Hamilton, 27 April 1846.
49. Curtin, *Image of Africa*, vol. 2, pp. 315–16.
50. Buckle to Hamilton, 27 April 1846.
51. The Niger Expedition was a government expedition intended primarily to establish trading posts and to obtain anti-slavery treaties with local chiefs, the object being to promote a legitimate and profitable trade in place of slavery. The expedition sailed in April 1841 but was forced to withdraw because of high mortality from fever. See Curtin, *Image of Africa*, vol. 2, chap. 12.
52. Buckle to Hamilton, 27 April 1846, F.534, GCRO.
53. Ibid.
54. Letter to *The Times*, 3 October 1845, from 'One who has suffered from fever'.
55. Letter from 'A Naval Officer' to *The Times*, undated, F.529, GCRO; Mark Harrison, 'An "Important and Truly National Subject": The West Africa Service and the Health of the Royal Navy in the Mid-Nineteenth Century', in S. Archer and D. Haycock (eds), *Health and Medicine at Sea*, London: Boydell & Brewer, 2010, 108–27.
56. Mr H.W. Macaulay to the Earl of Aberdeen, Saint Nicholas, Cape Verde Islands, 24 December 1846, F.534, GCRO.
57. Estcourt to Mr Peter Kenny, Cape Surgeon, Boa Vista, 1 September 1845, F.534, GCRO.
58. Capt. Gallway, H.M. Consul, Naples, to Earl of Aberdeen, 17 October 1845, F.546, GCRO.
59. Burnett to Secretary to the Admiralty, 14 November 1845, F.546, GCRO.
60. Lord Bathurst, Office of the Privy Council, to Secretary of the Commissioners of Customs, 21 October 1845, F.546, GCRO.
61. Burnett to Secretary to Admiralty, 14 November 1845, F.546, GCRO.
62. Aberdeen to the Hon. William Temple, 30 and 31 October 1845, F.546, GCRO.
63. Mr Addington to the Hon. W.L. Bathurst, Foreign Office, 10 November 1845, F.546, GCRO.
64. Baldwin, *Contagion*, p. 536.
65. Ibid., p. 532.
66. Aberdeen to Temple, 24 November 1845, F.546, GCRO.
67. Temple to Aberdeen, 14 November 1845; Aberdeen to Mr Abercromby, 24 November 1845.
68. Mr J. Rendall, HM Consul, Cape Verde Islands, to Aberdeen, 22 December 1845, F.546.
69. *Revista Universal Lisbonense*, vol. 5, ser.3, 22 January 1846, p. 361.
70. Dr Bernardino Antonio Gomes, 20 February 1846, Instituto de Investigação Científica Tropical, Cabo Verde, Archivo Histórico Colonial [AHC], Lisbon.
71. Rendall to Aberdeen, 22 December, F.546, GCRO.
72. Consul-General of Portugal, London, to the Lords of the Privy Council, 19 January 1846, F.546.

73. The British government eventually agreed to pay a sum of 1,000 guineas in compensation to the islanders of Boa Vista. See General Board of Health, *Second Report on Quarantine: Yellow Fever*, PP 1852 XX, p. 101.
74. Joaquim M. Franco, Provincial Surgeon, to Bernadino Antonio Gomes, President of the Naval Council of Health, 20 February 1846, Archivo Histórico Ultramarino [AHU], Lisbon.
75. Joaquim P. de Moraes to Gomes, 4 February 1846; J.M. Franco to Gomes, 14 December 1845, AHC.
76. Franco to Gomes, 14 December 1845, AHC.
77. Moraes to Gomes, 15 February 1846, AHU.
78. *Boletim Oficial do Governo de Cabo-Verde*, 116, 5 December 1845, p. 465.
79. Question from Captain Layard, M.P., 6 August 1846, Hansard, *Parl. Debates*, 3rd ser., LXXXVIII, col. 359.
80. See Chapter 2.
81. Question from Dr Bowring, 19 May 1846, Hansard, *Parl. Debates* 3rd ser., LXXVI, col. 878. McWilliam's opinion seems to have been shared by other surgeons involved in the expedition: *Papers Relative to the Expedition to the River Niger*, PP 1843 [472] XLVIII, e.g. report from W. Cook, Surgeon of the *Wilberforce*, to Lord Stanley, Secretary of State for the Colonies, 11 March 1843.
82. Mr Hume, Hansard, *Parl. Debates*, 3rd ser., LXXVI, col. 881.
83. Sir G. Clerk, ibid.
84. Harrison, 'An "Important and Truly National Subject"', pp. 113–19.
85. *Report on the Fever at Boa Vista by Dr. McWilliam*, London: T.R. Harrison, 1847, p. 17.
86. Ibid., p. 106.
87. Ibid., p. 110.
88. Ibid., p. 111.
89. Ibid.
90. Pym to Lords of Privy Council, undated, F.549, GCRO.
91. Pym to Lords of Privy Council, undated, F.549, GCRO.
92. Burnett to Secretary to the Admiralty, F.548, GCRO.
93. Report of Dr King on the Fever at Boa Vista, addressed to Sir William Burnett, 10 October 1847, F.550, GCRO.
94. Don Jose Miguel de Novonha, Governor-General of the Cape Verde Islands, to Portuguese Consul, Gibraltar, enclosure No. 3, *Report on the Fever at Boa Vista by Dr. McWilliam*.
95. Report of Dr King, F.550, GCRO, p. 3.
96. Pym to Lords of the Privy Council, 15 May 1848, F.551, GCRO.
97. See, for example, motion proposed by Dr Bowring and seconded by Mr Hume, 18 March 1847, Hansard, 3rd ser., XCI, cols 150–5.
98. Pelling, *Cholera*, p. 63.
99. *Report of the General Board of Health on Quarantine*, PP 1849, XXIV.
100. Pelling, *Cholera*, pp. 63–80.
101. The Board did not have time to begin its planned investigation into plague before the resignation of Chadwick in 1854 and its abolition in 1858.
102. General Board of Health, *Second Report on Quarantine*, pp. 89–90, 101.
103. Ibid., pp. 96.
104. Ibid., pp. 97.
105. Ibid., pp. 97–8.
106. Ibid., pp. 113–16.
107. Ibid., p. 136.
108. See, for example, G.F. Bone, *Inaugural Dissertation on Yellow Fever, and on the Treatment of that Disease with Saline Medicines*, London: Longman & Co., 1846; Alexander Bryson, *Report on the Climate and Principal Diseases of the African Station* London: HMSO, 1847; Daniel Blair, *Some Account of the Last Yellow-Fever Epidemic*

of British Guiana, London: Longman & Co., 1850. Bone was an assistant surgeon in the Army, Bryson a naval surgeon, and Blair – a highly respected figure – was Surgeon-General of British Guiana.
109. E.g. Richard Birtwhistle (a naval surgeon), 'Account of the Yellow Fever on Board the "Volage"', *Lancet*, 3 January 1846, 8–9.
110. John Wilbin and Alexander Harvey, 'An Account of Yellow Fever, as it occurred on board R.M.S. Ship "La Plata", in the Month of November, 1852', *Lancet*, 12 February 1853, 148–51.
111. T. Bacon Phillips, 'Yellow Fever as it occurred on board the R.M. Steamer "La Plata", on her homeward Voyage from St. Thomas, West Indies, in the Month of November last', *Lancet*, 26 March 1853, 293–9; William J. Cummins, 'The Yellow Fever in the West Indies', *Lancet*, 28 May 1853, 488–90; letter from Gavin Milroy to the *Lancet*, 9 April 1853, p. 350.
112. Gavin Milroy (1805–86), *DNB*.
113. General Board of Health, *Second Report on Quarantine*, p. 133.
114. Earls of St Germans and Malmesbury, 19 November 1852, Hansard, 3rd ser., CXXIII, cols 224–8.
115. J.C. McDonald, 'The History of Quarantine in Britain during the 19th Century', *Bulletin of the History of Medicine*, 25 (1951), 39.
116. Baldwin, *Contagion*, p. 150.
117. For example, the *Eclair* incident was mentioned by the Earl of St Germans when criticizing the imposition of a quarantine vessels from the Baltic in 1852. See 19 November 1852, Hansard, 3rd ser. CXXIII, col. 227.
118. See the *Lancet*, 16 July 1853, p. 50.
119. Letter from Gavin Milroy, *Lancet*, 6 August 1853, p. 121.
120. Cummins, 'Yellow Fever in the West Indies', p. 490.
121. Wiblin and Harvey, 'An Account of Yellow Fever'; T. Bacon Phillips, 'Yellow Fever'; news item in the *Lancet*, 22 January 1853, p. 99.
122. Dispatch from Sir W.M.G. Colebrook, Governor of Barbados, to Rt.-Hon. Sir John Packington, M.P., 15 January 1853, F.552, GCRO.
123. Capt. E.P. Halsted to Sir W.M.C. Colebrooke, 13 January 1853, F.552.
124. Letters from Gavin Milroy to the *Lancet*, 4 June 1853, 525–6; 1 October 1853, 324–5.
125. Baldwin, *Contagion* pp. 224–6.
126. E.g. the Bermuda epidemic of 1856. See letter from George Thomas Keele, *Lancet*, 8 November 1856; *Report on the British West Indian Conference on Quarantine*, Georgetown, Demerara: C.K. Jardine, 1888, p. 14; *Report of the Australian Sanitary Conference of Sydney, N.S.W.*, 1884, Sydney: Thomas Richards, 1884, pp. 41–2. These conferences argued for ten-day quarantines under humane conditions, using the Boa Vista epidemic as an illustration of the dangers of not introducing such measures.
127. Letter from T. Dunn, No. 2 Bt., 2/17 Regt., 8 May 1862, to his 'Dear Nicis', describing yellow fever outbreaks on Barbados in 1852 and Trinidad in 1853, and Barbados in 1855 [BRO] 29596; Henry C. Wilkinson, *Bermuda from Sail to Steam: A History of the Island from 1784 to 1901*, vol. 1, London: Oxford University Press, 1973, pp. 607–8, 662, 720.
128. Report of the Quarantine Committee, Royal College of Physicians, 9 May 1889, Call No. 2248/2 Royal College of Physicians, London [RCPL].
129. *Documents relating to British Guiana* (Demerara: Royal Gazette Office, 1848), p. iii; W.E. Riviere, 'Labour Shortage in the British West Indies after Emancipation', *Journal of Caribbean History*, 4 (1972), 1–30; William A. Green, *British Slave Emancipation: The Sugar Colonies and the Great Experiment 1830–1865*, Oxford: Clarendon Press, 1991 [1976]; Look Lai Walton, *Indentured Labor, Caribbean Sugar: Chinese and Indian Migrants to the British West Indies, 1838–1918*, Baltimore and London: Johns Hopkins University Press, 1993; Madhavi Kale, *Fragments of Empire: Capital, Slavery, and Indian Indentured Labor Migration in the British Caribbean*, Philadelphia: University of Pennsylvania Press, 1998.

Chapter 5: Yellow fever resurgent

1. Paul Gottheil, 'Historical Development of Steamship Agreements and Conferences in the American Foreign Trade', *Annals of the American Academy of Political and Social Science*, 55 (1914), 52.
2. Sidney Chalhoub, 'The Politics of Disease Control: Yellow Fever in Nineteenth-Century Rio de Janeiro', *Journal of Latin American Studies*, 25 (1993), 441–63; Teresa Meade, '"Civilizing Rio de Janeiro": The Public Health Campaign and the Riot of 1904', *Journal of Social History*, 20 (1986), 301–22.
3. John Duffy, *Sword of Pestilence: The New Orleans Fever Epidemic of 1853*, Baton Rouge: Louisiana State University Press, 1966.
4. Message from Mayor of New Orleans, 12 December 1854, p. 2; and Extract from 'Report of the Sanitary Commission on the Epidemic of Yellow Fever of 1853', p. 14, in E.H. Barton, *Care and Prevention of Yellow Fever at New Orleans and other Cities in America*, New York: H. Ballière, 1857.
5. 'The Lisbon Epidemic', *Lancet* 9 January 1858, p. 45.
6. J.B. Lyons, 'A Dublin Observer of the Lisbon Yellow Fever Epidémie', *Vesalius*, 1 (1995), 8–12.
7. 'The Lisbon Epidemic', p. 45.
8. *Lancet*, 2 January 1858, p. 24.
9. *Lancet*, 9 January 1858, p. 49.
10. 'The Lisbon Epidemic', p. 45.
11. *Lancet*, 13 March 1858, p. 281; 20 March 1858, p. 305.
12. 'Shutting the Stable-Door – Yellow Fever in Lisbon', *Lancet*, 10 April 1858, p. 378.
13. Letter from Gavin Milroy, *Lancet*, ii, 10 August 1861, pp. 146–7.
14. *Lancet*, 12 September 1861, p. 263.
15. Bertrand Hillemand, 'L Épidémie de fièvre jaune de Saint-Nazaire en 1861', *Histoire des sciences médicales*, 40 (2006), 23–36; *The Lancet*, ii, 28 September 1861, p. 310.
16. Coleman, *Yellow Fever* p. 135.
17. *Lancet*, 30 August 1862, p. 238.
18. *Lancet*, 4 October 1862, p. 378; 11 October 1862, p. 406.
19. *Lancet*, 21 June 1862, p. 679.
20. Letter from 'R.E.P.', West Indies, October 1862, *Lancet*, ii, 29 November 1862, p. 609.
21. *Lancet*, 6 December 1862, p. 635.
22. P.D. Meers, 'Yellow Fever in Swansea, 1865', *Journal of Hygiene*, 97 (1986), 185–91.
23. *Lancet*, 7 October 1865, p. 415.
24. Coleman, *Yellow Fever*, pp. 167–8.
25. *Lancet*, 17 November 1866, p. 550.
26. Ibid.
27. *Lancet*, 1 December 1866, p. 611.
28. Ibid.
29. *Lancet*, 8 December 1866, p. 641.
30. *Lancet*, 29 December 1866, p. 731.
31. *British Medical Journal*, i, 29 April 1871, p. 454.
32. Leandro Ruiz Moreno, *La Peste Historica de 1871: Fiebre Amarilla en Buenos Aires y Corrientes*, Parana: Neuva Impressora, 1949, p. 71.
33. J.M. Bustillo to J.M. Solsona, 15 December 1870; Solsona to Bustillo, 26 January 1871, reprinted ibid., pp. 156–7.
34. Ibid., p. 35.
35. *British Medical Journal*, 29 April 1871, p. 454.
36. *La Democracia*, 26 March 1871, reprinted in Moreno, *La Peste*, p. 40.
37. Ibid., pp. 37–8.
38. Cited in *British Medical Journal*, i, 20 May 1871, p. 537; see also Evergisto de Vergara, 'La epidemia de fiebre amarilla de 1871 en Buenos Aires', http://www.ieela.com.ar, accessed 14/12/10.

39. E.g. Moreno, *La Peste*, pp. 35–8.
40. *British Medical Journal*, 10 June 1871, p. 617.
41. *British Medical Journal*, 3 June 1871, p. 591.
42. 'Yellow Fever', *Encylopaedia Britannica* (1885), pp. 734–6.
43. *British Medical Journal*, 12 April 1873, p. 412.
44. Julyan G. Peard, *Race, Place, and Medicine: The Idea of the Tropics in Nineteenth-Century Brazilian Medicine*, Durham, NC and London: Duke University Press, 1999.
45. *Gazeta Medica da Bahia*, 15 February and 15 March 1973, cited in *British Medical Journal*, i, 3 May 1873, p. 485.
46. Jaime Larry Benchimol, *Pereira Passos: Um Haussmann Tropical*, Rio de Janeiro: Biblioteca Carioca, 1990, pp. 49–50.
47. Proust, *Essai sur l'hygiène*, p. 183.
48. Benchimol, *Pereira Passos*, pp. 65–75.
49. Chalhoub, 'The Politics of Disease Control', pp. 455–9.
50. *British Medical Journal*, 5 September 1874, p. 309.
51. *British Medical Journal*, 1 May 1875, p. 582.
52. *British Medical Journal*, 22 April 1876, pp. 515–16.
53. *British Medical Journal*, 6 May 1876, p. 572.
54. E.g. *British Medical Journal*, 30 September 1876, p. 434; ii, 24 November 1877, p. 738; i, 9 March 1878, p. 339.
55. *British Medical Journal*, 15 July 1876, p. 83; i, 12 May 1877, p. 588.
56. *British Medical Journal*, 13 April 1876, p. 540.
57. *British Medical Journal*, 24 November 1877, p. 738; i, 9 March 1878, p. 339.
58. Letter from Surg.-Maj. George A. Hutton, *British Medical Journal*, i, 8 June 1878, pp. 847–8.
59. *British Medical Journal*, 20 October 1877, p. 573.
60. Margaret Humphreys, *Yellow Fever and the South*, Baltimore and London: Johns Hopkins University Press, 1992, pp. 55–6.
61. Ibid., p. 61.
62. Ibid., p. 63.
63. Ibid., p. 69.
64. Editorial, *British Medical Journal*, 21 September 1878, p. 426.
65. Editorial, *British Medical Journal*, 2 November 1878, pp. 667–8; Katrina E. Towner, 'A History of Port Health in Southampton, 1825 to 1919', University of Southampton Ph.D. thesis, 2008.
66. Patterson, 'Yellow Fever Epidemics' p. 683.
67. Humphreys, *Yellow Fever*, pp. 77–111.
68. Joseph Holt, *Quarantine and Commerce: Their Antagonism destructive to the Prosperity of City and State. A Reconciliation an imperative Necessity. How this may be Accomplished*, New Orleans: L. Graham & Sons, 1884, p. 19.
69. E.g. J.M. Keating, *A History of the Yellow Fever: The Yellow Fever Epidemic of 1878, in Memphis, Tenn.*, Memphis: Howard Association, 1879, pp. 291, 325.
70. Humphreys, *Yellow Fever*, pp. 13–15.
71. E.g. H.D. Schmidt, *The Pathology and Treatment of Yellow Fever; with some Remarks upon the Nature of its Cause and its Prevention*, Chicago: Chicago Medical Press Association, 1881, pp. 231, 234, 236; John Gamgee, *Yellow Fever: A Nautical Disease. Its Origin and Prevention*, New York: D. Appleton & Co., 1879.
72. Robert C. Keith, *Baltimore Harbor*, Baltimore and London: Johns Hopkins University Press, 2005, pp. 7–8.
73. Quarantine record for 1881–1918, RG 19 S2, Box 29, Baltimore City Archives [BCA], sample of records for 1883 and 1884.
74. Gamgee, *Yellow Fever*, pp. 193–4.
75. State Dept. memorandum, 29 July 1880, RG 43, International Sanitary Conference, 1881, Box 1, National Archives and Records Administration [NARA], College Park, MD, USA.

76. Memorandum from Mr Frank, State Dept., 30 July 1880, RG 43, Box 1, NARA.
77. Protocol I and attachment, Proceedings of the International Sanitary Conference, Washington, DC, 5–24 January, 1881, RG 43, Box 1, NARA.
78. Humphreys, *Yellow Fever*, pp. 119–27.
79. Act March 27, 1890, *An Act to prevent the introduction of contagious diseases from one State to another and for the punishment of certain offences*.
80. *An Act granting additional quarantine powers and imposing additional duties upon the Marine-Hospital Service*, February 15 1893.
81. Statement by Dr A.H. Doty, New York, *Committee on Interstate and Foreign Commerce of the House of Representatives on Bills (H.R. 4363 and S. 2680) to Amend an Act entitled 'An Act granting additional Quarantine Powers and Imposing Duties upon the Marine-Hospital Service'*, Washington, DC: Govt. Printing Office, 1898, p. 4.
82. Curiously, Spooner represented the northern state of Wisconsin.
83. Statement by Dr H.B. Horlbeck, Health Officer, Charleston, SC, in *Committee on Interstate and Foreign Commerce.*, pp. 20–4.
84. Statement by Dr Joseph Y. Porter, ibid., pp. 49–50.
85. Alan M. Kraut, *Silent Travelers: Germs, Genes, and the 'Immigrant Menace'*, New York: Basic Books, 1994; idem, 'Plagues and Prejudice: Nativism's Construction of Disease in Nineteenth- and Twentieth-Century New York City', in D. Rosner (ed.), *Hives of Sickness: Public Health and Epidemics in New York City*, New Brunswick: Rutgers University Press, 1995, 65–94; Howard Markel, *Quarantine! East European Jewish Immigrants and the New York City Epidemics of 1892*, Baltimore: Johns Hopkins University Press, 1997.
86. Statement by Walter Wyman, in *Committee on Interstate and Foreign Commerce*, pp. 53–5.
87. Walter Wyman, 'Quarantine and Commerce', An Address delivered before the Commercial Club of Cincinnati, October 15th, 1898, unpublished pamphlet, RA 655 W9, Library of Congress [LOC], Washington, DC.
88. Mariola Espinosa, *Epidemic Invasions: Yellow Fever and the Limits of Cuban Independence, 1878–1930*, Chicago: University of Chicago Press, 2009.
89. Wyman, 'Quarantine and Commerce', pp. 9–11.
90. Espinosa, *Epidemic Invasions*.
91. Matthew Parker, *Panama Fever: The Battle to Build the Canal*, London: Hutchinson, 2007.
92. Paul S. Sutter, 'Nature's Agents or Agents of Empire? Entomological Workers and Environmental Change during the Construction of the Panama Canal', *Isis*, 98 (2007), 724–54; W.C. Gorgas, *Sanitation in Panama*, New York: Appleton, 1915.
93. Letter from MHS to John T. Morgan, United States Senate, 22 March 1905, Central File 1897–1923, File 1104 (Canal Zone), Box 99, RG 185, NARA.
94. Hugh V. Cumming to Surg.-Gen., PH and MHS, 11 November 1903, ibid.
95. File 14–A–X2, Colombian law establishing a quarantine station against yellow fever at Taboga, 1899, The Panama Canal – French Records, Box 168, RG 185, NARA. This quarantine station maintained a quarantine of six day' observation on all vessels arriving at Taboga.
96. Rear Adm. Henry Glass, Commander-in-Chief, Pacific Squadron, to Legation of the United States, Panama, 7 January 1904, File 1104, Box 99, RG 185, NARA.
97. Glass to W.I. Buchanan, US Envoy and Minister Plenipotentiary, Panama, 7 January 1904, ibid.
98. Buchanan to Sec. of State, 8 January 1904, ibid.
99. Sec. of State to Buchanan, 9 January 1904, ibid.
100. Claude Pierce, Asst.-Surg. US PH & MHS to Surg.-Genl. US PH & MHS, 18 January 1904, ibid.
101. Memorandum for the Secretary to the Treasury, undated [1913], ibid.
102. Memorandum for the Secretary to the Treasury, received 19 December, ibid.

103. Memorandum in connection with letter from the Secretary of the Treasury ... addressed to the President in regard to the Treasury Department handling the quarantine stations on the Panama Canal, 4 December 1913, Box 99, RG 185, NARA.
104. Executive Order, Woodrow Wilson, 14 August 1914, File 1104, Box 99, RG 185, NARA.
105. Consolidated Report of Quarantine Transactions at the Ports of Balboa-Panama and Colon-Cristobal for the three months ending 31 March 1918, Annual Health Report 1918, p. 41, Box 1, RG 185, NARA.
106. Marcos Cueto, *The Value of Health: A History of the Pan American Health Organization*, Washington, DC: Pan American Health Organization, 2001, pp. 18–19.
107. Surg.-Gen. R.H. Geel to S.B. Grubbs, Chief Quarantine Officer, Canal Zone, 11 February 1919; Chester Harding, Governor of the Panama Canal Zone, to Geel, 17 September 1919, File 1104, Box 99, RG 185, NARA.
108. *Second General International Sanitary Convention of the American Republics. Pan-American Sanitary Conference, 1905*, Washington, DC: Govt. Printing Office, 1907, pp. 1–2.
109. E.g. *Report of the Delegation from the United States of America to the Sixth International Sanitary Conference at Montevideo, December 12, 1920*, Washington, DC: Govt. Printing Office, 1920, p. 79.
110. Cueto, *The Value of Health*, p. 45.
111. Juan Guiteras, 'Remarks on the Washington Sanitary Convention of 1905, with special reference to Yellow Fever and Cholera', *American Journal of Public Health*, 2 (1912), p. 513.
112. Espinosa, *Epidemic Invasions*, pp. 103–7; Norman Howard-Jones, *The Pan American Health Organization: Origins and Evolution*, Geneva: WHO, 1981, p. 8.
113. Cueto, *The Value of Health*, p. 20.
114. Secretary of American Chamber of Commerce, Mexico, to Carter H. Glass, Sec. of the Treasury, 7 June 1919, Central File 15859, Box 762, RG 90, NARA.
115. Grubbs to Surg.-Genl. Rupert Blue, 16 April 1919, File 1104, Box 99, RG 185, NARA.
116. E.g. the British colonial government of Malaya. British Ambassador, Washington, DC, to Secretary of State, 23 July 1912, ibid.
117. S.P. James, 'The Protection of India from Yellow Fever', *Journal of Indian Medical Research*, 1 (1913), 213–57. The experience gained by James on this tour led to his appointment as a member and then as President of the Yellow Fever Committee of the Office International d'Hygiène Publique. He served on this committee for many years and later advised on means of protecting against yellow fever on air transport.
118. Espinosa, *Epidemic Invasions*, p. 121.
119. See Anne-Emanuelle Birn, *Marriage of Convenience: Rockefeller International Health and Revolutionary Mexico*, Rochester, NY: Rochester University Press, 2006, pp. 47–60.
120. Harding to Geel, 17 September 1919, File 1104, Box 99, RG 185, NARA.
121. Venezuelan Foreign Minister, Caracas, to Secretary of State, 23 March 1921, ibid.
122. W.C. Rucker, Chief Quarantine Officer, to Surg.-Gen., 4 January 1921, ibid.
123. S.B. Grubbs to Surg.-Gen., 23 June 1920, ibid.
124. Birn, *Marriage of Convenience*.

Chapter 6: A stranglehold on the East

1. Christopher Hamlin, *Cholera: The Biography*, Oxford: Oxford University Press, 2009, pp. 2–3.
2. Mark Harrison, 'A Question of Locality: The Identity of Cholera in British India, 1860–1890', in D. Arnold (ed.), *Warm Climates and Western Medicine: The Emergence of Tropical Medicine, 1500–1900*, Amsterdam: Rodopi, 1996, 133–59; Hamlin, *Cholera*, pp. 52–70.
3. Pelling, *Cholera*; Harrison, *Climates*, pp. 177–92.

4. Myron Echenberg, *Africa in the Time of Cholera: A History of Pandemics from 1817 to the Present*, Cambridge: Cambridge University Press, 2011, pp. 52-70; Gallagher, *Medicine and Power*, p. 44.
5. J. Netten Radcliffe, 'Memorandum on Quarantine in the Red Sea, and on the Sanitary Regulation of the Pilgrimage to Mecca', in *Ninth Annual Report of the Local Government Board 1879-80, Supplement containing Report and Papers submitted by the Medical Officer on the Recent Progress of Levantine Plague, and on Quarantine in the Red Sea*, London: George E. Eyre & William Spottiswoode, 1881, p. 92.
6. Amir A. Afkhami, 'Defending the Guarded Domain: Epidemics and the Emergence of an International Sanitary Policy in Iran', *Comparative Studies of South Asia, Africa and the Middle East*, 19, (1999), 130; David Arnold, 'The Indian Ocean as a Disease Zone, 1500-1950', *South Asia*, n.s. 14 (1991), 1-22.
7. Bayly, *Birth of the Modern World*.
8. See Martin H. Geyer and Johannes Paulmann (eds), *The Mechanics of Internationalism: Culture, Society, and Politics from the 1840s to the First World War*, Oxford: Oxford University Press, 2001.
9. Valeska Huber, 'The Unification of the Globe by Disease? The International Sanitary Conferences on Cholera, 1851-1894', *Historical Journal*, 49 (2006), p. 462; Arnold, 'Indian Ocean as a Disease Zone', pp. 17-18.
10. Norman Howard-Jones, *The Scientific Background to the International Sanitary Conferences, 1851-1938*, Geneva: World Health Organization, 1975; William F. Bynum, 'Policing Hearts of Darkness: Aspects of the International Sanitary Conferences', *History and Philosophy of the Life Sciences*, 15 (1993), 421-34.
11. Christopher Hamlin, 'Politics and Germ Theories in Victorian Britain: The Metropolitan Water Commissions of 1867-9 and 1892-3', in R. MacLeod (ed.), *Government and Expertise: Specialists, Administrators and Professionals, 1860-1919*, Cambridge: Cambridge University Press, 1988, 110-27.
12. A. H. Leith, *Abstract of the Proceedings and Reports of the International Sanitary Conference of 1866*, Bombay: Press of the Revenue, Financial and General Departments of the Secretariat, 1867, pp. 48-53.
13. Ibid., p. 94.
14. Ibid., pp. 66-85.
15. Ibid., pp. 92-105. For the sanitary aspects of the Haj, see also Mark Harrison, *Public Health in British India: Anglo-Indian Preventive Medicine 1859-1914*, Cambridge: Cambridge University Press, 1994, chap. 5; William R. Roff, 'Sanitation and Security: The Imperial Powers and the Nineteenth-Century Hajj', in R. Serjeant and R.L. Bidwell (eds), *Arabian Studies, VI*, Cambridge: Cambridge University Press, 1982, 143-60; Saurabh Mishra, *Pilgrimage, Politics, and Pestilence: The Haj from the Indian Subcontinent 1860-1920*, New Delhi: Oxford University Press, 2011; idem, 'Beyond the Bounds of Time? The Haj Pilgrimage from the Indian Subcontinent, 1865-1920', in B. Pati and M. Harrison (eds), *The Social History of Health and Medicine in Colonial India*, London: Routledge, 2009, pp. 31-44.
16. Extract from Proc. of Government of India (Home Dept.) Resolution of 21 January 1868, Proc. 75, Govt. of Bengal, Genl. Dept., February 1868, West Bengal State Archives, Kolkata [WBSA].
17. Memorandum on establishment of a quarantine station near Mokha; remarks on Dr Dickson's letter of 4th June to H.M. Ambassador, Proc. 77 (Medical), Govt. of Bengal, Genl. Dept., February 1868, WBSA.
18. Surg.-Maj. O. Turner, 20 March 1871 to Political Resident, Aden, Home (Agriculture, Revenue and Commerce) No. 3 (A), 15 July 1871, National Archives of India, New Delhi [NAI].
19. Government of Bombay to Secretary of State for India, 27 June 1871, Home (Agriculture, Revenue and Commerce), No. 3 (A), 15 July 1871, NAI.
20. Report by Asst. Political Resident, 12 September 1872, Proc. No. 7, Quarantine – General Department, R/20/A/399, APAC, BL.

21. Netten Radcliffe, 'Memorandum on Quarantine'.
22. Ibid., p. 103.
23. H.L. Harrison, Jr. Sec. to the Government of Bengal, to Capt. H. Howse, Officiating Master Attendant, port of Calcutta, 25 September 1867, No. 59, September 1867; Howse to Officiating Sec. to Government of Bengal, 20 August 1867, No. 57, September 1867, WBSA.
24. *Fourth Report of the Sanitary Commissioner with the Government of India 1867*, Calcutta: Office of the Superintendent of Govt. Printing, 1868.
25. Sheldon Watts, 'From Rapid Change to Stasis: Official Responses to Cholera in British-Ruled India and Egypt: 1860–c.1921', *Journal of World History*, 12 (2001), 349–51.
26. Harrison, 'A Question of Locality'.
27. Harrison, *Public Health*, chap. 3.
28. Ibid., pp. 107–8.
29. Despite numerous entreaties from Cuningham, the Government of India did not overturn the decision to impose military cordons in the Punjab in 1873. In this decision it was supported by both the Inspector-General of the Indian Medical Department and the Sanitary Commissioner with the Government of Bombay. See Proc. 511, Government of Bengal, Genl. Dept., 1873, Maharashtra State Archives, Mumbai [MSA].
30. See B.R. Tomlinson, *The Economy of Modern India, 1860–1970*, Cambridge: Cambridge University Press, 1993, chaps 2–3; Tapan Raychaudhuri, Dharma Kumar and Meghnad Desai (eds), *The Cambridge Economic History of India*, vol. 2, Cambridge: Cambridge University Press, 1980, pp. 572–7.
31. 'Quarantine at Aden', Government of Bombay No. 1223–43, 27 April 1874, Home (Sanitary), No. 31 (A), June 1874, NAI.
32. A good example is the reporting of cholera outbreaks in Nubia and the quarantines imposed at Suez against shipping from western Arabian ports in 1872. Information from the Upper Nile seems to have been incomplete or inaccurate, resulting in fluctuating quarantine periods. These were communicated to the Government of Bombay via the British consul in Alexandria and the Political Resident in Aden (mostly by telegraph). See Edward Stanton, Consul-General, Alexandria, to Brig.-Gen. J.W. Schneider, Political Resident, Aden, 6 September; Schneider to Stanton, 7 September; Schneider to Secretary to Government of Bombay, 10 September; Stanton to Schneider, 28 October; Schneider to Secretary to Government of Bombay, 29 October; Stanton to Schneider, 16 November; Schneider to Secretary to the Government of Bombay, 16 November, R/20/A/399, APAC, BL.
33. See marginal notes to memo. From Assistant-Resident, Aden, to Harbour Master, 17 July 1872, Proc. No. 7, Quarantine, R/20/A/399, APAC, BL.
34. Secretary of State for India to Government of India, forwarding request from Secretary of State for Foreign Affairs, 21 May 1874, Home (Sanitary), No. 7 (A), July 1874, NAI.
35. Mr H. Calvert, to Major-General Edward Stanton, 27 June 1874, Nos 31–2 (A) Home (Sanitary) September 1874, NAI.
36. A.C. Lyell, memo., 7 September 1874, Nos 31–2 (A) Home (Sanitary), Nos 31–2 Home (Sanitary), September 1874, NAI.
37. See Mike Davis, *Late Victorian Holocausts*, London: Verso, 2001.
38. Viceroy to Secretary of State, 19 September 1874, Nos 31–2 (A), Home (Sanitary), September 1874, NAI.
39. J.M. Cuningham, memo., 4 September 1874, Nos 31–2 (A) Home (Sanitary), September 1874, NAI.
40. Health Officer to Commissioner of Police, 12 January 1871, No. 430, Govt. of Bombay, Genl. Dept., Vol. 10, 1871, MSA.
41. Government of Bombay to Secretary of State, 26 October 1874, Nos 2–4 (A), Home (Sanitary), December 1874, NAI.
42. Port Surgeon to Political Resident Aden, 20 September 1872, Proc. No. 7; Resident to Port Surgeon, 'Quarantine – General Rules for Aden', 8 October 1872, R/20/A/399, APAC, BL.

43. Political Resident, Aden, to Sec. to Govt. of Bombay, 25 February 1873, Govt. of Bombay, Genl. Dept., Vol. 70, MSA.
44. Government of Bombay to Government of India, 19 August 1874, Nos 29–31 (A), Home (Sanitary), November 1874, NAI.
45. J.M. Cuningham, memo., 18 November 1874, Nos 2–4 (A), Home (Sanitary), December 1874, NAI.
46. Howard-Jones, *The Scientific Background of the International Sanitary Conferences*, pp. 35–40.
47. J.D. Tholozan, *Une Épidémie de peste en Mésopotamie en 1867*, Paris: Victor Masson & Fils, 1869.
48. Dispatch from Secretary of State, 17 June 1875, No. 22 (A), Home (Sanitary), October 1875, NAI.
49. Cuningham, memo. 19 August 1875, No. 22 (A), Home (Sanitary), October 1875, NAI.
50. Messrs Gray, Dawes & Co. to Secretary of State for Foreign Affairs, 12 May 1875, No. 23 (A), Home (Sanitary), October 1875, NAI.
51. Messrs Gray, Dawes & Co. to Secretary of State for Foreign Affairs, 25 May 1875, No. 23 (A) Home (Sanitary), October 1875, NAI.
52. Government of India to Secretary of State, 7 October 1875, No. 25 (A) Home (Sanitary), October 1875, NAI.
53. 'Proposed Quarantine Rules for Bombay' – response of J.M. Cuningham, 15 October 1875, No. 34 (A) Home (Sanitary), November 1875, NAI.
54. A.P. Howell, memos, 19 October and 23 October 1875; Howell to Secretary to Government of Bombay, 25 November 1875, No. 35 (A) Home (Sanitary), November 1875, NAI.
55. Hon. E.W. Ravenscroft to Officiating Secretary, Government of India, Home Department, 11 July 1876, No. 34 (A) Home (Sanitary), July 1876, NAI.
56. Government of Bombay Resoln., Government of Bombay, Home Department Proc. No. 8, March 1876, P/1106, APAC, BL.
57. J.M. Cuningham, memorandum on 'Plague and Quarantine', 15 June 1876, No. 47 (A) Home (Sanitary), July 1876, NAI.
58. Note by 'E.C.B.', 15 June 1876, No. 47 (A), July 1876, NAI.
59. Note by 'A.P.H.' (Arthur Howell), (Home Dept.), 17 June 1876, expressing Cuningham's concerns, No. 47 (A), July 1876, NAI.
60. Note by 'E.C.B.', 17 June 1876, No. 47 (A), July 1876, NAI.
61. Col. A.B. Kemball, H.M. Consul, Baghdad, to William Street H.M. Chargé d'Affaires, Constantinople, 28 September 1864, Foreign (Political), No. 267 (A), December 1864, NAI.
62. Extract from a Report by Her Majesty's Vice-Consul at Basra, 15 September 1864, Foreign (Political), No. 267 (A), December 1864, NAI.
63. W.P. Johnston, Basra, to C. Gonne, Secretary to the Government of Bombay, 16 January 1866; Notification, Foreign Department, No. 177, Fort William, 22 February 1866, Foreign (Political), Nos 151–2 (A), February 1866, NAI.
64. Foreign (General), No. 33 (B), December 1875, NAI.
65. Afkhami, 'Defending the Guarded Domain', pp. 131–2.
66. E.g. 'Establishment of quarantine at Bushire consequent on rumours of plague in Turkish Arabia', Foreign (General) August 1874, Nos 7–8, NAI; ditto Shiraz and Ispahan, Foreign (General), Nos 110–13 (B), January 1875, NAI.
67. Col. Herbert, Baghdad, 20 October 1870 to Charles Alison, Teheran, Foreign (Political), No. 244 (A), January 1871, NAI.
68. Charles Alison to Herbert, 25 October, Foreign (Political), No. 245 (A), January 1871; Alison to H. Ongley, 26 October, Foreign (Political), No. 246 (A), NAI.
69. Charles Alison, British Consul, Teheran, to Earl Granville, 30 October 1870, Foreign (Political), January 1871, No. 243 (A), NAI.
70. Afkhami, 'Defending the Guarded Domain'.

71. Col. J.P. Nixon, Political Agent and Consul-General in Turkish Arabia, to Secretary of State for Foreign Affairs, Baghdad, 15 March 1875, forwarding Baghdad Trade Report for Year ending 12 March 1875, Nos 191–5 (A) Foreign (Political), June 1875, NAI.
72. Cyril Elgood, *Medicine in Persia*, New York: Paul B. Hoeber, 1934, p. 76; Howard-Jones, *International Sanitary Conferences*, p. 40.
73. J.D. Tholozan, *Prophylaxie du choléra en orient: l'hygiène et la réforme sanitaire en Perse*, Paris: Victor Masson & Fils, 1869, pp. 32, 35, 43.
74. Ibid., p. 42.
75. Thomson to Viceroy, 16 April 1876, Government of Bombay Home (Sanitary), Proc. 218, May 1876, P/1106, APAC, BL.
76. Telegram from Foreign Secretary, Government of India, to Resident, Bushire, 20 April 1876, ibid.
77. Sir J. Dickson, memo., 14 April 1876; J. Ibrahim, Persian Minister for Foreign Affairs, to Mr Taylour Thomson, H.M. Legation, Tehran, 7 April 1876, No. 30 (A) Home (Sanitary), April 1876, NAI.
78. See James F. Stark, 'Industrial Illness in Cultural Context: *La maladie de Bradford* in Local, National and Global Settings, 1878–1919', University of Leeds Ph.D. thesis, 2011.
79. Letter from Political Resident, Aden, to Secretary to the Government of Bombay, Government of Bombay, Home Dept. (Sanitary), Proc. 286, July 1876, APAC, BL.
80. Commissioner of Customs, Bombay, 4 July 1876 and Government of Bombay Resolution, Proc. 287, Home (Sanitary), July 1876, APAC, BL.
81. Dr J.R.S. Dickson, 2 January 1877; Minutes of meeting of Tehran Board of Health, 14 January 1877, No. 9 (A) Home (Sanitary), January 1878, NAI.
82. Lt-Col. W.F. Prideaux to Secretary to Government of India, Foreign Department, 9 December 1876, No. 13 (A) Home (Sanitary), January 1878, NAI.
83. Duc Decazes to Count Wimffeu, Versailles, 30 October 1876, No. 76 (A) Home (Sanitary), October 1876, NAI.
84. Earl Derby, Foreign Secretary, to Count Buest, Austrian Ambassador, London, 13 July 1877, No. 78 Home (Sanitary), July 1876, NAI.
85. Derby to Buest, 7 December 1877, No. 96 (A) Foreign (General), April 1878, NAI.
86. Dr W.H. Colvill to Derby, 22 November 1877, Nos 61–100 (A) Foreign (General), April 1878, NAI.
87. Procès Verbaux du Conseil du Santé [PVCS], 26 June 1877, No 82 Foreign (General), April 1878, NAI.
88. PVCS, 3 July 1877, No. 83 (A) Foreign (General), April 1878, NAI.
89. W. Taylour Thomson to Secretary of State for Foreign Affairs, 28 January 1878; Mr Churchill, H.M. Consul at Resht, to Thomson, 24 December 1877, 7 January 1878, No. 11 (A) Home (Sanitary), April 1878, NAI.
90. Nixon to T.H. Thornton, Secretary to the Government of India, Foreign Department, 23 July 1877, No. 29 (A) Foreign (General), April 1878, NAI.
91. W.H. Colvill to H.M. Political Agent and Consul-General in Turkish Arabia, 16 March 1876, No. 37 (A) Foreign (General), April 1878, NAI.
92. Henry H. Calvert to Frank C. Lascelles, 17 September 1878, Foreign Office [FO] 881/4003, TNA.
93. Extracts from the *Procès-Verbal of the Consul Superieur de Santé*, 10 April and 17 April 1877, FO 881/3331, TNA.
94. Viceroy in Council to Salisbury, 4 January 1878, No. 26 (A) Home (Sanitary), April 1878, NAI.
95. Government of India to Secretary of State for India, 4 January 1878, No. 31 (A) Foreign (General), April 1878, NAI.
96. Towner, 'A History of Port Health in Southampton', (thesis), 2008; Anne Hardy, 'Cholera, Quarantine and the English Preventive System', *Medical History*, 37 (1993), 252–69.

97. Sir Louis Mallet, Under-Secretary of State for India, to Lord Tenterden, Under-Secretary of State for Foreign Affairs, 12 December 1877, No. 88 (A) Foreign (General), April 1878, NAI.
98. Sir Julian Pauncefote, Under-Secretary of State for Foreign Affairs to Mallet, 26 November 1877; Charles Cookson, Consul, Alexandria, to Derby, 9 November 1877, Nos 90–1 (A) Foreign (General), April 1878, NAI.
99. Henry Calvert to Cookson, 2 November 1877, No. 92 (A) Foreign (General), April 1878, NAI.
100. Cookson to Derby, 9 November 1877, No. 91 (A) Foreign (General), April 1878, NAI.
101. J.M. Cuningham, 6 September 1878, No. 5 (A) Home (Sanitary), December 1878, NAI.
102. Ravenscroft to Officiating Secretary to the Government of India, 23 September 1878, No. 8 (A) Home (Sanitary), December 1878, NAI.
103. Government of India to Secretary of State, 6 December 1878, No. 16 (A) Home (Sanitary), December 1878, NAI.
104. J.M. Cuningham, memo., 19 August 1878, No. 5 (A) Home (Sanitary), December 1878, NAI.
105. T.V. Lister, Foreign Office, to the Hon. C. Vivian, HM Consulate, Alexandria, 17 May 1878, No. 6 (A) Home (Sanitary), December 1878, NAI .
106 C. Vivian to Marquis of Salisbury, 18 August 1878, FO 881/4003, TNA; Louis Mallet to Under-Secretary of State for Foreign Affairs, 31 July 1878, No. 8 (A) Home (Sanitary), December 1878, NAI.
107. Copy of Home Department Resolution, 19 September 1878, No. 7 (A) Foreign (General), May 1879, NAI.
108. Secretary of State to Government of India, 20 February 1879, No. 7 (A) Foreign (General), May 1879, NAI.
109. 'C.B.,' memo., 11 March 1879; Government of India to Secretary of State, 16 June 1879, Nos 17–38 (A) Home (Sanitary) June 1879, NAI.
110. Appendices A and B, No. 7 (A) Foreign (General), May 1879, NAI.
111. Secretary to the Local Government Board to Lord Tenterden, Foreign Office, 18 June 1880, FO 881/4250X, TNA.
112. Akira Iriye, *Cultural Internationalism and World Order*, Baltimore and London: Johns Hopkins University Press, 1997.
113. Anthony Howe, 'Free Trade and Global Order: The Rise and Fall of a Victorian Vision', in Duncan Bell (ed.), *Victorian Visions of Global Order: Empire and International Relations in Nineteenth-Century Political Thought*, Cambridge: Cambridge University Press, 2007, pp. 26–46.
114. Howard-Jones, *International Sanitary Conferences*, p. 56.
115. *Protocoles et Procès-Verbaux de la Conférence Sanitaire Internationale de Rome inaugurée le 20 mai 1885*, Rome: Imprimerie du Ministère des Affairs Étrangères, 1885, pp. 17–18, 21–2.
116. Harrison, *Public Health*, pp. 126–31; Howard-Jones, *International Sanitary Conferences*, pp. 56–7, 62–4.

Chapter 7: Plague and the global economy

1. W.J.R. Simpson, *The Croonian Lectures on Plague*, London: Royal College of Physicians, 1907, p. 21.
2. Carol Benedict, *Bubonic Plague in Nineteenth-Century China*, Stanford, Calif.: Stanford University Press, 1996, chaps 1–2.
3. 'Warming up for a Plague Outbreak', *New Scientist*, 2566, 26 August 2006, 18.
4. HM Minister, Lisbon, to Foreign Office, 16 May 1894, *Correspondence relative to the Outbreak of Bubonic Plague at Hong Kong*, London: HMSO, 1894.
5. W.J. Simpson, *Report on the Causes and Continuance of Plague in Hongkong and Suggestions as to Remedial Measures*, London: Waterlow & Sons, 1903, p. 5.

6. Consul at Pakhoi to Colonial Secretary, 16 and 25 May 1894, *Correspondence relative to the Outbreak of Bubonic Plague*.
7. Kerrie L. MacPherson, 'Invisible Borders: Hong Kong, China and the Imperatives of Public Health', in M.L. Lewis and K.L. MacPherson (eds), *Public Health in Asia and the Pacific: Historical and Comparative Perspectives*, London: Routledge, 2008, pp. 14–15.
8. Sir William Robinson, Governor of Hong Kong, to Lord Ripon, 16 June 1894, *Correspondence relative to the Outbreak of Bubonic Plague*.
9. Extract from a report by Dr James A. Lowson, Govt. Civil Hospital, 16 May 1894, ibid.; Alexander Rennie, *Report on the Plague Prevailing in Canton during the Spring and Summer of 1894*, Shanghai: Chinese Imperial Maritime Customs, 1895, p. 5.
10. Robinson to Ripon, 23 May 1894, *Further Correspondence relative to the Outbreak of Bubonic Plague at Hong Kong*, London: HMSO, 1894.
11. Robinson to Ripon, 20 June 1894, ibid.
12. Ibid.
13. Ibid.
14. Robinson to Ripon 22 June 1894, ibid. On Kitasato and Yersin in Hong Kong, see: Edward Marriot, *The Plague Race: A Tale of Fear, Science and Heroism*, London: Picador, 2002; Mahito Fukuda, *Netsu to Makoto ga areba*, Kyoto: Minerva, 2008.
15. Robinson to Ripon, 20 June 1894, *Further Correspondence*.
16. Robinson to Ripon, 7 July 1894, ibid.
17. Robinson to Joseph Chamberlain, Colonial Secretary, 6 May 1896, *Hong Kong: Bubonic Plague*, London: HMSO, 1896, pp. 3–7.
18. Myron Echenberg, *Plague Ports: The Global Urban Impact of Bubonic Plague 1894–1901*, New York: New York University Press, 2007, p. 26.
19. E.g. the port of Chefoo (Yantai) on the Yellow Sea took no precautions despite the death of between 500,000 and a million people in neighbouring provinces during the cholera epidemic of 1888. However, some treaty ports did provide vaccination against smallpox, funded by private subscriptions. See China Imperial Maritime Customs, *Decennial Reports on the Trade, Navigation, Industries, etc., of the Ports open to Foreign Commerce in China and Corea, and on the Condition and Development of the Treaty Port Provinces, 1882–91*, Shanghai: Inspector General of Customs, 1893, p. 59.
20. Benedict, *Bubonic Plague*, p. 151. On the development of public health in Shanghai before 1894, see Kerrie L. MacPherson, *A Wilderness of Marshes: The Origins of Public Health in Shanghai, 1843–1893*, London: Oxford University Press, 1987.
21. China Imperial Maritime Customs, *Decennial Reports on the Trade, Navigation, Industries, etc., of the Ports open to Foreign Commerce in China and Corea, and on the Condition and Development of the Treaty Port Provinces, 1892–1901: I. Northern and Yangtze Ports*, Shanghai: Inspectorate General of Customs, 1904, pp. 472, 500–1.
22. Simpson, *Report*, p. 104.
23. Echenberg, *Plague Ports*, p. 46
24. Benedict, *Bubonic Plague*, pp. 152–5.
25. Ruth Rogaski, *Hygienic Modernity: Meanings of Health and Disease in Treaty-Port China*, Berkeley: University of California Press, 2004, p. 188.
26. E.g. the possibility of infection from endemic plague centres in the Himalayas. See Snow, *Report*, pp. 1–2.
27. Echenberg, *Plague Ports*, p. 70.
28. For the impact of the franchise on modernization in Bombay, see Sandip Hazareesingh, *The Colonial City and the Challenge of Modernity: Urban Hegemonies and Civic Contestations in Bombay (1900–1925)*, Hyderabad: Orient Longman, 2007.
29. Mridula Ramanna, *Western Medicine and Public Health in Colonial Bombay*, Hyderabad: Orient Longman, 2002, chap. 3.
30. R. Nathan, *The Plague in India, 1896, 1897*, Simla: Govt. Central Printing Office, 3 vols (1898), vol. 1, p. 355.
31. Harrison, *Public Health*, pp. 134–5.

32. Snow, *Report*, pp. 1–3; Nathan, *The Plague*, vol. 1, p. 356.
33. Snow, *Report*, p. 4.
34. See Ramanna, *Western Medicine and Public Health in Colonial Bombay*.
35. Snow, *Report*, pp. 8–9.
36. Nathan, *The Plague*, vol. 1, pp. 409–10.
37. Ibid., pp. 413, 416.
38. Harrison, *Public Health*, p. 134.
39. Nathan, *The Plague*, vol. 1, p. 419.
40. Harrison, *Public Health*, pp. 142–3.
41. Mark Harrison, 'Towards a Sanitary Utopia? Professional Visions and Public Health in India, 1880–1914', *South Asia Research*, 10 (1990), 26–7.
42. See David Arnold, 'Touching the Body: Perspectives on the Indian Plague, 1896–1900', in R. Guha (ed.), *Subaltern Studies V: Writings on South Asian History and Society*, Delhi: Oxford University Press, 1987, 55–90; R. Chandarvarkar, 'Plague Panic and Epidemic Politics in India, 1896–1914', in T. Ranger and P. Slack (eds), *Epidemics and Ideas*, Cambridge: Cambridge University Press, 1992, 203–40.
43. I.J. Catanach, 'Plague and the Tensions of Empire: India, 1896–1918', in D. Arnold (ed.), *Imperial Medicine and Indigenous Societies*, Manchester: Manchester University Press, 1988, 149–71.
44. Harrison, *Public Health*, p. 135.
45. Ibid., pp. 148–9.
46. Ibid., p. 135.
47. Nathan, *The Plague*, vol. 1, pp. 436–7.
48. A.W. Wakil, *The Third Pandemic of Plague in Egypt: Historical, Statistical and Epidemiological Remarks on the First Thirty-Two Years of its Prevalence*, Cairo: Egyptian University, 1932, p. 20.
49. Frederico Viñas y Cusí y Rosendo de Grau, *La Peste Bubónica Memoria sobre la Epidemia occurrida en Porto en 1899, por Jaime Ferran*, Barcelona: F. Sanchez, 1907, pp. 242–3.
50. Echenberg, *Plague Ports*, p. 80.
51. J. Lane Notter, 'International Sanitary Conferences of the Victorian Era', *Transactions of the Epidemiological Society of London*, 17 (1897–8), 13–14.
52. Marc Armand Ruffer, 'Measures taken at Tor and Suez against Ships coming from the Red Sea and the Far East', *Transactions of the Epidemiological Society of London*, 19 (1899–1900), 25–47.
53. *British Medical Journal*, 11 March 1899, p. 626.
54. *British Medical Journal*, 17 June 1899, p. 1506.
55. Wakil, *The Third Pandemic of Plague in Egypt*, pp. 29–32, 130.
56. Ibid., p. 28; F.M. Sandwith, *The Medical Diseases of Egypt*, London: Henry Kimpton, 1905, p. 168.
57. *British Medical Journal*, 27 May 1899, p. 1304.
58. *British Medical Journal*, 3 June 1899, p. 1351.
59. *British Medical Journal*, 20 May 1899, p. 1294.
60. Echenberg, *Plague Ports*, pp. 86–91.
61. Wakil, *The Third Pandemic of Plague in Egypt*, p. 85.
62. *British Medical Journal*, 3 June 1899, p. 1351.
63. *British Medical Journal*, 24 June 1899, p. 1552; 12 August 1899, p. 435.
64. *British Medical Journal*, 3 June 1899, p. 1351.
65. *British Medical Journal*, 1 July 1899, p. 94.
66. *British Medical Journal*, 8 July 1899, p. 102; 16 September 1899, p. 745.
67. Wakil, *The Third Plague Pandemic in Egypt*, p. 86; Sandwith, *The Medical Diseases of Egypt*, pp. 170–3.
68. De Grau, *La Peste*, p. 79.
69. A. Shadwell, 'Plague at Oporto', *Transactions of the Epidemiological Society of London*, 19 (1899–1900), p. 51.

70. A. Calmette and A.T. Salimberi, 'La Peste Bubonique étude de l'Épidémie d' Oporto en 1899', Sérothérapie', *Annales de l'Institut Pasteur*, 13 (December 1899), 865–936; pp. 868–70.
71. Calmette and Salimberi, 'La Peste Bubonique', pp. 870–1.
72. *British Medical Journal*, 19 August 1899, p. 498.
73. José Verdes Montenegro, *Bubonic Plague: Its Course and Symptoms and Means of Prevention and Treatment, according to the latest Scientific Discoveries; including notes on Cases in Oporto*, trans. W. Munro, London: Ballière, Tindall & Cox, 1900, pp. 49, 51.
74. Ibid., pp. 45, 49.
75. Ibid., p. 54.
76. *British Medical Journal*, 7 October 1899, p. 954.
77. *British Medical Journal*, 19 August 1899, p. 498; 2 September 1899, p. 621.
78. Echenberg, *Plague Ports*, p. 112.
79. *British Medical Journal*, 7 October 1899, p. 954.
80. *British Medical Journal*, 16 September 1899, p. 745; 23 September 1899, p. 808.
81. *British Medical Journal*, 23 September 1899, p. 808; 30 September 1899, p. 877.
82. *British Medical Journal*, 14 October 1899, p. 1039.
83. Shadwell, 'Plague at Oporto', p. 50.
84. *British Medical Journal*, 21 October 1899, p. 1127.
85. *British Medical Journal*, 9 September 1899, p. 681.
86. *British Medical Journal*, 14 October 1899, p. 1040.
87. *British Medical Journal*, 21 October 1899, p. 1127.
88. Echenberg, *Plague Ports*, p. 128.
89. *British Medical Journal*, 23 September 1899, p. 809.
90. Peter Curson and Kevin McCracken, *Plague in Sydney: The Anatomy of an Epidemic*, Kensington: University of New South Wales Press, 1989, pp. 33–4.
91. *British Medical Journal*, 23 September, p. 808.
92. *British Medical Journal*, 29 July 1899, p. 292.
93. Simpson, *Plague*, p. 21. On plague in Cape Town see: Molly Sutphen, 'Striving to be Separate? Civilian and Military Doctors in Cape Town during the Anglo-Boer War', in R. Cooter, M. Harrison and S. Sturdy (eds), *War, Medicine and Modernity*, Stroud: Sutton, 1998, 48–64; M.W. Swanson, 'The Sanitation Syndrome: Bubonic Plague and Native Policy in the Cape Colony, 1900–1909', *Journal of African History*, 18 (1977), 387–410.
94. *British Medical Journal*, 2 September 1899, p. 64.
95. Krista Maglen, 'A World Apart: Geography, Australian Quarantine, and the Mother Country', *Journal of the History of Medicine and Allied Sciences*, 60 (2005), 196–217; Katherine Foxhall, 'Disease at Sea: Convicts, Emigrants, Ships and the Ocean in the Voyage to Australia, c.1830–1860', University of Warwick Ph.D. thesis, 2008.
96. Warwick Anderson, *The Cultivation of Whiteness: Science, Health and Racial Destiny in Australia*, Melbourne: Melbourne University Press, 2002, pp. 93–4; Alison Bashford, *Imperial Hygiene: A Critical History of Colonialism, Nationalism and Public Health*, Basingstoke: Macmillan, 2004, pp. 115–36.
97. Katherine Foxhall, 'Fever, Immigration and Quarantine in New South Wales, 1837–1840', *Social History of Medicine*, 24 (2011), 624–43; Krista Maglen, '"In this Miserable Spot Called Quarantine": The Healthy and Unhealthy in Nineteenth-Century Australian and Pacific Quarantine Stations', *Science in Context*, 19 (2006), 317–36; Alan Mayne, '"The Dreadful Scourge": Responses to Smallpox in Sydney and Melbourne', in R. MacLeod and M. Lewis (eds), *Disease, Medicine, and Empire: Perspectives on Western Medicine and the Experience of European Expansion*, London: Routledge, 1988, 219–41.
98. John Ashburton Thompson, 'A Contribution to the Aetiology of Plague', *Journal of Hygiene*, 1 (1901), p. 154.
99. Curson and McCracken, *Plague in Sydney*, pp. 116–45; John Ashburton Thompson, *Report of the Board of Health on a Second Outbreak of Plague at Sydney, 1902*, Sydney: William Applegate Gullick, 1903, p. 4.

100. Milton J. Lewis, 'Public Health in Australia from the Nineteenth to the Twenty-First Century', in M.J. Lewis and K.L. MacPherson (eds), *Public Health in Asia and the Pacific; Historical and Comparative Perspectives*, London: Routledge, 2008, p. 224.
101. Curson and McCracken, *Plague in Sydney*, pp. 146–56.
102. Thompson, *Report*, p. 3.
103. Curson and McCracken, *Plague in Sydney*, p. 157.
104. Thompson, *Report*, pp. 10–11. On the outbreaks in Queensland and Western Australia, see Milton J. Lewis, *The People's Health: Public Health in Australia, 1788-1950*, Westport, Conn.: Praeger, 2003, pp. 129–31.
105. Thompson, *Report*, p. 4.
106. Echenberg, *Plague Ports*; Patrick Zylberman, 'Civilizing the State: Borders, Weak States and International Health in Modern Europe', in A. Bashford (ed.), *Medicine at the Border: Disease, Globalization and Security, 1850 to the Present*, Basingstoke: Palgrave Macmillan, 2006, p. 24.
107. Convention Sanitaire Internationale, Chap.II, Sec.III, Art. 20, *Bulletin de l'Office International d'Hygiène Publique* [*BOIHP*], 1 (1909), p. 16.
108. Ibid., p. 47.
109. *The International Sanitary Convention of Paris, 1903*, trans. Theodore Thomson, London: HMSO, 1904, p. 10.
110. Ibid., p. 32.
111. Acting-Governor, H. Bryan, introductory remarks, *Gold Coast. Report for 1908*, London: HMSO, 1909, pp. 3–5.
112. *British Medical Journal*, 14 March 1908, p. 634; *Lancet*, 29 February 1908, p. 580.
113. Bryan, introductory remarks, *Gold Coast*, pp. 5–6.
114. Ryan Johnson, '*Mantsemei*, Interpreters, and the Successful Eradication of Plague: The 1908 Plague Epidemic in Colonial Accra', in R. Johnson and A. Khalid (eds), *Public Health in the British Empire: Intermediaries, Subordinates, and the Practice of Public Health, 1850-1960*, London: Routledge, 2011, 135–53.
115. Myron Echenberg, *Black Death, White Medicine: Bubonic Plague and the Politics of Public Health in Colonial Senegal, 1914-1945*, Oxford: James Currey, 2002.
116. 'Agreement Internationale ... pour la création a Paris d'un Office Internationale d'Hygiène Publique', *BOIHP*, 1 (1909), pp. 62–6.
117. E.g. 'Assainissement Général, Prophylaxie: La Dératisation', *BOIHP*, 2 (1910), pp. 542–603.
118. 'Convention Sanitaire ... Washington le 14 Octobre 1905', Chap.I, Sec. I, Art. 1; Chap. II, Sec. I, Art. 10 and Sec.II, Art. 11, *BOIHP*, 1 (1909), 251–5.
119. Most of the differences related to arrangements for yellow fever. See Howard-Jones, *Pan American Health Organization*, p. 9.
120. *BOIHP*, 1 (1909), p. 256.
121. *Second International Sanitary Convention of the American Republics. Pan American Sanitary Conference, 1905*, Washington, DC: Govt. Printing Office, 1907, pp. 1–2.
122. Guiteras, 'Remarks on the Washington Sanitary Convention of 1905'; Cueto, *The Value of Health*, p. 48.
123. See Marilyn Chase, *The Barbary Plague: The Black Death in Victorian San Francisco*, New York: Random House, 2004; James C. Mohr, *Plague and Fire: Battling Black Death and the 1900 Burning of Honolulu's Chinatown*, New York: Oxford University Press, 2005; Kraut, *Silent Travelers*.
124. Hugh V. Cumming, San Francisco Quarantine Station, to Surgeon-General, US PHMHS, 11 November 1903, File 1104, Box 99 (Canal Zone), RG 185, NARA.
125. Claude Pierce, Office of Medical Officer in Command, Panama, to Surgeon-General US PHMHS, 18 January 1904.
126. Annual Health Report, vol. 1 (1906–17), Box 1, RG 185, NARA.
127. On the plague in Peru, see Marcos Cueto, *The Return of Epidemics: Health and Society in Peru during the Twentieth Century*, Aldershot: Ashgate, 2001, pp. 6–28.
128. M.C. Guthrie, Chief Quarantine Officer, Panama Canal, to Surgeon-General, US PHMHS, 15 December 1914, File 1104, Box 99 (Canal Zone), RG 185, NARA.

129. Hugh de Valin, Quarantine Officer, Honolulu, to Surgeon-General, 10 July 1926, File 0520/17, Box 91 (Quarantine Stations, Hawaii), RG 90, NARA.
130. H.S. Cumming to Budget Officer, Treasury Department, 9 December 1933, File 0135/14, Box 94 (Hawaiian Islands, Plague Station), RG 90, NARA.
131. *Report of the Delegation from the United States of America to the Sixth International Sanitary Conference at Montevideo, December 12–20 1920*, Washington, DC: Govt. Printing Office, 1920, pp. 34–5.
132. S.B. Grubbs, Medical Director and Chief Quarantine Officer, Honolulu, to Surgeon-General, 2 August 1932; Grubbs to Surgeon-General, 1 June 1932, File 0425/183, Box 94 (Hawaiian Islands, Plague Station), RG 90, NARA.
133. Grubbs to Surgeon-General, 20 February 1933, File 1850/95, Box 94 (Honolulu Quarantine), RG 90, NARA.
134. Infringements of arrangements, File 0425/32, Box 92 (Honolulu Quarantine), RG 90, NARA.
135. Lien-Teh Wu, 'A Short History of the Manchurian Plague Prevention Service', in L.-T. Wu (ed.), *Manchurian Plague Prevention Service. Memorial Volume 1912–1932*, Shanghai: National Quarantine Service, 1934, p. 1.
136. E.H. Hankin, 'On the Epidemiology of Plague', *Journal of Hygiene*, 5 (1905), 49.
137. See Steven G. Marks, *Road to Power: The Trans-Siberian Railroad and the Colonization of Asian Russia, 1850–1917*, London: I.B. Tauris, 1991.
138. G.D. Gray, 'The History of the Spread of Plague in North China', in *Report of the Inter-National Plague Conference held at Mukden, April, 1911*, Manila: Bureau of Printing, 1912, pp. 31–3.
139. Lien-Teh Wu, 'Inaugural Address delivered at the International Plague Conference, Mukden', in Wu (ed.), *Manchurian Plague*, pp. 13–19; idem, 'The Second Pneumonic Plague Epidemic in Manchuria, 1920–21', ibid., 51–78.
140. Mark Gamsa, 'The Epidemics of Pneumonic Plague in Manchuria 1910–1911', *Past and Present*, 190 (2006), 147–83.
141. *Forty-Second Annual Report of the Local Government Board 1912–13*, London: HMSO, 1913 p. 43; Lien-Teh Wu, 'First Report of the North Manchuria Plague Prevention Service', *Journal of Hygiene*, 13 (1913), 276.
142. Wu, 'Short History', p. 1.
143. *Thirty-Seventh Annual Report of the Local Government Board, 1907–08*, London: HMSO, 1909, p. 229; Imperial Maritime Customs, *Decennial Reports . . ., 1892–1901. Vol. I: Northern and Yangtze Ports*, p. 25.
144. Wu, 'Short History', p. 4.
145. Ibid., p. 153; Wu, 'The Second Pneumonic Plague Epidemic', p. 53.
146. Lien-Teh Wu, J.W.H. Chun and R. Pollitzer, 'The Cholera Epidemic of 1926', in Wu (ed.), *Manchurian Plague*, pp. 268–9; Robert J. Perrins, 'Doctors, Disease and Development: Emergency Colonial Public Health in Southern Manchuria, 1905–1926', in M. Low (ed.), *Building a Modern Japan: Science, Technology, and Medicine in the Meiji Era and Beyond*, Basingstoke: Palgrave, 2005.
147. Arthur Stanley, 'Quarantine Measures at Shanghai against Northern Ports infected with Pneumonic Plague', *Report of the Inter-National Plague Conference held at Mukden*, pp. 306–7.
148. 'Discussion', ibid., p. 310.
149. S.B. Grubbs, Chief Quarantine Officer, Panama Canal, to Surgeon-General, 4 June 1920, File 1104, Box 99 (Canal Zone), RG 185, NARA.
150. *Report of the Delegation from the United States of America to the Sixth International Sanitary Conference at Montevideo*, p. 80.
151. Warwick Anderson, *Colonial Pathologies: American Tropical Medicine, Race, and Hygiene in the Philippines*, Durham, NC: Duke University Press, 2006, pp. 61–3.
152. *Thirty-Seventh Annual Report of the Local Government Board 1907–08*, p. 230; *Forty-First Annual Report of the Local Government Board 1911–12*, London: HMSO, 1912, p. 47.

153. Informal Report of Medical Inspection of Japan and China for the US Public Health Service, V.G. Heiser, p. 1, Papers of Victor George Heiser, B.H357, APS.
154. Ibid., p. 4.
155. Ibid., pp. 5–6.
156. V.G. Heiser, 'International Aspects of Disease', Papers of Victor George Heiser, B.H357, APS.
157. *The International Sanitary Convention of Paris, 1911–12*, trans. R.W. Johnstone, London: HMSO, 1919, pp. 4–5, 7–9.
158. 'Revision of the International Sanitary Convention of 1912', *British Medical Journal*, 8 July 1922, p. 50.
159. Heiser, 'International Aspects of Disease', pp. 7–9.
160. Lenore Manderson, 'Wireless Wars in the Eastern Arena: Epidemiological Surveillance, Disease Prevention and the Work of the Eastern Bureau of the League of Nations Health Organization', in Weindling (ed.), *International Health Organizations*, pp. 134–53; Alison Bashford, 'Global Biopolitics and the History of World Health', *History of the Human Sciences*, 19 (2006), 67–88.
161. F. Norman White, *The Prevalence of Epidemic Disease and Port Health Organization and Procedure in the Far East*, Geneva: League of Nations, 1923, pp. 13–23.
162. Ibid., p. 23.
163. Anne Sealey, 'Globalizing the 1926 International Sanitary Conference', *Journal of Global History*, 6 (2011), p. 437.
164. White, *The Prevalence of Epidemic Disease*, p. 111.
165. Ibid., pp. 105–6.
166. *The Pan American Sanitary Code, International Convention signed at Habana, Cuba, November 14, 1924*, Washington, DC: Govt. Printing Office, 1925, p. 3.
167. Ibid., pp. 5–7, 14–15.
168. Sealey, 'Globalizing the 1926 International Sanitary Conference'.
169. *League of Nations – Health Organization*, Geneva: League of Nations Information Section, 1931, p. 8.
170. Echenberg, *Plague Ports*.
171. BOIHP, 21 (1929), 1606–13. Exceptions include the epidemics in West Java 1930–4 and Dakar 1944. See Echenberg, *Black Death*; Terence H. Hull, 'Plague in Java', in N.G. Owen (ed.), *Death and Disease in Southeast Asia: Explorations in Social, Medical and Demographic History*, Singapore: Oxford University Press, 1987, pp. 210–34.
172. *Convention amending the International Sanitary Convention (1938)*, hhttp://iea.uoregon.edu/texts/1938-Amendment-1926-Sanitary.EN.htm
173. *Final Act of the XII Pan American Sanitary Conference, Caracas, Venezuela, January 12–24, 1947*, Washington, DC: Pan American Union, 1947; Cueto, *The Value of Health*, chaps 3–4.
174. *XXII Pan American Sanitary Conference. XXXVIII Meeting of the Regional Committee of the World Health Organization for the Americas, Washington D.C. 22–27 September 1986. Verbatim Records*, Washington, DC: Pan American Health Organization, 1987.

Chapter 8: Protection or protectionism?

1. Emmanuel Le Roy Ladurie, 'A Concept: The Unification of the Globe by Disease', in his *Mind and Method of the Historian*, Brighton: Harvester, 1981, 28–83.
2. Roger de Herdt, *Bijdrage tot de Geschiedenis van de Vetteelt in Vlaanderen, inzonderheid tot de Geschiedenis van de Rundveepest*, Louvain: Belgisch Centrum voor Landelijke Geschiedenis, 1970, p. 7.
3. Ibid., p. 8.
4. Lyon Playfair, *The Cattle Plague in its Relation to Past Epidemics and to the Present Attack*, Edinburgh: Edmonston & Douglas, 1865, pp. 9–16; Lise Wilkinson, *Animals & Disease: An Introduction to the History of Comparative Medicine*, Cambridge: Cambridge University Press, 1992, p. 42.

5. Dominick Hünniger, 'Policing Epizootics: Legislation and Administration during Outbreaks of Cattle Plague in Eighteenth-Century Northern Germany as Continuous Crisis Management', in K. Brown and D. Gilfoyle (eds), *Healing the Herds: Disease, Livestock Economies, and the Globalization of Veterinary Medicine* Athens, Ohio: Ohio University Press, 2010, pp. 76–91.
6. De Herdt, *Geschiedenis van de Vetteelt in Vlaanderen*, pp. 11–13, 36–40; Wilkinson, *Animals & Disease*, p. 60; Peter A. Koolmees, 'Epizootic Diseases in the Netherlands, 1713–2002: Veterinary Science, Agricultural Policy, and Public Response', in K. Brown and D. Gilfoyle (eds), *Healing the Herds: Disease, Livestock Economies, and the Globalization of Veterinary Medicine*, Athens, Ohio: Ohio University Press, 2010, pp. 23–4.
7. Playfair, *The Cattle Plague*, p. 17; Budd, *Siberian Cattle Plague*, p. 6; de Herdt, *Geschiedenis van de Vetteelt in Vlaanderen*, p. 3.
8. Much of the trade was conducted by steam-liner companies: see C. Knick Harley, 'Steers Afloat: The North Atlantic Meat Trade, Liner Predominance, and Freight Rates, 1870–1913', *Journal of Economic History*, 68 (2008), 1028–58.
9. John Gamgee, 'Report of Professor Gamgee on the Lung Plague', in Horace Capron, *Report of the Commissioner of Agriculture on the Diseases of Cattle in the United States* Washington, DC: Government Printing Office, 1871, pp. 3–15.
10. Ibid., pp. 11, 15.
11. Ralph Whitlock, *The Great Cattle Plague: An Account of the Foot-and-Mouth Epidemic of 1967–8*, London: John Baker, 1968, pp. 9–11.
12. Playfair, *The Cattle Plague*, pp. 19–20.
13. Ibid., p. 20.
14. John Gamgee, *The Cattle Plague; with Official Reports of the International Veterinary Congresses, held in Hamburg, 1863, and in Vienna, 1865*, London: Robert Hardwicke, 1866, pp. 465–76.
15. Ibid., pp. 561–2.
16. Budd, *The Siberian Cattle Plague*, pp. 5–6.
17. Playfair, *The Cattle Plague*, pp. 31, 34, 38.
18. E.g. Josiah Bateman, *The Mighty Hand of God: A Sermon preached in the Parish Church, Margate, on Friday, March 9th, 1866, being the Day of Humiliation for the Cattle Plague*, London: Simpkin, Marshall & Co., 1866, p. 1.
19. W.W. Clarke, *The Cattle Plague: A Judgment from God for the Sins of the Nation*, London: Simpkins, Marshall & Co., 1866, p. 6.
20. William McCall, *God's Judgment and Man's Duty*, London: William Macintosh, 1866, pp. 9–10.
21. Charles Bell, *Remarks on Rinderpest*, London: Robert Hardwicke, 1866, pp. 8–9.
22. Playfair, *The Cattle Plague*, pp. 62–3.
23. Gamgee, *The Cattle Plague*, p. 134.
24. Ibid., p. 142.
25. Ibid., p. 141; Terrie M. Romano, 'The Cattle Plague of 1865 and the Reception of the "Germ Theory" in Mid-Victorian Britain', *Journal of the History of Medicine and Allied Sciences*, 52 (1997), 51–80.
26. Gamgee, *The Cattle Plague*, pp. vi–vii, 143–5.
27. Ibid., p. 132.
28. Ibid., p. 134.
29. Whitlock, *The Great Cattle Plague*, pp. 12–13.
30. Michael Worboys, 'Germ Theories of Disease and British Veterinary Medicine, 1860–1890', *Medical History*, 35 (1991), 308–27.
31. George Foggo, *The Policy of Restrictive Measures, or Quarantine, as applied to Cholera and Cattle Plague*, London: Head, Noble & Co., 1872, p. 3.
32. Ibid., p. 8.
33. Whitlock, *The Great Cattle Plague*, pp. 11–13.

34. *Report of the Commissioners appointed to Inquire into the Origin, Nature, etc, of the Indian Cattle Plagues*, Calcutta: Office of the Superintendent of Government Printing, 1871, pp. 764–71.
35. *Proceedings of the First Meeting of Veterinary Officers in India, held at Lahore on the 24th March 1919 and following days*, Calcutta: Superintendent of Government Printing, 1919, pp. 33–4; *Proceedings of the Second Meeting of Veterinary Officers in India, held at Calcutta from 26th February to 2nd March, 1923*, Calcutta: Superintendent of Government Printing, 1924, pp. 99–101.
36. Daniel F. Doeppers, 'Fighting Rinderpest in the Philippines, 1886–1941', in Brown and Gilfoyle (eds), *Healing the Herds*, pp. 108–28.
37. Pule Phoofolo, 'Epidemics and Revolutions: The Rinderpest Epidemics in Late Nineteenth-Century Southern Africa', *Past and Present*, 138 (1993), 112–43.
38. Thomas P. Ofcansky, 'The 1889–97 Rinderpest Epidemic and the Rise of British and German Colonialism in Eastern and Southern Africa', offprint from *Journal of African Studies*, 8 (1981) Rhodes House Library, University of Oxford, RHO 740.14.49 (9).
39. Institute for Animal Health, 'Disease Facts – Rinderpest', http://www.iah.bbsrc.ac.uk/disease/rinderpest1.shtml, accessed 15/09/2010.
40. Capron, *Report of the Commissioner of Agriculture*, p. 1.
41. Quoted in John Gamgee, 'Report of Professor Gamgee on the Splenic or Periodic Fever of Cattle', in Capron, *Report of the Commissioner*, p. 82.
42. Ibid., p. 124.
43. Ibid., pp. 85–7, 116.
44. Ibid., pp. 126–7.
45. Ibid., pp. 116–17.
46. Justin Kastner, et al., 'Scientific Conviction amidst Scientific Controversy in the Transatlantic Livestock and Meat Trade', *Endeavour*, 29 (2005), 78–83; US Ambassador, London, to Secretary of State, 5 April 1890, Box 1, RG 17, NARA.
47. Carl N. Tyson, 'Texas Fever', http://digital.library.okstate.edu/encyclopedia/entries/T/TE022.html, accessed 08/09/2010.
48. J.M. Rush to Secretary of State, 18 February 1890, Reports and Correspondence, Letter-book, p. 96, Box 1, RG 17, NARA.
49. Prince de Chimay to Edwin H. Terrell, US Ambassador, Brussels, 18 November 1890, Letter-book, pp. 191–2, Box 1, RG 17, NARA.
50. John L. Gignilliat, 'Pigs, Politics, and Protection: The European Boycott of American Pork, 1879–1891', *Agricultural History*, 35 (1961), 3–12.
51. Alfred Vagts, *Deutschland und die Vereinigten Staaten in der Weltpolitik*, New York: Macmillan, 1935; Otto zu Stolberg-Wernigerode, *Germany and the United States of America during the Era of Bismarck*, Reading, Penn.: Henry Janssen Foundation, 1937; Louis I. Snyder, 'The American–German Pork Dispute, 1879–1891', *Journal of Modern History*, 17 (1945), 16–28; Gignilliat, 'Pigs, Politics, and Protection'.
52. US Ambassador, Berlin, to Hon. James Blaine, Secretary of State, 24 January 1891, Letter-book, p. 212, Box 1, RG 17, NARA.
53. Cameron and Neal, *Concise Economic History of the World*, p. 298.
54. Suellen Hoy and Walter Nugent, 'Public Health or Protectionism? The German–American Pork War, 1880–1891', *Bulletin of the History of Medicine*, 63 (1989), 198–224.
55. G.W. Pope, 'Some Results of Federal Quarantine against Foreign Live-Stock Diseases', *Yearbook of the United States Department of Agriculture*, Washington, DC: Government Printing Office, 1919, p. 243.
56. Robert D. Leigh, *Federal Health Administration in the United States*, New York: Harper, 1927, pp. 35–6.
57. 'Creation de l'Office International des Épizooties', *BOIE*, 1 (1927–8), 8–13.
58. *BOIE*, 1 (1927), p. 3.

59. 'Procès-Verbaux de la Prèmiere Reunion du Comité de l'Office International des Épizooties', 8 March 1927, *BOIE*, 1 (1927), 14–15.
60. E.g. H.C.L.E. Berger, 'L'Immunisation anti-aphteuse des animaux exportés', *BOIE*, 1 (1927), 534–6; M. Burgi, 'Les Méthodes Générales de la prophylaie de la fièvre aphteuse', ibid., 537–77.
61. E. Leclainche, 'La Standardisation des bulletins sanitaires', ibid., 577–83; S. Arán, 'Unifaction des certificats sanitaires pour le commerce international des animaux, viands et produits carnés', *BOIE*, 2 (1928–9), 589–98.
62. 'Rapport du Comité d'experts en matière de mesures de police vétérinaire', ibid., pp. 584–8.
63. 'Rapport du Comité économique de la Société des Nations sur sa 2me Session', 6–11 May 1929, *BOIE*, 3 (1929–30), pp. 4–5.
64. 'Rapport général du Sous-Comité d'experts en matière de Police vétérinaire', *BOIE*, 4 (1930–31), p. 359.
65. 'Rapport du Comité consultative sur "application," des recommendations de la Conférence économique internationale', *BOIE*, 4 (1930–31), p. 410.
66. *BOIE*, 5 (1931–32), p. 291.
67. 'Projets de Conventions vétérinaires', ibid., pp. 292–303.
68. US Department of Agriculture, *Regulations Governing the Inspection and Quarantine of Livestock and other Animals offered for Importation (except from Mexico)*, Pamphlet SF 623 B3 1945, Washington, DC: Government Printing Office, 1945, pp. 2–4.
69. Manuel A. Machado, *Aftosa: A Historical Survey of Foot-and-Mouth Disease and Inter-American Relations*, Albany: State University of New York Press, 1969, pp. 20–8.
70. See James T. Critchell, *A History of the Frozen Meat Trade*, London: Constable, 1912.
71. Abigail Woods, *A Manufactured Plague? The History of Foot and Mouth Disease in Britain*, London: Earthscan, 2004, pp. 53–63.
72. 'Short history of the OIE', http://www.oie.int/Eng/OIE/en_histoire.htm
73. Statement by Lord Soulsby in L.A. Reynolds and E.M. Tansey (eds), *Foot and Mouth Disease: The 1967 Outbreak and its Aftermath. The Transcript of a Witness Seminar held by the Wellcome Trust Centre for the History of Medicine at UCL, on 11 December 2001*, London: The Wellcome Trust, 2003, p. 7.
74. Whitlock, *The Great Cattle Plague*, pp. 31–58, 94–5.
75. Woods, *A Manufactured Plague?*, pp. 109–30.
76. See proceedings of Asian regional conference in *BOIE*, 63 (1965).
77. 'OIE's role in the pandemic influenza H1N1 2009', http://www.oie.int/eng/edito/en_lastedito.htm
78. Machado, *Aftosa*, p. 116.
79. B.G. Cané, L.F. Leanes and L.O. Mascitelli, 'Emerging Diseases and their Impact on Animal Commerce: The Argentine Lesson', *Annals of the New York Academy of Sciences*, 1026 (2004), 1–7.
80. James S. Donnelly, *The Great Irish Potato Famine*, Stroud: Sutton, 2001.
81. Christy Campbell, *Phylloxera: How Wine was Saved for the World*, London: Harper Perennial, 2004, p. 173.
82. Andreas Dix, 'Phylloxera', in S. Krech III, J.R. McNeill and C. Merchant (eds), *Encyclopedia of World Environmental History*, London: Routledge, 2004, p. 1002.
83. J.F.M. Clark, *Bugs and the Victorians*, London and New Haven: Yale University Press, 2009, pp. 132–53.
84. K. Starr Chester, *Nature and Prevention of Plant Diseases*, Toronto: The Blakiston Company, 1950, pp. 4–5.
85. *Plant Quarantine in Asia and the Pacific: Report of an APO Study Meeting 17th–26th March, 1992, Taipei, Taiwan*, Tokyo: Asia Productivity Assoc., 1993, pp. 2, 99.
86. 'A Constructive Criticism of the Policies Governing the Establishment and Administration of Quarantines against Horticultural Products, June 1925,' p. 3, Pamphlet SB 987.C6, LOC.

87. Ibid., p. 10.
88. Ibid., pp. 32–3.
89. Ibid., p. 11.
90. Ibid., p. 17.
91. Harvey Smith and Ralph Lattimore, 'The search for non tariff barriers: fire blight of apples', Department of Economics and Marketing Discussion Paper no. 44 (1997), International Trade Policy Research Centre, Department of Economics and Marketing, Lincoln University, Canterbury, researcharchive.lincoln.ac.nz/dspace/bitstream/.../1/cd_dp_44.pdf, accessed 14/09/09.
92. Linda Calvin and Barry Krissoff, 'Resolution of the US–Japan apple dispute: new opportunities for trade', United States Department of Agriculture, www.ers.usda.gov., accessed 14/09/09.
93. American Phytopathological Society, 'Plant diseases plague world trade', 13 August 2001, http://www.apsnet.org/media/press/global.asp, accessed 22/01/08.
94. John Knight, 'Underarm bowling and Australia–New Zealand trade', 18 July 2005, http://www.australianreview.net/digest/2005/07/knight.html, accessed 23/02/06.
95. Anna Marie Kimball, *Risky Trade: Infectious Disease in the Era of Global Trade*, Aldershot: Ashgate, 2006, p. 9.
96. 'Standards and safety', World Trade Organization, http://www.wto.org/english/thewto_e/whatis_e/tif_e/agrm4_e.htm
97. SPS Agreement Training Module: Chapter 4: Implementation – the SPS Committee', 5.1: 'Dispute Settlement', http://www.wto.org, accessed 21/12/11.
98. 'SPS Agreement Principles: Dispute Settlement', http://www.aphis.usda.gov/is/sps/mod1/1disput.html, accessed 28/02/06.
99. Ibid.
100. Joost Pauwelyn, 'The WTO Agreement on Sanitary and Phytosanitary (SPS) Measures as Applied in the First Three SPS Disputes', *Journal of International Economic Law*, 4 (1999), 641–64.
101. Quoted in Knight, 'Underarm bowling'.
102. 'WTO to rule over apple ban', http://www.dawn.com/2008/01/22/ebr13.htm, accessed 04/09/09.
103. 'Aussies block WTO probe into New Zealand apple ban', *New Zealand Herald*, 19 December 2007, http://www.nzherald.co.nz/trade/news/print.cfm?c_id=96&objectid=10483112, accessed 04/09/09.
104. New Zealand Ministry of Foreign Affairs, 'New Zealand involvement in WTO disputes – Australia apples', http://www.mfat.govt.nz/Treaties-and-International-Law/02, accessed 04/09/09.
105. 'NZ apple ban costs $20m a year, says grower', *New Zealand Herald*, 1 July 2009, http://www.nz.herald.co.nz/nz-exports/news/print.cfm?c_id=193&objectid=10581756, accessed 04/09/09.
106. Reuters, 'Update 2 – WTO condemns Australian ban on New Zealand apples', 9 August 2010, www.reuters.com, accessed 01/10/10.
107. Maglen, 'A World Apart'.
108. Joint Statement, Federal Minister for Agriculture, Fisheries and Forestry, Warren Truss, Federal Minister for Trade, Mark Vaile, 'Quarantine is helping Australia reach the world's markets', 29 August 2001, Australian Ministry for Agriculture, Fisheries and Forestry, http://www.maff.gov.au/releases/01/01237wtj.html, 28/06/05.
109. Bridges Weekly Trade News Digest, DSU Update, 'Philippines to launch case on bananas and paw-paws', http://www.ictsd.org/weekly/03-07-17/story5.htm, accessed 28/02/06.
110. Peter Lewis, 'Banana industry anxiously awaits import decision', 23 June 2002, http://www.abc.net.au/landline/stories/s586538.htm, accessed 23/02/06; Natasha Simpson, 'Philippines takes Aust to WTO over banana ban', 15 July 2003, http://www.abc.net.au/am/content/2003/s902217.htm, accessed 23/02/06.

111. *Horticulture and Home Pest News*, Plant Pathology, 'Banana – black sigatoka disease', http://www.ipm.iastate.edu/ipm/hortnews/2005/3-23-2005/banana.html, accessed 23/02/06.
112. 'Protecting an Industry – What It Takes', *National Marketplace News*, December/January, 2003/04, p. 46.
113. Simpson, 'Philippines takes Aust to WTO'.
114. 'Philippine banana growers hail gov't move to ban Australian beef', *AsiaPulse News*, 10 December 2002, http://www.highbeam.com/doc/1G1-97014491.html, accessed 04/09/09.
115. The multinational banana industry has been seen by many as typical of the iniquities of globalization. See Steve Striffler and Mark Moberg (eds), *Banana Wars: Power, Production, and History in the Americas*, Durham, NC: Duke University Press, 2003.
116. Sarah C. Fajardo, 'Philippines–Australia agricultural trade dispute: the case of quarantine on Philippine fruit exports', International Gender and Trade Network, Monthly Bulletin, June 2002, vol. 2, www.genderandtrade.net/Regions/Asia.html, 23/02/06.
117. Peter Gallagher, 'Bringing quarantine barriers to account', *Australian Financial Review*, 10 April 2003, http://www.inquit.com/article/228/bringing-quarantine-barriers-to-account, accessed 23/02/06.
118. 'The good news: RP bananas back in Australian market soon?', *Official Gazette of the Republic of the Philippines* http://www.gov.ph/index.php?option=com_content&task=view&id=19263&19263&Itemid=2, accessed 04/09/09.
119. Josyline Javelosa and Andrew Schmitz, 'Costs and Benefits of a WTO Dispute: Philippine Bananas and the Australian Market', *Estey Centre Journal of International Law and Trade Policy*, 7 (2006), 78–83.
120. Robert Fagan, 'Globalization, the WTO and the Australia–Philippines "Banana War"', in N. Fold and B. Pritchard (eds), *Cross-Continental Agro-Food Chains: Structures, Actors and Dynamics in the Global Food System*, London: Routledge, 2006, 207–22.
121. Javelosa and Schmitz, 'Costs and Benefits'.
122. Joint Ministerial Statement, Australian Ministers for Foreign Affairs and Trade, 11–12 August 2005, 'Inaugural Philippines–Australia Ministerial Meeting', http://www.foreignminister.gov.au/releases/2005/joint_philippines_120805.html, accessed 28/02/06.
123. Bernama, 'Australian farmers fight to keep ban on Philippine bananas', 5 November 2008, http://www.bernama.com/bernama/v5/newsindex.php?id=369508, accessed 04/09/09.
124. Brad Miller, 'Aerial spraying issue turns seesaw court battle', 29 November 2007, http://ipsnews.net/news.asp?idnews=40264, accessed 15/07/09.
125. Allan Nawal, Jeffrey M. Tupas and Joselle Badilla, 'Banana firms to contest aerial spraying ban at CA', *Inquirer*, http://services.inquirer.net/print/print.php?article_id=20070924-90391, accessed 04/09/2009; 'Banana exec vows all-out war vs. Critics', *Durian Post*, 31 July 2009, http://durianpost.wordpress.com/2009/07/31/, accessed 04/09/09.
126. Dario Agnote, 'Calls to scrap aerial spraying of bananas in Philippines', 16 June 2010, www.dirtybananas.org, accessed 01/10/10.
127. Bernama, 'Australian Farmers'.
128. Daniel Palmer, 'Agreement to import bananas from the Philippines leads to grower fears and IGA ban', *Australian Food News*, 5 March 2009, http://www.ausfoodnews.com.au/2009/03/05, accessed 04/09/09.
129. Aurelio A. Pena, 'Australia's farmers scared of losing market to Philippine bananas', *Philpress*, http://goldelyonn.wordpress.com/2009/06/11, accessed 04/09/09.
130. Tom Bicknell, 'Japan grabs Philippine banana supplies', http://www.fruitnet.com/content.aspx?cid=4190&ttid=17, accessed 04/09/09.
131. 'CFBF Comments on US–Australia Free Trade Agreement Feb. 2003', California Farm Bureau Federation, http://www.cfbf.com//issues//trade//us_aus_fta.cfm, accessed 23//02//06.

132. Australia–United States Free Trade Agreement, Sanitary and Phytosanitary Measures, Department of Food, Agriculture and Trade, Australia http:/www.dfat.gov.au/trade/negotiations/us_fta/outcomes/17_sanitary_phytosanitary.html, accessed 28/06/05.
133. Sally Kingsland, submission to the Senate Select Committee on the Free Trade Agreement, 30 April 2004, http://www.aph.gov.au/senate_freetrade/submissions/sublist.html, accessed 28/06/05.
134. John Landos, 'Where is the SPS agreement going?', http://www.apec.org.au/docs/landos.pdf, accessed 28/11/10.

Chapter 9: Disease and globalization

1. Francis Fukuyama, *The End of History and the Last Man*, New York: Avon Books, 1992.
2. Hopkins (ed.), *Globalization in World History*.
3. See, for example: Joseph E. Stiglitz, *Globalization and its Discontents*, London: Penguin, 2002; Jerry Mander and Edward Goldsmith (eds), *The Case Against the Global Economy and For a Turn towards the Local*, Berkeley: Sierra Club Books, 1996.
4. E.g. Lal, *Reviving the Invisible Hand*; Jagdish Bhagrati, *In Defense of Globalization*, New York and Oxford: Oxford University Press, 2004.
5. The exposure of cattle to toxins, for example: see Mark Purdey, 'An explanation of mad cow disease', http://www.ourcivilisation.com/madcow/madcow.htm, accessed 15/06/10.
6. BBC News, 'UK BSE Timeline', http://news.bbc.co.uk/1/hi/uk/218676.stm, accessed 15/09/10.
7. 'France refuses to lift British beef ban', BBC News, 12 March 2002, http://news.bbc.co.uk/1/hi/world/europe/1869367.stm, accessed 24/03/09.
8. John R. Fisher, 'Cattle Plagues Past and Present: The Mystery of Mad Cow Disease', *Journal of Contemporary History*, 33 (1998), p. 225.
9. Christian Durcot, Mark Arnold, Aline de Koeijer, Dagmar Heim and Didier Calvas, 'Review on the epidemiology and dynamics of BSE epidemics', *Veterinary Research*, 39 (2008), http://www.vetres.org, accessed 12/09/10.
10. 'USA-BSE hits exports hard', 27 December 2003, http://www.medicalnewstoday.com/articles/5017.php, accessed 24/03/09; 'Major importers ban US beef', *Guardian*, 24 December 2003, http://guardian.co.uk/world/2003/dec/usa.bse2, accessed 24/03/09.
11. Biotechnology Working Party, 'First cases of BSE in USA and Canada: risk assessment of ruminant materials originating from USA and Canada', European Medicines Agency, London, 21 July 2004, www.emea.europa.eu/docs/en_GB/document.../WC500003607.pdf, accessed 28/11/10.
12. 'U.S. Statement at WTO on Japan's beef import ban', United States Embassy, Tokyo, http://tokyo.usembassy.gov/e/p/tp-20050309-77.html, accessed 24/03/09.
13. 'New US beef import ban in Japan', BBC News, 20 January 2006, http:news.bbc.uk/2/hi/business/4631580.stm, accessed 24/03/09.
14. 'Japan lifts US beef import ban imposed against Mad Cow Disease', 28 July 2007, http://www.ens-newswire.com/ens/jul2006/2006-07-28-02.asp, accessed 24/03/09.
15. Song Jung-a, 'South Korea Removes Ban on US Beef', *Financial Times*, 30 May 2008.
16. 'The bulldozer boss and the beef crisis', editorial, *Financial Times*, 16 June 2008.
17. 'Q&A: S Korea beef protests', BBC News, 25 June 2008, http://news.bbc.co.uk/1/hi/world/asia-pacific/7457087.stm, accessed 24/03/09.
18. John Sudworth, 'Political price paid in beef row', BBC News, 5 June 2008, http://news.bbc.co.uk/1/hi/world/asia-pacific/7436914, accessed 28/05/08.
19. 'S Korean leader in beef apology', BBC News, 19 June 2008, http://news.bbc.co.uk/1/hi/world/asia-pacific/7462776.stm, accessed 28/06/08; 'S Korean leader replaces top aides', BBC News, 20 June 2008, http://news.bbc.co.uk/1/hi/world/asia-pacific/7464906.stm, accessed 28/06/08; John Sudworth, 'S Korea head admits beef failings', 7 July 2008, http://news.bbc.co.uk/1/hi/world/asia-pacific/7492562.stm, accessed 24/03/09.

20. Interview: Korea's summer of discontent, *International Socialism*, 120, 2 October 2008, www.isj.org.uk/index.php4?id=480=120, accessed 24/03/09.
21. Ser Myo-ja, 'Religious Groups Join Protests against Beef Deal', *Joong Ang Daily*, 2 July 2008, p. 1.
22. Kim Yon-se, 'President Appeals for End to Beef Row', *Korean Times*, 26 June 2008, p. 1
23. Park Sang-woo and Ser Myo-ja, 'Despite Warning, Thousands Strike', *Joong Ang Daily*, 3 July 2008, p. 1; 'S Korean car workers join protests', https:lists.resist.ca/pipermail/onthebarricades/2008/00505.html, accessed 24/03/09.
24. Ser Myo-ja, 'Teachers' Union Joins Protests to Prevent Imports of US Beef', *Joong Ang Daily*, 4 July 2008, p. 1.
25. Ser, 'Religious Groups Join Protests'.
26. Oh Byung-sang, 'Keep Church and State Separate', *Joong Ang Daily*, 2 July 2008, p. 10; Kim Jong-soo, 'Spaghetti Westerns Come to Seoul', *Joong Ang Daily*, 3 July 2008, p. 10; Ser, 'Teachers' Union Joins Protests'.
27. Lim Mi-jin and Moon Byung-joo, 'Sales of US Beef Strong for A-Meat', *Joong Ang Daily*, 3 July 2008, p. 3.
28. 'South Korean doctors hold US beef-eating event to dispel Mad Cow Disease fears', 9 July 2008, http://www.dailymail.co.uk/news/worldnews/article-1033673/South-Korean-doctors, accessed 24/03/09.
29. Kim So-hyun, 'Expat, Korean Coalition plans Counter-Demonstration in Seoul', *Korea Herald*, 4 July, 2008, p. 3.
30. Peter M. Beck, 'Candlelight Protests: Finding a Way Forward', *Korea Herald*, 4 July 2008, p. 4.
31. Lee Hee-ok, 'Spirit of the Times', *Joong Ang Daily*, 4 July 2008, p. 10; interviews with author, *Kyunghyang*, July 2008, p. 6 and *Hankyoreh*, 8 July 2008, p. 28.
32. James Cogan, 'South Korean government turns to repression to curb protests', 3 July 2008, http://www.wsws.org/articles/2008/jul2008/skor-j03.shtml, accessed 24/03/09.
33. 'Thousands in S Korea beef protest', 5 July 2008, http://news.bbc.co.uk/2/hi/asia-pacific/7491482.stm; Park Chan-kyong, 'New mass protest against S. Korean government', http://english.chosun.com/w21data/html/news/200807/200807180020.html, accessed 24/03/09.
34. 'Protestors rally against US beef imports in central Seoul', http://www.news.com.au/heraldsun/story/0,21985,24086463-5005963-5005961,00.html, accessed 24/03/09.
35. Thomas Abraham, *Twenty-First Century Plague*, Baltimore: Johns Hopkins University Press, 2004, pp. 14–36.
36. 'Timeline: SARS Virus', BBC News, 7 July 2004, http://news.bbc.co.uk/1/hi/world/asia-pacific/2973415.stm, accessed 28/06/05; Deborah Davis and Helen Siu, 'Introduction', in Deborah Davis and Helen Siu (eds), *SARS: Reception and Interpretations in Three Chinese Cities*, London and New York: Routledge, 2007, pp. 1–2.
37. Elizabeth Fee and Daniel M. Fox (eds), *AIDS: The Making of a Chronic Disease*, Berkeley and Los Angeles: California University Press, 1992; Pater Baldwin, *Disease and Democracy: The Industrialized World Faces AIDS*, Berkeley and Los Angeles: California University Press, 2005.
38. John Iliffe, *The African AIDS Epidemic*, Oxford: James Currey, 2006; Toyin Falola and Matthew M. Heaton, *HIV/AIDS, Illness, and African Well-being*, Rochester, NY: University of Rochester Press, 2007, especially Part IV.
39. Anarfi Asamoa-Baah, 'Can new infectious diseases be stopped? Lessons from SARS and avian Influenza', *OECD Observer*, 243, May 2004, http://www.oecdobserver.org.news.printpage.php/aid/1284, accessed 28/11/10.
40. Deborah Davis and Helen Siu, 'Introduction', in Davis and Siu (eds), *SARS: Reception and Interpretations in Three Chinese Cities*, London and New York: Routledge, 2007, p. 3; 'China may execute SARS quarantine violators (no ifs or buts)', 15 May 2003, http://www.hypocrites.com/article11925.html, accessed 28/06/2005; 'China widens SARS quarantine', BBC News, 24 April 2003, http://news.bbc.co.uk/1/hi/world/asia-pacific/2974739.stm, accessed 28/06/05; Hannah Beech, 'The Quarantine Blues', *Time*

Asia, http://www.time.com/time/asia/magazine/article/0,13673,501030519-451009,00. html, accessed 28/06/05.
41. Philip Thornton, 'WTO Warns SARS Could Halt Revival of International Trade', *Independent*, 24 April 2003, www.Independent.co.uk/news/business/news/wto-warns-sars-could-halt-revival-of-international-trade-595505.html, accessed 24/03/2009.
42. 'SARS weighs on Asia's economies', BBC News, 7 April 2003, http://newsvote.bbc.co.uk/mpapps/pagetools/print/news/bbc.co.uk/2/hi/business/2924557.stm, accessed 24/03/09.
43. 'SARS crisis: potential threat to Asian economy', *Hindu*, 28 April 2003, http://www.theHindu.com/2003/04/28/stories/2003042800170200.htm, accessed 24/03/09.
44. 'US economy starts to feel effects of SARS', 24 April 2003, *Sydney Morning Herald*, http://www.smh.com.au/articles/2003/04/24/1050777337621.html, accessed 24/03/09.
45. Philip Browning, 'Asian currencies: how SARS could cause a trade war', *Herald Tribune*, 15 May 2003, http://www.iht.com/articles/2003/05/15/edbow_ed3_.php, accessed 24/03/09.
46. 'Bush order allows SARS quarantine', 4 April 2003, http://www.cnn.com/2003/HEALTH/04/04/sars.bush/, accessed, 27/10/2005; Centers for Disease Control [CDC], 'Fact Sheet on Isolation and Quarantine', 3 May 2005, http://www.cdc.gov/ncidod/sars/isolationquarantine.htm, accessed 28/06/05.
47. 'MPs call for SARS quarantine', 24 April 2003, http://news.bbc.co.uk/1/hi/health/2972109.stm, accessed 28/06/05.
48. Tim Harcourt, 'The economic effects of SARS: what do we know so far?' 1 May 2003. http://www.austrade.gov.au/Default.aspx?PrintFriendly=True&ArticleID=6039, accessed 24/03/09.
49. 'Experts: SARS damage to China's foreign trade is limited', Xinhua News Agency, 30 May 2003, http://www.china.org.cn/english/features/SARS/65833.htm, accessed 24/03/09.
50. Jong-Wha Lee and Warwick J. McKibbin, 'Globalization and disease: the case of SARS', *Brookings*, 20 May 2003, http://www.brookings.edu/papers/2003/0520development_lee.aspx?p=1, accessed 24/03/09.
51. Davis, 'Introduction', in Davis and Siu (eds), *SARS*, p. 4
52. Christine Loh and Jennifer Welker, 'SARS and the Hong Kong Community', in C. Low (ed.), *At the Epicentre: Hong Kong and the SARS Outbreak*, Hong Kong: Hong Kong University Press, 2004, 215–34.
53. Yun Fan and Ming-chi Chen, 'The Weakness of a Post-authoritarian Democratic Society: Reflections upon Taiwan's Societal Crisis during the SARS Outbreak', in Davis and Siu (eds), *SARS*, pp. 147–64.
54. Tseng Yen-Fen and Wu Chia-Ling, 'Governing Germs from Outside and Within Borders', in A.K.C. Leung and C. Furth (eds), *Health and Hygiene in Chinese East Asia*, Durham, NC and London: Duke University Press, 2010, 225–72.
55. 'Update 70 – Singapore removed from list of areas with local SARS transmission', *Inquirer* 30 May 2003; 'Singapore offers beach resort for SARS quarantine', 13 May 2003, *Inquirer*, http://www.inq7.net/wnw/2003/may/13/wnw_10-1.htm, accessed 28/06/05.
56. WHO, 'Update 70 – Singapore removed from the list of areas with local SARS transmission', 30 May 2003, http://www.who.int/csr/don/2003_05_30a/en/, accessed 27/10/05.
57. 'Official defends 10-day SARS quarantine', 7 June 2003, http://www.cbc.ca/stories/2003/06/07/sars_quarantine030607, accessed 28/06/05; 'Ontario offers SARS quarantine compensation', 13 June 2003, http://www.cbc.ca/stories/2003/06/13/ont_sarscomp 030613, accessed 28/06/05.
58. 'Efficiency of quarantine during an epidemic of Severe Acute Respiratory Syndrome – Beijing, China, 2003', 31 October 2003, http://www.cdc.gov/mmwr/preview/mmwrhtml/mm5243a2.htm, accessed 28/06/05.
59. Georgeo C. Benjamin, 'Afterword', in Tim Brookes, *Behind the Mask: How the World Survived SARS, the First Epidemic of the Twenty-First Century*, Washington, DC: American Public Health Association, 2005, p. 236.

60. Kimball, *Risky Trade*, pp. 163–5, 174–5.
61. See Howard Phillips and David Killingray (eds), *The Spanish Influenza Pandemic of 1918–19: New Perspectives*, London: Routledge, 2003.
62. David Gratzer, 'SARS 101', *National Review*, 19 May 2003, http://www.manhattan-institute.org/cgi-bin/apMI/print.cgi, accessed 24/03/09.
63. Jong-Wha Lee and Warwick J. McKibbin, 'Globalization and Disease: The Case of SARS', *Brookings*, 20 May 2003, http://www.brookings.edu/papers/2003/0520 development_lee.aspx?p=1.
64. Kimball, *Risky Trade*, pp. 2–3.
65. Meeting of APEC Ministers Responsible for Trade, Khon Kaen, Thailand, 2–3 June 2003, 'Ministerial Statement on Severe Acute Respiratory Syndrome (SARS)', http://www.apec.org/apec/ministerial_statements/sectoral_ministerial/trade/2003_trade/stm, accessed 24/03/09; 'Asia unites against SARS', BBC News, 28 June 2003, http://news.bbc.co.uk/1/hi/world/asia-pacific/3027994.stm, accessed 24/03/09.
66. Takuro Nozawa, 'Impact of SARS: business with China remains stable thus far', Centre for Strategic Studies, *Japan Watch*, 16 July 2003, 1–2; 'Animal diseases "threaten humans"', BBC News, 13 January 2004, http://news.bbc.co.uk/1/hi/health/3391899.stm, accessed 24/03/09; S.M.A. Kazmi, 'SARS gone, effect lingers on Indo-China annual trade', http://www.indianexpress.com/oldstory.php?storyid=27793, accessed 24/03/09; Cortlan Bennett, 'Alarm as China's wild animal trade is blamed for "new case of SARS"', 3 January 2004, http://www.telegraph.co.uk/news/worldnews/asia/china/1450911, accessed 24/03/09; Amanda Katz and Sarah Wahlert, 'SARS prompts crackdown on wildlife trade', http://www.buzzle.com/editorials/5-26-2003-40813.asp, accessed 24/03/09; Martin Enserink and Dennis Normille, 'Search for SARS Origins Stalls', *Science*, 31, 302 (October 2003), 766–7.
67. Stephen Brown, 'The Economic Impact of SARS', in C. Loh (ed.), *At the Epicentre: Hong Kong and the SARS Outbreak*, Hong Kong: Hong Kong University Press, 2004, pp. 179–93.
68. E.g. Ilona Kickbusch and Evelyne de Leeuw, 'Global Public Health: Revisiting Healthy Public Policy at the Global Level', *Health Promotion International*, 14 (1999), 285–88; Ilona Kickbusch, 'The Development of International Health Policies – Accountability Intact?', *Social Science and Medicine*, 51 (2000), 979–89.
69. See Susan Peterson, 'Epidemic Disease and National Security', *Security Studies*, 12 (2002), 43–81.
70. Gratzer, 'SARS 101'.
71. See Peter Washer, *Emerging Infectious Diseases and Society*, Basingstoke: Palgrave, 2010.
72. Kimball, *Risky Trade*, p. xv; Meredith T. Mariani, *The Intersection of International Law: Agricultural Biochemistry, and Infectious Disease*. Leiden: Brill, 2007, p. 31.
73. David P. Fidler, 'Germs, Governance, and Global Public Health in the Wake of SARS', *Journal of Clinical Investigation*, 113 (2004), 799–804. See also Richard Dodgson, Kelley Lee and Nick Drager, 'Global Health Governance', Discussion Paper No. 1, Centre on Global Change and Health, London School of Hygiene and Tropical Medicine, in John J. Kirton (ed.), *Global Health*, Aldershot: Ashgate, 2009, pp. 439–62.
74. Terrence O'Sullivan, 'Globalization, SARS and other catastrophic disease risks: the little known international security threat', http://allacademic.com/meta/p73646_index.html, accessed 24/03/09.
75. Neil V. Mugas, 'Laguna hospital placed in SARS quarantine', *The Manila Times*, 7 January 2004, http://www.manilatimes.net/national/2004/jan/07/yehey/top_stories/20040107top5.html, accessed 24/03/09; Bill Andrews, '500 quarantines in new SARS outbreak', *Edinburgh Evening News*, 26 April 2004, http://news.scotsman.com/print.cfm?id=470902004, accessed 23/03/09.
76. David Bell, Philip Jenkins and Julie Hall, 'World Health Organization Global Conference on Severe Acute Respiratory Syndrome', CDC Emerging Infectious Diseases, http://www.cdc.gov/ncidod?EID/vol9no9/03-0559.htm, accessed 27/10/05.

77. WHO, 'H5N1 avian influenza: a chronology of key events', http://www.int./influenza/resources/documents/chronology/en/index.html, accessed 29/12/11.
78. Asamo-Baah, 'Can new infectious diseases be stopped?'
79. Lester Haines, 'Bird flu pandemic inevitable, says WHO', *Science*, 8 September 2005, http://www.theregister.co.uk/2005/09/08/bird_flu_pandemic/print.html, accessed 27/10/05.
80. 'Military quarantine for bird flu', 4 October 2005, http://www.cbsnews.com/stories/2005/10/04/health/main910425.shtml, accessed 27/10/05
81. Kimball, *Risky Trade*, pp. 54–5.
82. Elisabeth Rosenthal, 'Bird flu is linked to global trade in poultry', *International Herald Tribune*, 12 February 2007, http://www.iht.com/bin/print.php?id=4568849, accessed 28/11/10.
83. Quoted in Kimball, *Risky Trade*, p. 57.
84. Compassion in World Farming, 'The role of the intensive poultry production industry in the spread of avian influenza', February 2007, http://www.apeiresearch.net/document_file/document_20070706104415-1.pdf, accessed 21/12/11.
85. Indonesia was the country worst hit by avian influenza and had the highest human mortality rate. See 'Bird flu situation in Indonesia critical', FAO Newsroom, 18 March 2008 www.fao.org/newsroom/en/news/200/1000813/index.html, accessed 21/12/11.
86. Compassion in World Farming 'The role of intensive poultry production . . . ', p. 11.
87. 'Mexico confirms H1N1 flu not from pigs on Smithfield farm', 15 May 2009, National Park Producers Council, http://www.nppc.org/News/DocumentsPrint.aspx?DocumentID=24726, accessed 04/09/09; Edwin D. Kilbourne, 'A Virologist's Perspective on the 1918–19 Pandemic', in H. Phillips and D. Killingray (eds), *The Spanish Influenza Pandemic of 1918–19: New Perspectives*, London: Routledge, 2003, p. 34.
88. Jo Tuckman and Robert Booth, 'Four-year-old Could Hold Key Clue in Mexico's Search for Ground Zero', *Guardian*, 28 April 2009, pp. 4–5.
89. Ian Traynor, 'EU Official Accused of "Alarmism" for Telling Travellers to Avoid Americas', *Guardian*, 28 April 2009, p. 4.
90. Chris McGreal, Severin Carrell and Patrick Wintour, 'Swine flu – Global Threat Raised', *Guardian*, 28 April 2009; http://www.guardian.co.uk/world/2009/apr/27/swine-flu-race-to-contain-outbreak, accessed 16/04/2012; Severin Carrell and Patrick Wintour, 'Minister Urges Calm as Virus Reaches Britain', ibid., p. 5.
91. 'Swine flu: "inevitable flu pandemic" would be fourth in a century', *Daily Telegraph*, 27 April 2009, http://www.telegraph.co.uk/health/swine-flu/5230062, accessed 04/09/09.
92. 'Trade tensions over pork', http://www.foxnews.com/story/0,2933,518935,00.html, accessed 04/09/09.
93. WHO, 'The international response to the influenza pandemic: WHO responds to the critics', http://www.who.int/csr/diseases/swineflu/notes/briefing_20100610/en, accessed 29/12/11.
94. Owen Bowcott, 'Swine flu could kill 65,000 in UK, warns chief medical officer', *Guardian*, 16 July 2009, http://www.guardian.co.uk/world/2009/jul/16/swine-flu-pandemic-warning-helpline, accessed 22/01/11.
95. Daniel Martin, 'Sir Liam Donaldson quits the NHS . . . but critics say resignation is two years too late', *Daily Mail*, 16 December 2009, http://www.dailymail.co.uk/article-1236179/Sir-Liam-Donaldson-quits-NHS-critics-say-resignation-years-late.html, accessed 22/01/11.
96. 'This is serious: I'm not crying wolf about swine flu', *Independent*, 2 May 2009, pp. 6–7; Robert McKie, 'Swine Flu is Officially a Pandemic. But Don't Worry . . . Not Yet, Anyway', *Observer*, 14 June 2009, p. 22.
97. 'China's quarantine measures "proper and necessary"', *China View*, 5 May 2009, http://news.xinhuanet.com/english/2009–05/05/content_1137190.htm, accessed 04/09/09; 'US Citizens quarantined in China over swine flu fears', Reuters, 5 May 2009, http://www.foxnews.com/story/0,2933,518935,00.html, accessed 04/09/09.
98. 'Swine flu fears cause multiple countries to ban US, Mexican pork imports', *Area Development*, 27 April 2009, http://www.areadevelopment.com/newsitems/4–27–2009, accessed 04/09/09.

99. 'Russian port ban over swine flu "unjustified": EU official', http://www.eubusiness. com/news-eu/1241513222.7/, accessed 04/09/09.
100. 'Pork producers contend with swine flu fallout', http://www.cbc.ca/canada/calgary/ story/2009/05/04/pork-market.html; Geena Teel, 'Ottawa threatens trade action if pork ban isn't lifted', *National Post*, 5 May 2009, http://www.nationalpost.com/m/ story.html?id=1564824, accessed 04/09/09.
101. Editorials from the Director General, 'OIE's role in the pandemic influenza H1N1 2009', 5 November 2009, http://www.oie.int/eng/edito/en_lastedito.htm, 04/03/10; David Fidler, 'The swine flu outbreak and international law', *Insights*, 13, 27 April 2009, http://www.asil.org/insights090427.cfm.
102. Daniel Workman, 'Swine flu infects pork trade policies in China', 5 May 2009, http:// world-trade-organization.suite101.com/article.cfm/swine_fluinfects_pork_trade, accessed 04/09/09.
103. 'FM: China's bans on pigs, pork imports in line with WTO rules', 6 May 2009, *English DBW* http://english.dbw.cn/system/2009/05/06/000130444.shtml, accessed 04/09/09.
104. 'Russia lifts flu-lined pork ban on Wisconsin, Ontario', *China Post*, 20 July 2009, http://www.chinapost.com.tw/business/europe/2009/07/20/216955/Russia-lifts.htm, accessed 04/09/09.
105. Nataliya Vasilyeva, 'US: Lifting pork imports could boost Russia's WTO talks', http:blog.targana.com/n/us-lifting-pork-imports-could-boost-russias-wto-talks-73287/, accessed 04/09/09.
106. Workman, 'Swine flu'; 'Trade: WTO SPS Committee discusses trade responses to swine flu', Third World Network, 26 June 2009, http://www.twnside.org.sg/title2/wto. info/2009/twninfo20090705.htm, accessed 04/09/09.
107. Michael Greger, 'CDC confirms ties to virus first discovered in US pig factories', *The Humane Society of the United States*, 28 August 2009, http://www.humane.society.org/ news/2009/04/swine_flu_virus_origin_1998_042909.html, accessed 21/11/10; idem, 'Industrial Animal Agriculture's Role in the Emergence and Spread of Disease', in J. D'Silva and J. Webster (eds), *The Meat Crisis: Developing More Sustainable Production and Consumption*, London: Earthscan, 2010, p. 166.
108. Debora MacKenzie, 'Pork industry is blurring the science of swine flu', *New Scientist*, 30 April 2009, http://www.newscientist.com/blogs/shortsharpscience/2009/04/why-the-pork-industry-hates-th.html, accessed 21/11/10.
109. 'WTO SPS Committee discusses trade responses to swine flu'.
110. 'China requests WTO panel to probe US poultry import ban', *People's Daily Online*, 21 July 2009, http://english.peoplesdaily.com.cn/90001/90776/90884/6705246.html, accessed 04/09/09; 'WTO to investigate US/China poultry import ban', 4 August 2009, WATTAgNET, http://www.wattagnet.com/10105.html, accessed 04/09/09.
111. On the tendency of recession to reduce international governance, see John Gray, *False Dawn: The Delusions of Global Capitalism*, London: Granta, 2009, p. xxiv.
112. WTO Agreement on the Application of Sanitary and Phytosanitary Measures, www.wto.org, accessed 28/02/06.
113. Compassion in World Farming, 'The role of intensive poultry production . . .'.

Conclusion: Sanitary pasts, sanitary futures

1. Ladurie, 'A Concept: The Unification of the Globe by Disease', in his *Mind and Method of the Historian*, pp. 28–83.
2. John McDermott and Delia Grace, 'Agriculture-associated diseases: adapting agriculture to improve human health', International Livestock Research Institute Policy Brief, February 2011, http://www.ilri.org, accessed 21/12/11.
3. WTO Committee of Sanitary and Phytosanitary Measures, 'Review of the operation and implementation of the SPS Agreement', p. 21; 'Chronological list of disputes', http://www.wto.org, accessed 22/12/11.

4. David P. Fidler, 'International Law and the *E. coli* Outbreaks in Europe', *Insights*, 15, 6 June 2011.
5. Compassion in World Farming, 'The role of intensive poultry production in the spread of avian influenza', February 2007, http://www.apeiresearch.net/document_file/document_20070706104415-1.pdf, accessed 21/12/11.
6. Tseng Yen-Fen and Wu Chia-Ling, 'Governing Germs from Outside and Within Borders', in A.K.C. Leung and C. Furth (eds), *Health and Hygiene in Chinese East Asia*, Durham, NC and London: Duke University Press, 2010, pp. 255–72.
7. 'Putting Meat on the Table: Industrial Farm Animal Production in America', Pew Commission on Industrial Farm Animal Production, 2008, www.pewtrusts.org, accessed 21/12/11.
8. *Grain*, April 2009, http://www.grain.org/articles/?id=48, accessed 21/11/10.
9. Webster quoted in Greger, 'Industrial Animal Agriculture's Role', p. 165.
10. McDermott and Grace, 'Agriculture-associated diseases'.
11. E.g. WHO, FAO & OIE, 'Global Early Warning and Response System for Major Animal Diseases, including Zoonoses (GLEWS)', February 2006; 'The FAO–OIE–WHO Collaboration: Sharing Responsibilities and Coordinating Global Activities to Address Health Risks at the Animal–Human–Ecosystems Interfaces. A Tripartite Concept Note', April 2010. http://www.who.int/zoonoses/outbreaks/glews/en/, accessed, 08/04/12.
12. 'Emerging diseases drive human, animal health alliance', 20 July 2007, American Government Archive, http://www.america.gov/st/develop-english/2007/July/20070720135159lcnirell0.28751, accessed 07/09/2009.
13. Washer, *Emerging Infectious Diseases*, chap. 5.
14. C.M. Delgado, M. Rosegrant, H. Steinfeld, S. Ehui and C. Courbois, *Livestock to 2020: The Next Food Revolution*, Washington, DC: International Food Policy Research Institute, 1999.
15. Tom Levitt, 'Asian factory farming boom spreading animal diseases like avian influenza', *Ecologist*, 11 February 2011, http://www.theecologist.org, accessed 21/12/11.
16. K.-H. Zessin, 'Emerging Diseases: A Global and Biological Perspective', *Journal of Veterinary Science*, 53 (2006), 7–10.
17. Greger, 'Industrial Animal Agriculture's Role', p. 167; Peter Cowen and Roberta A. Morales, 'Economic and Trade Implications of Zoonotic Diseases', in T. Burroughs, S. Knobler and J. Lederberg (eds), *The Emergence of Zoonotic Diseases: Understanding the Impact on Animal and Human Health*, Washington, DC: National Academy Press, 2001, 20–5, pp. 24–5.

Bibliography

Primary sources

Archives

India

Maharashtra State Archives, Mumbai
General Department Proceedings.

National Archives of India, New Delhi
Foreign (General) Proceedings.
Foreign (Political) Proceedings.
Home (Agriculture, Revenue and Commerce) Proceedings.
Home (Sanitary) Proceedings.

West Bengal State Archives, Kolkata
Home (Medical) Proceedings.

Portugal

Lisbon
Archivo Histórico Colonial.
Archivo Histórico Ultramarino.

United Kingdom

Bristol City Record Office
Dunn correspondence, BRO 29596.

British Library
Aden Residency files.
Government of Bombay, Home Department (Sanitary).

Chester and Cheshire Archives
Correspondence from Privy Council to Mayor and Aldermen of Chester, ZM/L/4.

Derbyshire County Records Office
Correspondence from Office of the Privy Council to Lord Lieutenant in Ireland, D 3135.
Correspondence from W. Sharpe to Lord George Sackville, D 3155/C1271.
Order in Council, D 3155/C1018.

Gloucestershire County Records Office
Sotherton-Estcourt Papers.

National Maritime Museum
Office of Sick and Wounded Seamen to Admiralty Board, ADM/F/27.

Royal College of Physicians of Edinburgh
Papers of Sir John Pringle.

Royal College of Physicians, London
Report of the Quarantine Committee, Royal College of Physicians, 1889.

The National Archives, London
Foreign Office.
Privy Council.
State Papers.
Treasury.

University of Oxford
Extrait des Régistres de Parlement, BOD Vet. E3d.83 (53), Bodleian Library, Special Collections Reserve.
Thomas P. Ofcansky, offprint from *Journal of African Studies*, 8 (1981), RHO 740.14.49 (9), Rhodes House Library.

Wellcome Library for the History and Understanding of Medicine, London
Balthasar de Aperregui, 'Orderes relatives a sanidad y lazarettos en el Puerto de Barcelona, con motivo de la peste, en el año de 1714, y siguientes', Western MS.963.
Nicolas Chervin, Certified copies and original documents relating to yellow fever in Guadeloupe and the West Indies, Western MS. Amer. 113.
Correspondence from Thomas Mayne to Sir Charles Hamilton, Western MS.7313.
Thursday meeting book, Kingston-upon-Hull Corporation, Western MS.3109.
Maj.-Gen. Sir Charles Phillips, 'Letters and Instructions to the Officers during the Plague at Corfu, 1816', Western MS.3883.
A.D. Vasse St. Ouen, French Consul at Larnaca, Cyprus, to A.R. Roussin, French Ambassador at Constantinople, Western MS.4911.

West Sussex Record Office
Goodwood Papers, MS.1451.
Petition from John March, Deputy Governor of the Levant Company, n.d. [*c*.1763], PHA/35.

United States of America

American Philosophical Society, Philadelphia
Stephan Girard Collection.
Papers of Victor George Heiser.
Letters and Papers of David Hosack, 1795–1835.

Baltimore City Archives
Quarantine record for 1881–1918, RG 19 S2.

Library of Congress, Washington, DC
Walter Wyman, ' "Quarantine and Commerce", An Address delivered before the Commercial Club of Cincinnati, October 15th, 1898', unpublished pamphlet, RA 655 W9.
'A Constructive Criticism of the Policies Governing the Establishment and Administration of Quarantines against Horticultural Products, June 1925', Pamphlet SB 987.C6.

National Archives and Records Administration, College Park, Maryland
RG 17 Records of the Bureau of Animal Industry.
RG 43 International Sanitary Conference, 1881.
RG 90 Quarantine Stations (Hawaii).
RG 185 Panama Canal Zone.

Published official documents

China

China Imperial Maritime Customs, *Decennial Reports on the Trade, Navigation, Industries, etc., of the Ports open to Foreign Commerce in China and Corea, and on the Condition and Development of the Treaty Port Provinces, 1882-91*. Shanghai: Inspector General of Customs, 1893.

China Imperial Maritime Customs, *Decennial Reports on the Trade, Navigation, Industries, etc., of the Ports open to Foreign Commerce in China and Corea, and on the Condition and Development of the Treaty Port Provinces, 1892-1901: I. Northern and Yangtze Ports*, Shanghai: Inspectorate General of Customs, 1904.

France

L' Ordre public pour la ville de Lyon, pendant la maladie contagieuse. Lyons: A. Valancol, 1670.

International Sanitary Conferences and Offices

Procès-Verbaux de la conférence sanitaire internationale, ouverte a Paris le 27 juillet 1851, 2 vols. Paris: Imprimerie Nationale, 1851.

Protocoles et Procès-Verbaux de la Conférence Sanitaire Internationale de Rome inaugurée le 20 mai 1885. Rome: Imprimerie du Ministère des Affairs Étrangères, 1885.

Second International Sanitary Convention of the American Republics. Pan American Sanitary Conference, 1905. Washington, DC: Govt. Printing Office, 1902.

The International Sanitary Convention of Paris, 1903, trans. Theodore Thomson. London: HMSO, 1904.

Bulletin de l'Office International d'Hygiène Publique. Paris: OIHP, 1909-29.

Report of the Inter-National Plague Conference held at Mukden, April, 1911. Manila: Bureau of Printing, 1912.

The International Sanitary Convention of Paris, 1911-12, trans. R.W. Johnstone. London: HMSO, 1919.

Report of the Delegation from the United States of America to the Sixth International Sanitary Conference at Montevideo, December 12-20, 1920. Washington, DC: Govt. Printing Office, 1920.

The Pan American Sanitary Code, International Convention signed at Habana, Cuba, November 14, 1924. Washington, DC: Govt. Printing Office, 1925.

Bulletin de l' Office International des Épizooties. Paris: BOIE, 1927-65.

League of Nations-Health Organization. Geneva League of Nations Information Section, 1931.

Final Act of the XII Pan American Sanitary Conference, Caracas, Venezuela, January 12-24, 1947. Washington, DC: Pan American Union, 1947.

Organization for the Americas, Washington, DC September 22-27, 1986. Verbatim Records. Washington, DC: Pan American Health Organization, 1987.

XXII Pan American Sanitary Conference. XXXVIII Meeting of the Regional Committee of the World Health Organization for the Americas, Washington, DC 22-27 September 1986. Verbatim Records. Washington, DC: Pan American Health Organization, 1987.

Plant Quarantine in Asia and the Pacific: Report of an APO Study Meeting 17th-26th March, 1992, Taipei, Taiwan. Tokyo: Asia Productivity Assoc., 1993.

Portugal and colonies
Boletim Oficial do Governo de Cabo-Verde.

United Kingdom and colonies
Hansard, *Parliamentary Debates.*
Report from the Select Committee appointed to consider the validity of the Doctrine of Contagion in the Plague, 14 June 1819, PP 449, 1819.
Second Report of the Select Committee appointed to consider the means of improving and maintaining the Foreign Trade of the Country, Quarantine. PP 417, 1824.
Correspondence respecting the Operation of the Commercial Treaty with Turkey, of August 16, 1838, PP 341, 1838.
Collective Note of the Representatives of Austria, France, Great Britain, Prussia, and Russia at Constantinople, to the Porte, July 27, 1839, PP 205, 1839.
Convention of Commerce and Navigation between Her Majesty, and the Sultan of the Ottoman Empire, PP 157, 1839.
Copy of the Tariff agreed upon by the Commissioners appointed under the Seventh Article of the Convention of Commerce and Navigation between Turkey and England, PP 549, 1839.
Correspondence relative to the Affairs of the Levant, Part III, PP 337, 1841.
Convention between Great Britain, Austria, France, Prussia, Russia, and Turkey respecting the Straits of the Dardanelles and of the Bosphorous, PP 350, 1842.
Correspondence relative to the Contagion of Plague and the Quarantine Regulations of Foreign Countries, 1836–1843, PP 475, 1843.
Papers Relative to the Expedition to the River Niger, PP 472, 1843.
Correspondence on the Subject of the 'Eclair' and the Epidemic which broke out on the said Vessel, PP 707, 1846.
Correspondence respecting the Quarantine Laws since the Correspondence last presented to Parliament, PP 718, 1846.
Report on the Fever at Boa Vista by Dr. McWilliam, PP 116, London: T.R. Harrison, 1847.
Documents relating to British Guiana. Demerara: Royal Gazette Office, 1848.
General Board of Health, Report on Quarantine, PP 1070, 1849.
General Board of Health, Second Report on Quarantine, PP 1473, 1852.
Statistical Reports on the Health of the Navy, for the Years 1837–43: Part III. North Coast of Spain Station, West Coast of Africa Station, Packet Service, Home Station, Ships employed variously. London: HMSO, 1854.
Papers respecting Quarantine in the Mediterranean, London: Harrison & Sons, 1860.
Fourth Report of the Sanitary Commissioner with the Government of India 1867. Calcutta: Office of the Superintendent of Government Printing, 1868.
Report of the Commissioners appointed to Inquire into the Origin, Nature, etc, of the Indian Cattle Plagues. Calcutta: Office of the Superintendent of Government Printing, 1871.
Report of the Australian Sanitary Conference of Sydney, N.S.W., 1884. Sydney: Thomas Richards, 1884.
Report on the British West Indian Conference on Quarantine. Georgetown, Demerara: C.K. Jardine, 1888.
Correspondence relative to the Outbreak of Bubonic Plague at Hong Kong. London: HMSO, 1894.
Further Correspondence relative to the Outbreak of Bubonic Plague at Hong Kong. London: HMSO, 1894.
Hong Kong: Bubonic Plague. London: HMSO, 1896.
Gold Coast. Report for 1908. London: HMSO, 1909.
Thirty-Seventh Annual Report of the Local Government Board, 1907–08. London: HMSO, 1909.
Forty-First Annual Report of the Local Government Board, 1911–12. London: HMSO, 1912.
Forty-Second Annual Report of the Local Government Board, 1912–13. London: HMSO, 1913.

Proceedings of the First Meeting of Veterinary Officers in India, held at Lahore on the 24th March 1919 and following days. Calcutta: Superintendent of Government Printing, 1919.

Proceedings of the Second Meeting of Veterinary Officers in India, held at Calcutta from 26th February to 2nd March, 1923. Calcutta: Superintendent of Government Printing, 1924.

United States of America

Philadelphia, Commonwealth of, *An Act for the Establishing an Health Office, for Securing the City and Port of Philadelphia, from the Introduction of Pestilential and Contagious Diseases.* Philadelphia: True American, 1799.

Baltimore, City Health Department, *The First Thirty-Five Annual Reports.* Baltimore: The Commissioner of Health of Baltimore, Maryland, 1853.

Act March 27, 1890, An Act to prevent the introduction of contagious diseases from one State to another and for the punishment of certain offences.

An Act granting additional quarantine powers and imposing additional duties upon the Marine-Hospital Service, February 15 1893.

Committee on Interstate and Foreign Commerce of the House of Representatives on Bills (H.R. 4363 and S. 2680) to Amend an Act entitled 'An Act granting additional Quarantine Powers and Imposing Duties upon the Marine-Hospital Service'. Washington, DC: Govt. Printing Office, 1898.

Second General International Sanitary Convention of the American Republics. Pan-American Sanitary Conference, 1905. Washington, DC: Govt. Printing Office, 1907.

Report of the Delegation from the United States of America to the Sixth International Sanitary Conference at Montevideo, December 12, 1920. Washington, DC: Govt. Printing Office, 1920.

US Department of Agriculture, *Regulations Governing the Inspection and Quarantine of Livestock and other Animals offered for Importation (except from Mexico)*, Pamphlet SF 623 B3 1945. Washington, DC: Government Printing Office, 1945.

Newspapers and periodicals
American Journal of Public Health
Annales de l'Institut Pasteur
British Medical Journal
Daily Telegraph
Financial Times
Gentleman's Magazine
Gloucester Journal
Gloucestershire Chronicle
Guardian
Hankyoreh
Japan Watch
Joong Ang Daily
Journal of Hygiene
Journal of Indian Medical Research
Korea Herald
Korean Times
Kyunghyang
Lancet
Medical Repository
Medico-Chirurgical Review
Nautical Magazine and Naval Chronicle
Naval Intelligencer
New Scientist
Philosophical Transactions of the Royal Society of London
Revista Universal Lisbonense

The Times
Transactions of the Epidemiological Society of London

Other published works

Ahmud, Syud (ed.), *Tuzuk-i-Jahangiri*, Aligarh: Private Press, 1864.
Alpini, Prosper, *De medicina Aegyptiorum*. Paris: G. Pele & I. Duval, 1646.
Anon., *Traité de la peste*. Paris: Guillaume Caelier, 1722.
Anon., *Della peste ossia della cura per preservarsene, e guarire da questo fatalismo morbo*. Venice: Leonardo & Giammaria, 1784.
Anon., 'Some Account of the Life and Writings of the late Dr. Richard Mead,' *Gentleman's Magazine*, 24 (1754), 512.
Anon., 'Quarantaines', *Dictionnaire encyclopédie des sciences medicalés*. Paris: P. Asselin & G. Masson, 1874.
Arán, S., 'Unifaction des certificats sanitaires pour le commerce international des animaux, viands et produits carnés'. *Bulletin de l' Office International des Épizooties*, 2 (1928-29), 589-98.
Arnoul, Martin, *Histoire de la derniere peste de Marseilles, Aix, Arles, et Toulon. Avec plusiers avantures arrives pendant la contagion*. Paris: Paulus du Mesnil, 1732.
Assalini, P., *Observations on the Disease called The Plague, on the Dysentery, The Opthalmy of Egypt, and on the Means of Prevention, with some Remarks on the Yellow Fever of Cadiz*, trans. A. Neale. New York: T.J. Swords, 1806.
Astruc, Jean, *Dissertation sur l'origine des maladies epidemiques et principalement sur l'origine de la peste*. Montpellier: Jean Martel, 1721.
Ayre, Joseph, *A Letter addressed to the Right Honourable Lord John Russell, M.P., Secretary of State for the Home Department, on the evil Policy of those Measures of Quarantine, and Restrictive Police, which are employed for arresting the Progress of Asiatic Cholera*. London: Longman, Orme, Brown, Green & Longman, 1837.
Bacon Phillips, T., 'Yellow Fever as it occurred on board the R.M. Steamer "La Plata", on her homeward Voyage from St. Thomas, West Indies, in the Month of November last'. *Lancet*, 26 March 1853, 293-9.
Bally, Victor, *Du typhus d'Amérique ou fièvre jaune*. Paris: Smith, 1814.
Bancroft, Edward N., *An Essay on the Disease called Yellow Fever, with Observations concerning Febrile Contagion, Typhus Fever, Dysentery, and the Plague, partly delivered as the Gulstonian Lectures, before the College of Physicians, in the Years 1806 and 1807*. London: T. Cadell & W. Davies, 1811.
——, *A Sequel to the Essay on Yellow Fever; principally intended to prove, by incontestable Facts and important Documents, that the Fever, called Bulam, or Pestilential, has no Existence as a Distinct, or a Contagious Disease*. London: J. Callow, 1817.
Barton, E.H., *Care and Prevention of Yellow Fever at New Orleans and other Cities in America*, New York: H. Baillière, 1857.
Bateman, Josiah, *The Mighty Hand of God: A Sermon preached in the Parish Church, Margate, on Friday, March 9th, 1866, being the Day of Humiliation for the Cattle Plague*. London: Simpkin, Marshall & Co., 1866.
Bell, Charles, *Remarks on Rinderpest*. London: Robert Hardwicke, 1866.
Berger, H.C.L.E., 'L'Immunisation anti-aphteuse des animaux exportés'. *Bulletin de l' Office International des Épizooties*, 1 (1927), 534-6.
Bertrand, Jean-Baptiste, *A Historical Relation of the Plague at Marseilles, in the Year 1720*, trans. A. Plumptre. London: Mawman, 1973; reprint of 1805 edn.
Birtwhistle, Richard, 'Account of the Yellow Fever on Board the "Volage"'. *Lancet*, 3 January 1846, 8-9.
Blair, Daniel, *Some Account of the Last Yellow-Fever Epidemic of British Guiana*. London: Longman & Co., 1850.
Blane, Gilbert, *Observations on the Diseases Incident to Seamen*. London: Joseph Cooper, 1785.

Bone, G.F., *Inaugural Dissertation on Yellow Fever, and on the Treatment of that Disease with Saline Medicines*. London: Longman & Co., 1846.

Bowring, John, *Observations on the Oriental Plague, and on Quarantine as a Means of Arresting its Progress*. Edinburgh: W. Tait, 1838.

Brownrigg, William, *Considerations on the Means of Preventing the Communication of Pestilential Contagion and of Eradicating it in Infected Places*. London: Lockyer Davis, 1771.

Bryson, Alexander, *Report on the Climate and Principal Diseases of the African Station*. London: W. Clowes & Sons, 1847.

Budd, William, *The Siberian Cattle Plague, or, the Typhoid Fever of the Ox*. Bristol: Kerslake & Co., 1865.

Bullein, William, 'A Dialogue both pleasant and pietyful' (1564), in Rebecca Totaro (ed.), *The Plague in Print: Essential Elizabethan Sources, 1558–1603*. Pittsburgh: Duquesne University Press, 2010.

Burgi, M., 'Les Methodes générales de la prophylaie de la fièvre aphteuse'. *Bulletin de l' Office International des Épizooties*, 1 (1927), 537–77.

Butterfield, L.H. (ed.), *Letters of Benjamin Rush*, Princeton, NJ: Princeton University Press, 1951, vol. 1.

Caldwell, Charles, *Anniversary Oration on the Subject of Quarantines, delivered to the Philadelphia Medical Society, on the 21st of January, 1807*. Philadelphia: Fry & Kammerer, 1807.

Calmette, A. and Salimberi, A.T., 'La Peste bubonique étude de l'épidémie d' Oporto en 1899; Sérothérapie', *Annales de l'Institut Pasteur*, 13 (December 1899), 865–936.

Cané, B.G., Leanes, L.F. and Mascitelli, L.O., 'Emerging Diseases and their Impact on Animal Commerce: The Argentine Lesson'. *Annals of the New York Academy of Sciences*, 1026 (2004), 1–7.

Capron, Horace (ed.), *Report of the Commissioner of Agriculture on the Diseases of Cattle in the United States*. Washington, DC: Government Printing Office, 1871.

Careri, J.F.G., 'A Voyage Round the World by Dr. John Francis Gernelli Careri, Containing the Most Remarkable Things he saw in Indostan', in J.P. Guha (ed.), *India in the Seventeenth Century*. New Delhi: Associated Publishing House, 1976, vol. 2.

Chisholm, Colin, *An Essay on the Malignant Pestilential Fever introduced into the West Indian Islands from Boullam, on the Coast of Guinea, as it appeared in 1793 and 1794*. London: C. Dilly, 1795.

———. *An Essay on the Malignant Pestilential Fever, introduced into the West Indian Islands from Boullam, on the Coast of Guinea, as it appeared in 1793, 1794, 1795, and 1796. Interspersed with Observations and Facts, tending to Prove that the Epidemic existing at Philadelphia, New-York, &c. was the same Fever introduced by Infection imported from the West Indian Islands: and Illustrated by Evidences found on the State of those Islands, and Information of the most eminent Practitioners residing on them*. London: J. Mawman, 1801.

———. *A Letter to John Haygarth, M.D. F.R.S., London and Edinburgh, &c., from Colin Chisholm, M.D. F.R.S., &c., Author of An Essay on the Pestilential Fever: Exhibiting farther Evidence of the Infectious Nature of this Fatal Distemper in Grenada, during 1793, 4, 5, and 6; and in the United States of America, from 1793 to 1805: in order to correct the pernicious Doctrine promulgated by Dr Edward Miller, and other American Physicians, relative to this destructive Pestilence*. London: Joseph Newman, 1809.

Clark, James, *A Treatise on the Yellow Fever, as it appeared in the Island of Dominica, in the Years 1793-4-5*. London: J. Murray & S. Highley, 1797.

Clark, John, *Observations on the Diseases in Long Voyages to Hot Countries, and Particularly to those which prevail in the East Indies*. London: D. Wilson & G. Nicol, 1773.

Clarke, W.W., *The Cattle Plague: A Judgment from God for the Sins of the Nation*. London: Simpkins, Marshall & Co., 1866.

Clot-Bey A.-B., *De la Peste observée en Égypte; recherches et considerations sur cette maladie*. Paris: Fortin, Masson, 1840.

Cowdrey, Jonathan, 'A Description of the City of Tripoli, in Barbary; with Observations on the Local Origin and Contagiousness of the Plague'. *Medical Repository*, 3 (1806), 154–9.

Crausius, Rudolphus Guillelmus, *Excerpta quaedam ex observation in nupera peste Hambugensi*. Jena: I. F. Beerwinckel, 1714.
Croissante, Pichaty de, 'Some Account of the Plague at Marseilles in the Year 1720'. *Gentleman's Magazine*, 24 (1754), 32–6.
Cummins, William J., 'The Yellow Fever in the West Indies'. *Lancet*, 28 May 1853, 488–90.
Curtis, Charles, *An Account of the Diseases of India*. Edinburgh: W. Laing, 1807.
Davidson, George, 'Practical and Diagnostic Observations on Yellow Fever, as it occurs in Martinique'. *Medical Repository*, 2 (1805), 244–52.
De Grau, Frederico Viñas y Cusí y Rosendo, *La Peste Bubónica Memoria sobre la Epidemia occurrida en Porto en 1899, por Jaime Ferran*. Barcelona: F. Sanchez, 1907.
De la Brosse, Guy, *Traité de la Peste*. Paris: L. & C. Perier, 1623.
De Rochefort, Charles, *Histoire naturelle et morale des Antilles de l'Amerique*, 2nd edn. Rotterdam: Arnout Leers, 1665.
Defoe, Daniel, *A Journal of the Plague Year*. London: Penguin, 2003 [1722].
Devèze, Jean, *Memoire sur la fièvre jaune afin de la proven non-contagieuse*. Paris: Ballard, 1821.
Dupeyron, De Segur, *Rapport adressé a son exc. le ministre du commerce, chargé de procéder a une enquête sur les divers régimes sanitaires de la Mediterranée*. Paris: L'Imprimerie Royale, 1834.
Ferro, Paskal Joseph, *Untersuchung der Pestanstekung, nebst zwei Aufsätzen von der Glaubwürdigkeit der meisten Pestberichte aus der Moldau und Wallachia, unter der Schädlichkeit der bisherigen Contumanzen von D. Lange and Fronius*. Vienna: Joseph Edlen, 1787.
Ffirth, S., 'Practical Remarks on the Similarity of American and Asiatic Fevers'. *Medical Repository*, 4 (1807), 21–7.
Frank, Johann Peter, *A System of Complete Medical Police*, ed. E. Lesky. Baltimore: J.H.V. Press, 1976; trans. E. Vlim from 3rd edn, Vienna, 1786.
Foggo, George, *The Policy of Restrictive Measures, or Quarantine, as applied to Cholera and Cattle Plague*. London: Head, Noble & Co., 1872.
Gamgee, John, *The Cattle Plague; with Official Reports of the International Veterinary Congresses, held in Hamburg, 1863, and in Vienna, 1865*. London: Robert Hardwicke, 1866.
——, 'Report of Professor Gamgee on the Lung Plague', in Horace Capron, *Report of the Commissioner of Agriculture on the Diseases of Cattle in the United States*. Washington, DC: Government Printing Office, 1871, 3–15.
——, *Yellow Fever: A Nautical Disease. Its Origin and Prevention*. New York: D. Appleton & Co., 1879.
A Gentleman, *A Journal Kept on a Journey from Bassora to Bagdad; over the Little Desert, to Cyprus, Rhodes, Zante, Corfu; and Otranto, in Italy; in the Year 1779*. Horsham: Arthur Lee, 1784.
Gorgas, W.C., *Sanitation in Panama*. New York: Appleton, 1915.
Granville, A.B., *A Letter to the Right Honble. W. Huskisson, M.P., President of the Board of Trade, on the Quarantine Bill*. London: J. Davy, 1825.
Gray, G.D., 'The History of the Spread of Plague in North China', in *Report of the Inter-National Plague Conference held at Mukden, April, 1911*. Manila: Bureau of Printing, 1912, 31–3.
Guiteras, Juan, 'Remarks on the Washington Sanitary Convention of 1905, with special reference to Yellow Fever and Cholera', *American Journal of Public Health*, 2 (1912), 506–14.
Halford, Henry, Blane, Gilbert, et al., *Cholera Morbus: Its Causes, Prevention, and Cure; with Disquisitions on the Contagious or Non-Contagious Nature of this dreadful Malady, by Sir Henry Halford, Sir Gilbert Blane, and eminent Birmingham Physicians, and the Lancet, and Medical Gazette, together with ample Directions regarding it, by the College of Physicians and Board of Health*. Glasgow: W.R. M'Phan, 1831.
Hankin, E.H., 'On the Epidemiology of Plague', *Journal of Hygiene*, 5 (1905), 48–83.

Heiser, V.G., 'International Aspects of Disease', in *The International Sanitary Convention of Paris, 1911-12*, trans. R.W. Johnstone. London: HMSO, 1919.

Hillary, William, *Observations on the Changes of the Air, and the Concomitant Epidemical Diseases in the Island of Barbadoes. To which is added, a Treatise on the Putrid Bilious Fever, commonly called the Yellow Fever; and such other Diseases as are indigenous or endemial in the West India Islands or in the Torrid Zone*. London: C. Hitch & L. Hawes, 1759.

Hirsch, August, *Handbook of Geographical and Historical Pathology*, vol. 1. London: The Sydenham Society, 1883.

Hirst, L. Fabian, 'Plague Fleas, with Special Reference to the Milroy Lectures, 1924'. *Journal of Hygiene*, 24 (1925), 1-16.

Holroyd, Arthur T., *The Quarantine Laws, their Abuses and Inconsistencies. A Letter, addressed to the Rt. Hon. Sir John Cam Hobhouse, Bart., M.P., President of the Board of Control*. London: Simpkins, Marshall & Co., 1839.

Holt, Joseph, *Quarantine and Commerce: Their Antagonism destructive to the Prosperity of City and State. A Reconciliation an imperative Necessity. How this may be Accomplished*. New Orleans: L. Graham & Sons, 1884.

Howard, John, *An Account of the Principal Lazarettos in Europe*. Warrington: William Eyres, 1789.

Huxham, John, *Essay on Fevers*. London: S. Austen, 1750.

Ingram, Dale, *An Historical Account of the Several Plagues that have appeared in the World since the Year 1346 with An Enquiry into the Present prevailing Opinion, that the Plague is a Contagious Disease, capable of being transported in Merchandize, from one Country to Another*. London: R. Baldwin, 1755.

Jacobi, Ludwig Friedrich, *De Peste*. Erfurt H. Grochius, 1708.

James, S.P., 'The Protection of India from Yellow Fever', *Journal of Indian Medical Research*, 1 (1913), 213-57.

Johnson, J.E., *An Address to the Public on the Advantages of a Steam Navigation to India*. London: D. Sidney & Co., 1824.

Kanold, Johannes, *Einiger Medicorus Schreiben von der in Preussen an. 1708, in Dantzig an. 1709, in Rosenberg an. 1708, und in Fraustadt an. 1709 grassireten Pest, wie auch von der wahren Beschaffenheir des Brechens, des Schweisses under der Pest-Schwären, sonderlich der Beulen*. Breslau: Fellgiebel, 1711.

Keating, J.M., *A History of the Yellow Fever: The Yellow Fever Epidemic of 1878, in Memphis, Tenn*. Memphis: Howard Association, 1879.

Khan, 'Inayat, *The Shah Jahan Nama of 'Inayat Khan*, trans. A.R. Fuller. New Delhi: Oxford University Press, 1990.

Khawajah, Hamdan ibn 'Uthman, *Ithaf al-munsifinwa-al-udaba fi al-ihtiras an al-waba*. Algiers: al-Sharikah, 1968.

Laitner, Christiano, *De febrius et morbis acutis*. Venice: H. Albricium, 1721.

Lane Notter, J., 'International Sanitary Conferences of the Victorian Era'. *Transactions of the Epidemiological Society of London*, 17 (1897-8).

Lange, Martin, *Rudimenta doctrinae de peste*. Vienna: Rudolph Graeffer, 1784.

Le Brun, Corneille, *Voyages de Corneille le Brun au Levant, c'est-à-dire, dans les principaux endroits de l'Asie Mineure, dans les isles de Chio, Rhodes, Chypres, etc.* Paris: P. Gosse & J. Neautme, 1732.

Leclainche, E., 'La Standardisation des bulletins sanitaires'. *Bulletin de l' Office International des Épizooties*, 1 (1927), 577-83.

Leith, A. H., *Abstract of the Proceedings and Reports of the International Sanitary Conference of 1866*. Bombay: Press of the Revenue, Financial and General Departments of the Secretariat, 1867.

Lind, James, *An Essay on Diseases incidental to Europeans in Hot Climates with the Method of Preventing their Fatal Consequences*. London: J. & J. Richardson, 1808 [1768].

Mabit, J., *Essai sur les maladies de l'armée de St.-Domingue en l'an XI, et principalement sur la fièvre jaune*. Paris: Ecole de Médecine, 1804.

McCabe, James, *Observations on the Epidemic Cholera of Asia and Europe*. Cheltenham: G.A. Williams, 1832.

McCall, William, *God's Judgment and Man's Duty*. London: William Macintosh, 1866.

M'Gregor [McGrigor], James *Medical Sketches of the Expedition to Egypt*. London: John Murray, 1804.

Mackenzie, Mordach, 'Extracts of Several Letters of Mordach Mackenzie, M.D. concerning the Plague at Constantinople', *Philosophical Transactions of the Royal Society*, 47 (1752), 385.

——, 'A Further Account of the late Plague at Constantinople, in a Letter of Dr. Mackenzie from thence', *Philosophical Transactions of the Royal Society*, 47 (1752), 515.

Maclean, Charles, *To the Inhabitants of British India*. Calcutta: s.n., 1798.

——, *The Affairs of Asia considered in their Effects on the Liberties of Britain*. London: s.n., 1806.

——, *A View of the Consequences of Laying Open the Trade to India, to Private Ships*. London: Black, Perry & Co., 1813.

——, *Results of an Investigation respecting Epidemic and Pestilential Diseases: Including Researches in the Levant concerning the Plague*, 2 vols. London: Thomas & George Underwood, 1817.

——, *Specimens of Systematic Misrule*. London: H. Hay, 1820.

——, *Evils of Quarantine Laws, and Non-Existence of Pestilential Contagion; deduced from the Phenomena of the Plague of the Levant, the Yellow Fever of Spain, and the Cholera Morbus of Asia*. London: T.& G. Underwood, 1824.

M'Lean, Hector, *An Enquiry into the Nature, and Causes of the Great Mortality among the Troops at St. Domingo: with Practical Remarks on the Fever of that Island, and Directions, for the Conduct of Europeans on their first Arrival in Warm Climates*. London: T. Cadell, 1797.

MacPherson, John, *Annals of Cholera: From the Earliest Periods to the Year 1817*. London: Ranken & Co., 1872.

M'William, James O., *Medical History of the Expedition to the Niger during the Years 1841–2, comprising an Account of the Fever which led to its abrupt Termination*, London: HMSO, 1843.

Makittrick, Jacobus, *Dissertatio medica inauguralis de febre indiae occidentalis maligna flava*. Edinburgh: A. Donaldson, 1764.

Manget, Jean Jacques, *Traité de la peste*. Geneva: Philippe Planche, 1721.

Manningham, Richard, *A Discourse concerning the Plague and Pestilential Fevers*. London: Robinson, 1758.

Martel, Jean, *Dissertation sur l'origine des maladies epidemiques et principalement sur l'origine de la peste*. Montpellier: Imprimeur Ordinaire du Roy, 1721.

Mead, Richard, *Of the Power of the Sun and Moon on Humane Bodies; and of the Diseases that Rise from Thence*. London: Richard Wellington, 1712.

——, *A Short Discourse concerning Pestilential Contagion, and the Methods used to Prevent it*. London: S. Buckley, 1720.

Miles, S. (ed.), *Medical Essays and Observations relating to the Practice of Physic and Surgery*. London: S. Birt, 1745.

Milroy, Gavin, *Quarantine and the Plague: Being a Summary of the Report on these Subjects recently addressed to the Royal Academy of Medicine in France*. London: Samuel Highley, 1846.

Montenegro, José Verdes, *Bubonic Plague: Its Course and Symptoms and Means of Prevention and Treatment, according to the latest Scientific Discoveries; including notes on Cases in Oporto*, trans. W. Munro. London: Baillière, Tindall & Cox, 1900.

Moseley, Benjamin, *A Treatise on Tropical Diseases*. London: T. Cadell, 1789.

Murray, John, *The Plague and Quarantine. Remarks on some Epidemic and Endemic Diseases; (including the Plague of the Levant) and the Means of Disinfection: with a Description of the Preservative Phial. Also a Postscript on Dr. Bowring's Pamphlet*, 2nd edn. London: John Murray, 1839.

Nathan, R., *The Plague in India, 1896, 1897*. Simla: Govt. Central Printing Office, 3 vols, 1898.
Netten Radcliffe, J., 'Memorandum on Quarantine in the Red Sea, and on the Sanitary Regulation of the Pilgrimage to Mecca', in *Ninth Annual Report of the Local Government Board 1879–80, Supplement containing Report and Papers submitted by the Medical Officer on the Recent Progress of Levantine Plague, and on Quarantine in the Red Sea*. London: George E. Eyre & William Spottiswoode, 1881, 98–103.
Norman White, F., *The Prevalence of Epidemic Disease and Port Health Organization and Procedure in the Far East*. Geneva: League of Nations, 1923.
O'Halloran, Thomas, *Remarks on the Yellow Fever of the South and East Coasts of Spain*. London: Callow & Wilson, 1823.
Papon, J.P., *De la Peste, ou les époques mémorables de ce fléau, et les moyen de s'en preserver*, 2nd vols. Paris: Lavillette, 1799.
Pariset, Étienne, *Observations sur la fièvre jaune, faites a Cadiz*. Paris: Audot, 1820.
Philadelphia, Academy of Medicine of, *Proofs of the Origin of Yellow Fever, in Philadelphia & Kensington, in the Year 1797, from Domestic Exhalation; and from the Foul Air of the Snow Navigation, from Marseilles: and from that of the Ship Huldah, from Hamburgh, in two Letters addressed to the Governor of the Commonwealth of Philadelphia*. Philadelphia: T. & S.F. Bradford, 1798.
A Philadelphian, *Occasional Essays on the Yellow Fever*. Philadelphia: John Ormorod, 1800.
Playfair, Lyon, *The Cattle Plague in its Relation to Past Epidemics and to the Present Attack*. Edinburgh: Edmonston & Douglas, 1865.
Pope, G.W., 'Some Results of Federal Quarantine against Foreign Live-Stock Diseases', *Yearbook of the United States Department of Agriculture*. Washington, DC: Government Printing Office, 1919.
Proust, Adrien, *Essai sur l'hygiène internationale ses applications contre la peste, la fièvre jaune et le cholera asiatique*. Paris: G. Masson, 1873.
Pym, William, *Observations upon the Bulam Fever, which has of late Years prevailed in the West Indies, on the Coast of South America, at Gibraltar, Cadiz, and other parts of Spain: with a Collection of Facts proving it to be a highly Contagious Disease*. London: J. Callow, 1815.
Rennie, Alexander, *Report on the Plague Prevailing in Canton during the Spring and Summer of 1894*. Shanghai: Chinese Imperial Maritime Customs, 1895.
Ruffer, Marc Armand, 'Measures taken at Tor and Suez against Ships coming from the Red Sea and the Far East'. *Transactions of the Epidemiological Society of London*, 19 (1899–1900), 25–47.
Rush, Benjamin, *An Account of the Bilious Remitting Yellow Fever*. Philadelphia: Thomas Dobson, 1794.
Russell, Alexander, *The Natural History of Aleppo, and Parts Adjacent, containing a Description of the City, and the Principal Natural Productions in its Neighbourhood, together with an Account of the Climate, Inhabitants, and Diseases; particularly of the PLAGUE, with the Methods used by Europeans for their Prevention*. London: A. Millar, 1756.
Russell, Patrick, *A Treatise of the Plague*. London: G.G.J. & J. Robinson, 1791.
Samoilowitz, Daniel (Samoilovich), *Mémoire sur la peste, qui, en 1771, ravage l'Empire Russe, sur-tout Moscou, la Capitale*. Paris: Leclerc, 1783.
Sandwith, F.M., *The Medical Diseases of Egypt*. London: Henry Kimpton, 1905.
Schmidt, H.D., *The Pathology and Treatment of Yellow Fever; with some Remarks upon the Nature of its Cause and its Prevention*. Chicago: Chicago Medical Press Association, 1881.
Senac, Jean Baptiste, *Traité des causes des accidens, et de la cure de la peste*. Paris: P.-J. Mariette, 1744.
Shadwell, A., 'Plague at Oporto'. *Transactions of the Epidemiological Society of London*, 19 (1899–1900).
Simpson, W.J.R., *Report on the Causes and Continuance of Plague in Hongkong and Suggestions as to Remedial Measures*. London: Waterlow & Sons, 1903.
——, *The Croonian Lectures on Plague*. London: Royal College of Physicians, 1907.

Snow, P.C.H., *Report on the Outbreak of Bubonic Plague in Bombay, 1896–7*. Bombay: Times of India Steam Press, 1897.
Starr, Chester K., *Nature and Prevention of Plant Diseases*. Toronto: The Blakiston Company, 1950.
Taylor, John, *Travels from England to India, in the Year 1789*, 2 vols. London: S. Low, 1799.
Tholozan, J.D., *Une Épidémie de peste en Mésopotamie en 1867*. Paris: Victor Masson & Fils, 1869.
——, *Prophylaxie du choléra en Orient: l'hygiène et la réforme sanitaire en Perse*. Paris: Victor Masson & Fils, 1869.
Thompson, John Ashburton, 'A Contribution to the Aetiology of Plague'. *Journal of Hygiene*, 1 (1901), 153–67.
——, *Report of the Board of Health on a Second Outbreak of Plague at Sydney, 1902*, Sydney: William Applegate Gullick, 1903.
Tommasini, J. *Recherches pathologiques sur la fièvre de Livorne de 1804, sur la fièvre jaune d'amérique*. Paris: Arthus-Bertrand, 1812.
Untainted Englishman, *The Nature of a Quarantine, as it is performed in Italy: To Guard against... the Plague: with Important Remarks on the Necessity of Laying Open the Trade to the East Indies*. London: J. Williams, 1767.
Wade, John P., *A Paper on the Prevention and Treatment of the Disorders of Seamen and Soldiers in Bengal*. London: J. Murray, 1793.
Wakil, A.W., *The Third Pandemic of Plague in Egypt: Historical, Statistical and Epidemiological Remarks on the First Thirty-Two Years of its Prevalence*. Cairo: Egyptian University, 1932.
Warren, Henry, *A Treatise concerning the Malignant Fever in Barbadoes, and the Neighbouring Islands: with an Account of the Seasons there, from the Year 1734 to 1738. In a Letter to Dr. Mead*. London: Fletcher Gyles, 1740.
Weir, T.S., 'Report from Brigade-Surgeon-Lieutenant-Colonel T.S. Weir, Executive Health Officer, Bombay, in P.C.H. Snow (ed.), *Report on the Outbreak of Bubonic Plague in Bombay, 1896–97*. Bombay: Times of India Steam Press, 1897, 237–90.
White, Norman, F., *The Prevalence of Epidemic Disease and Port Health Organization and Procedure in the Far East*. Geneva: League of Nations, 1923.
White, William, *The Evils of Quarantine Laws, and Non-existence of Pestilential Contagion*. London: Effingham Wilson, 1837.
Wilbin, John and Harvey, Alexander, 'An Account of Yellow Fever, as it occurred on board R.M.S. Ship "La Plata", in the Month of November, 1852'. *Lancet*, 12 February 1853, 148–51.
Wu, Lien-teh, 'First Report of the North Manchuria Plague Prevention Service', *Journal of Hygiene*, 13 (1913), 237–90.
—— (ed.), *Manchurian Plague Prevention Service. Memorial Volume 1912–1932*. Shanghai: National Quarantine Service, 1934.

Secondary sources

Published works

Abraham, Thomas, *Twenty-First Century Plague*. Baltimore: Johns Hopkins University Press, 2004.
Ackerknecht, Erwin, 'Anticontagionism between 1821 and 1861'. *Bulletin of the History of Medicine*, 22 (1948), 561–93.
Ackroyd, Marcus, Brockliss, Laurence, Moss, Michael, Retford, Kate and Stevenson, John, *Advancing with the Army: Medicine, the Professions, and Social Mobility in the British Isles 1790–1850*. Oxford: Oxford University Press, 2006.
Afkhami, Amir, A., 'Defending the Guarded Domain: Epidemics and the Emergence of an International Sanitary Policy in Iran'. *Comparative Studies of South Asia, Africa and the Middle East*, 19 (1999), 122–34.

Alden, D. and Miller, J.C., 'Out of Africa: The Slave Trade and the Transmission of Smallpox to Brazil, 1560–1831', in R.I. Rotberg (ed.), *Health and Disease in Human History: A Journal of Interdisciplinary History Reader*. Cambridge, Mass.: MIT Press, 2000, pp. 203–30.

Alexander, John T., *Bubonic Plague in Early Modern Russia: Public Health and Urban Disaster*. Oxford: Oxford University Press, 2003.

Ali, Harris and Keil, Roger (eds), *Networked Disease: Emerging Infections in the Global City*. Oxford: Blackwell, 2002.

Allen, Charles E., 'World Health and World Politics'. *International Organization*, 4 (1950), 27–43.

al-Sayyid Marsot, Ataf Lufti, *Egypt in the Reign of Muammad Ali*. Cambridge: Cambridge University Press, 1994.

Anderson, Warwick, *The Cultivation of Whiteness: Science, Health and Racial Destiny in Australia*. Melbourne: Melbourne University Press, 2002.

——, *Colonial Pathologies: American Tropical Medicine, Race, and Hygiene in the Philippines*. Durham, NC: Duke University Press, 2006.

Appleby, Andrew, 'The Disappearance of Plague: A Continuing Puzzle'. *English Historical Review*, 33 (1980), 161–73.

Arnold, David, 'Touching the Body: Perspectives on the Indian Plague, 1896–1900', in R. Guha (ed.), *Subaltern Studies V: Writings on South Asian History and Society*. Delhi: Oxford University Press, 1987, 55–90.

——, 'The Indian Ocean as a Disease Zone, 1500–1950'. *South Asia*, n.s. 14 (1991), 1–22.

Arrizabalaga, Jon, Henderson, John and French, Roger, *The Great Pox: The French Disease in Renaissance Europe*. New Haven and London: Yale University Press, 1997.

Baldwin, Peter, *Contagion and the State in Europe 1830–1930*. Cambridge: Cambridge University Press, 1999.

——, *Disease and Democracy: The Industrialized World Faces AIDS*. Berkeley and Los Angeles: California University Press, 2005.

Barrett, R., Kuzawa, C.W., McDade, T. and Armelagos, G.J., 'Emerging and Re-emerging Infectious Diseases: The Third Epidemiological Transition'. *Annual Review of Anthropology*, 27 (1998), 247–71.

Barry, Stephane and Gualde, Norbert, 'La Peste noir dans l'Occident chrétien et muslaman, 1347–1353', *Canadian Bulletin of Medical History*, 25 (2008), 461–98.

Bashford, Alison, *Imperial Hygiene: A Critical History of Colonialism, Nationalism and Public Health*. Basingstoke: Macmillan, 2004.

——, 'Global Biopolitics and the History of World Health'. *History of the Human Sciences*, 19 (2006), 67–88.

—— (ed.), *Medicine at the Border: Disease, Globalization and Security, 1850 to the Present*. Basingstoke: Palgrave Macmillan, 2006.

—— and Claire Hooker (eds), *Contagion: Historical and Cultural Studies*. London: Routledge, 2001.

Bayly, C.A., *The Birth of the Modern World 1780–1914*. Oxford: Blackwell, 2004.

Beck, Peter M., 'Candlelight Protests: Finding a Way Forward'. *Korea Herald*, 4 July 2008, 4.

Benchimol, Jaime Larry, *Pereira Passos: Um Haussmann Tropical*. Rio de Janeiro: Biblioteca Carioca, 1990.

Benedict, Carol, *Bubonic Plague in Nineteenth-Century China*. Stanford, Calif.: Stanford University Press, 1996.

Benedictow, Ole J., *The Black Death 1346–1353: The Complete History*. Woodbridge: Boydell Press, 2004.

Bhagrati, Jagdish, *In Defense of Globalization*. New York and Oxford: Oxford University Press, 2004.

Biraben, J.-N., *Les Hommes et la peste en France et dans les pays européens et méditerranées*, vol. 1. Paris: Mouton, 1975.

——, 'Les Routes maritimes des grandes épidémies au moyen âge', in C. Buchet (ed.), *L'Homme, la santé et la mer*. Paris: Champion, 1997, 23-37.
Birn, Anne-Emanuelle, *Marriage of Convenience: Rockefeller International Health and Revolutionary Mexico*. Rochester, NY: Rochester University Press, 2006.
Black, Jeremy, *European Warfare 1660-1815*. London and New Haven: Yale University Press, 1994.
——, *The British Seaborne Empire*. New Haven and London: Yale University Press, 2004.
Blackburn, Robin, *The Making of New World Slavery: From the Baroque to the Modern 1492-1800*. London: Verso, 1997.
Booker, John, *Maritime Quarantine: The British Experience, c.1650-1900*. Aldershot: Ashgate, 2007.
Borsch, Stuart J., *The Black Death in Egypt and England: A Comparative Study*. Austin: University of Texas Press, 2005.
Boxer, C.R. *The Dutch Seaborne Empire 1600-1800*. London: Penguin, 1990.
Bozhong, Li, 'Was there a "fourteenth-century turning point"? Population, Land, Technology and Farm Management', in P. J. Smith and R. von Glahn (eds), *The Song-Yuan Transition in Chinese History*. Cambridge, Mass.: Harvard University Press, 2003, 134-75.
Braudel, Fernand, *Civilization and Capitalism, 15th-18th Century: Volume III, The Perspective of the World*. London: Collins, 1988.
Bridge, F.R. and Bullen, Roger, *The Great Powers and the European States System 1815-1914*. London: Longman, 1980.
Brockliss, Laurence and Jones, Colin, *The Medical World of Early Modern France*. Oxford: Clarendon Press, 1997.
Brookes, Tim, *Behind the Mask: How the World Survived SARS, the First Epidemic of the Twenty-First Century*. Washington, DC: American Public Health Association, 2005.
Brooks, Francis J., 'Revising the Conquest of Mexico', in R.I. Rotberg (ed.), *Health and Disease in Human History: A Journal of Interdisciplinary History Reader*. Cambridge, Mass.: MIT Press, 2000, 15-28.
Brown, Michael, 'From Foetid Air to Filth: The Cultural Transformation of British Epidemiological Thought, ca. 1780-1848'. *Bulletin of the History of Medicine*, 82 (2008), 515-44.
Brown, Karen and Gilfoyle, Daniel (eds), *Healing the Herds: Disease, Livestock Economies, and the Globalization of Veterinary Medicine*. Athens, Ohio: Ohio University Press, 2010.
Brown, Stephen, 'The Economic Impact of SARS', in C. Loh (ed.), *At the Epicentre: Hong Kong and the SARS Outbreak*. Hong Kong: Hong Kong University Press, 2004, 179-93.
Buckley, Roger N., *The British Army in the West Indies: Society and the Military in the Revolutionary Age*. Gainesville: University Press of Florida, 1998.
Bullen, Roger, *Palmerston, Guizot and the Collapse of the Entente Cordiae*. London: Athlone Press, 1974.
Bushnell, O.A., *The Gifts of Civilization: Germs and Genocide in Hawai'i*. Honolulu: University of Hawaii Press, 1993.
Bynum, W.F., 'Cullen and the Study of Fevers in Britain, 1760-1820', in W.F. Bynum and V. Nutton (eds), *Theories of Fever from Antiquity to the Enlightenment*, Medical History Supplement, No. 1. London: Wellcome Institute for the History of Medicine, 1981, 135-48.
——, 'Policing Hearts of Darkness: Aspects of the International Sanitary Conferences'. *History and Philosophy of the Life Sciences*, 15 (1993), 421-34.
Byrne, Joseph P., *The Black Death*. Westport, Conn.: Greenwood Press, 2004.
Cain, P.J. and Hopkins, A.G., *British Imperialism, 1688-2000*. London: Routledge, 2001.
Cameron, Rondo and Neal, Larry, *A Concise Economic History of the World*. New York and Oxford: Oxford University Press, 2003.
Campbell, Christy, *Phylloxera: How Wine was Saved for the World*. London: Harper Perennial, 2004.
Cantor, David (ed.), *Reinventing Hippocrates*. Aldershot: Ashgate, 2002.

Cantor, Norman, *In the Wake of Plague: The Black Death and the World it Made*. London: Simon & Schuster, 2001.
Carmichael, Ann G., 'Plague Legislation in the Italian Renaissance'. *Bulletin of the History of Medicine*, 57 (1983), 519–25.
——, *Plague and the Poor in Renaissance Florence*. Cambridge and New York: Cambridge University Press, 1986.
——, 'Universal and Particular: The Language of Plague, 1348–1500', in Nutton (ed.), *Pestilential Complexities*, 17–52.
Carrell, Severin and Wintour, Patrick, 'Minister Urges Calm as Virus Reaches Britain'. *Guardian*, 28 April 2009, 5.
Catanach, I.J., 'Plague and the Tensions of Empire: India, 1896–1918', in D. Arnold (ed.), *Imperial Medicine and Indigenous Societies*. Manchester: Manchester University Press, 1988, 149–71.
Chalhoub, Sidney, 'The Politics of Disease Control: Yellow Fever in Nineteenth-Century Rio de Janeiro'. *Journal of Latin American Studies*, 25 (1993), 441–63.
Chandarvarkar, R., 'Plague Panic and Epidemic Politics in India, 1896–1914', in T. Ranger and P. Slack (eds), *Epidemics and Ideas*. Cambridge: Cambridge University Press, 1992, 203–40.
Chapman, Tim, *The Congress of Vienna: Origins, Processes and Results*. London: Routledge, 1998.
Chase, Marilyn, *The Barbary Plague: The Black Death in Victorian San Francisco*. New York: Random House, 2004.
Chaudhury, S. and Morineau, M. (eds), *Merchants, Companies and Trade: Europe and Asia in the Early Modern Era*. Cambridge: Cambridge University Press, 1999.
Christensen, Peter, '"In These Perilous Times": Plague and Plague Policies in Early Modern Denmark', *Medical History*, 47 (2003), 413–50.
Cipolla, Carlo M., *Fighting the Plague in Seventeenth-Century Italy*. Madison: University of Wisconsin Press, 1981.
Clark, J.F.M., *Bugs and the Victorians*. London and New Haven: Yale University Press, 2009.
Clemow, Frank G., *The Geography of Disease*. Cambridge: Cambridge University Press, 1903.
Cohn Jr., Samuel K., *The Black Death Transformed: Disease and Culture in Early Renaissance Europe*. London: Arnold, 2002.
——, 'Epidemiology of the Black Death and Successive Waves of Plague', in Nutton (ed.), *Pestilential Complexities*, 74–100.
——, *Cultures of Plague: Medical Thinking at the End of the Renaissance*. Oxford: Oxford University Press, 2010.
Coleman, William, *Yellow Fever in the North: The Methods of Early Epidemiology*. Madison: University of Wisconsin Press, 1987.
Conrad, Lawrence I., 'Epidemic Disease in Formal and Popular Thought in Early Islamic Society', in T. Ranger and P. Slack (eds), *Epidemics and Ideas: Essays on the Historical Perception of Pestilence*. Cambridge: Cambridge University Press, 1992, 77–100.
——, 'A Ninth-Century Muslim's Scholar's Discussion of Contagion', in L.I. Conrad and D. Wujastyk (eds), *Contagion: Perspectives from Pre-Modern Societies*. Aldershot: Ashgate, 2000, 163–78.
Cook, Alexandra Parma and Cook, David Noble *The Plague Files: Crisis Management in Sixteenth-Century Seville*. Baton Rouge: Louisiana State University Press, 2009.
Cook, David Noble, *Born to Die: Disease and the New World Conquest, 1492–1650*. New York: Cambridge University Press, 1998.
Cooper, Donald B., *Epidemic Disease in Mexico City 1761–1813: An Administrative, Social, and Medical Study*. Austin: University of Texas Press, 1965.
Cooper, Frederick, *Colonialism in Question: Theory, Knowledge, History*. Berkeley: University of California Press, 2005.
Cooter, Roger, 'Anticontagionism and History's Medical Record', in P. Wright and A. Treacher (eds), *The Problem of Medical Knowledge: Examining the Social Construction of Medicine*. Edinburgh: Edinburgh University Press, 1982, 87–108.

Coulter, J.L.S. and Lloyd, C., *Medicine and the Navy 1200–1900. Vol. IV: 1815–1900*. Edinburgh and London: E. & S. Livingstone, 1963.
Cowen, Peter and Morales, Roberta A., 'Economic and Trade Implications of Zoonotic Diseases', in T. Burroughs, S. Knobler and J. Lederberg (eds), *The Emergence of Zoonotic Diseases: Understanding the Impact on Animal and Human Health*. Washington, DC: National Academy Press, 2001, 20–5.
Critchell, James T., *A History of the Frozen Meat Trade*. London: Constable, 1912.
Crosby, Alfred W., *The Columbian Exchange: The Biological and Cultural Consequences of 1492*. Westport, Conn.: Greenwood Press, 1974.
——, *Ecological Imperialism: The Biological Expansion of Europe, 900–1900*. Cambridge: Cambridge University Press, 1986.
——, 'Hawaiian Depopulation as a Model for the Amerindian Experience', in P. Slack and T. Ranger (eds), *Epidemics and Ideas: Essays on the Historical Perception of Pestilence*. Cambridge: Cambridge University Press, 1992, 175–202.
Cueto, Marcos, *The Return of Epidemics: Health and Society in Peru during the Twentieth Century*. Aldershot: Ashgate, 2001.
——, *The Value of Health: A History of the Pan American Health Organization*. Washington, DC: Pan American Health Organization, 2001.
Curson, Peter and McCracken, Kevin, *Plague in Sydney: The Anatomy of an Epidemic*. Kensington: University of New South Wales Press, 1989.
Curtin, Philip D., '"The White Man's Grave": Image and Reality, 1750–1850'. *Journal of British Studies*, 1 (1961), 94–110.
——, *The Image of Africa: British Ideas and Action, 1780–1850*. Madison: University of Wisconsin Press, 1964.
——, *The Atlantic Slave Trade: A Census*. New York: Norton, 1981.
——, *Death by Migration: Europe's Encounter with the Tropical World in the Nineteenth Century*. Cambridge: Cambridge University Press, 1989.
——, *The Rise and Fall of the Plantation Complex*, 2nd edn. Cambridge: Cambridge University Press, 1998.
Darwin, John, *After Tamerlane: The Global History of Empire since 1405*. London: Allen Lane, 2007.
——, *The Empire Project: The Rise and Fall of the British World System 1830–1970*. Oxford: Oxford University Press, 2009.
Das Gupta, Ashin, 'The Merchants of Surat, c. 1700–50', in E. Leach and S.N. Mukherjee (eds), *Elites in South Asia*. Cambridge: Cambridge University Press, 1970, 201–22.
——, 'Indian Merchants and the Trade in the Indian Ocean, c. 1500–1750', in T. Raychaudhuri (ed.), *The Cambridge Economic History of India, Volume I: c.1200–c.1750*. Cambridge: Cambridge University Press, 1982, 407–33.
Davis, Deborah and Siu, Helen (eds), *SARS: Reception and Interpretations in Three Chinese Cities*. London and New York: Routledge, 2007.
Davis, Mike, *Late Victorian Holocausts*. London: Verso, 2001.
De Herdt, Roger, *Bijdrage tot de Geschiedenis van de Vetteelt in Vlaanderen, inzonderheid tot de Geschiedenis van de Rundveepest*. Louvain: Belgisch Centrum voor Landelijke Geschiedenis, 1970.
De Vries, Jan, 'The Industrial Revolution and the Industrious Revolution'. *Journal of Economic History*, 54 (1994), 240–70.
De Vries, Jan and van de Woude, Adriaan, *The First Modern Economy: Success, Failure and the Perseverance of the Dutch Economy*. Cambridge: Cambridge University Press, 1997.
Delgado, C.M., Rosegrant, M., Steinfeld, H., Ehui, S. and Courbois, C., *Livestock to 2020: The Next Food Revolution*. Washington, DC: International Food Policy Research Institute, 1999.
Diamond, Jared, *Guns, Germs and Steel: A Short History of Everybody for the Last 13,000 Years*. London: Vintage, 1998.
Dix, Andreas, 'Phylloxera', in S. Krech III, J.R. McNeill and C. Merchant (eds), *Encyclopedia of World Environmental History*. London: Routledge, 2004, 1,002.

Dodgson, Richard, Lee, Kelley and Drager, Nick, 'Global Health Governance', Discussion Paper No. 1, Centre on Global Change and Health, London School of Hygiene and Tropical Medicine, in John J. Kirton (ed.), *Global Health*. Aldershot: Ashgate, 2009, 439–62.
Doeppers, Daniel F., 'Fighting Rinderpest in the Philippines, 1886–1941', in K. Brown and D. Gilfoyle (eds), *Healing the Herds: Disease, Livestock Economies, and the Globalization of Veterinary Medicine*. Athens, Ohio: Ohio University Press, 2010, 108–28.
Dols, Michael, *The Black Death in the Middle East*. Princeton, NJ: Princeton University Press, 1977.
Donnelly, James S., *The Great Irish Potato Famine*. Stroud: Sutton, 2001.
Doughty, Edward, *Observations and Inquiries into the Nature and Treatment of the Yellow, or Bulam Fever, in Jamaica and at Cadiz; particularly in what regards its primary Cause and assigned Contagious Powers*. London: Highley & Son, 1816.
Drancourt, Michel, et al., 'Detection of 400-Year Old *Yersinia pestis* DNA in Human Dental Pulp'. *Proceedings of the National Academy of Sciences of the USA*, 95 (1998), 12637–40.
Duffy, John, *Sword of Pestilence: The New Orleans Fever Epidemic of 1853*. Baton Rouge: Louisiana State University Press, 1966.
——, *A History of Public Health in New York City 1625–1866*. New York: Russell Sage Foundation, 1968.
Durey, Michael, *Return of the Plague: British Society and the Cholera of 1831–2*. London: Macmillan, 1979.
Echenberg, Myron, *Black Death, White Medicine: Bubonic Plague and the Politics of Public Health in Colonial Senegal, 1914–1945*. Oxford: James Currey, 2002.
——, *Plague Ports: The Global Urban Impact of Bubonic Plague 1894–1901*. New York: New York University Press, 2007.
——, *Africa in the Time of Cholera: A History of Pandemics from 1817 to the Present*. Cambridge: Cambridge University Press, 2011.
Elgood, Cyril, *Medicine in Persia*. New York: Paul B. Hoeber, 1934.
Enserink, Martin and Normille, Dennis, 'Search for SARS Origins Stalls'. *Science*, 31, 302 (October 2003), 766–7.
Espinosa, Mariola, *Epidemic Invasions: Yellow Fever and the Limits of Cuban Independence, 1878–1930*. Chicago: University of Chicago Press, 2009.
Evans, Richard, *Death in Hamburg: Society and Politics in the Cholera Years 1830–1910*. Oxford: Clarendon Press, 1987.
——, 'Epidemics and Revolutions: Cholera in Nineteenth-Century Europe', in P. Slack and T. Ranger (eds), *Epidemics and Ideas*. Cambridge: Cambridge University Press, 1992, 149–74.
Fagan, Robert, 'Globalization, the WTO and the Australia–Philippines "Banana War"', in N. Fold and B. Pritchard (eds), *Cross-Continental Agro-Food Chains: Structures, Actors and Dynamics in the Global Food System*. London: Routledge, 2006, 207–22.
Falola, Toyin and Heaton, Matthew M., *HIV/AIDS, Illness, and African Well-being*. Rochester, NY: University of Rochester Press, 2007.
Fan, Yun and Chen, Ming-chi, 'The Weakness of a Post-authoritarian Democratic Society: Reflections upon Taiwan's Societal Crisis during the SARS Outbreak', in D. Davis and H.F. Siu (eds), *SARS: Reception and Interpretation in Three Chinese Cities*. London: Routledge, 2007, 147–64.
Fee, Elizabeth and Fox, Daniel M. (eds), *AIDS: The Making of a Chronic Disease*. Berkeley and Los Angeles: California University Press, 1992.
Fidler, David P., 'Germs, Governance, and Global Public Health in the Wake of SARS'. *Journal of Clinical Investigation*, 113 (2004), 799–804.
Findlay, G.M., 'The First Recognized Epidemic of Yellow Fever'. *Transactions of the Royal Society of Tropical Medicine and Hygiene*, 35 (1941), 143–54.
Findlay, Ronald and O'Rourke, Kevin H., *Power and Plenty: Trade, War, and the World Economy in the Second Millennium*. Princeton, NJ: Princeton University Press, 2007.
Finger, Simon, 'An Indissoluble Union: How the American War for Independence Transformed Philadelphia's Medical Community and Created a Public Health Establishment'. *Pennsylvania History*, 77 (2010), 37–72.

Fischer, W. and McInnis, R.M. (eds), *The Emergence of a World Economy 1500-1914*, 2 vols. Wiesbaden: Franz Steiner Verlag, 1986.
Fisher, John R., 'Cattle Plagues Past and Present: The Mystery of Mad Cow Disease'. *Journal of Contemporary History*, 33 (1998), 215-28.
Flinn, M.W., 'Plague in Europe and the Mediterranean Countries'. *Journal of European Economic History*, 8 (1979), 131-48.
Foucault, Michel, 'The Politics of Health in the Eighteenth Century', in C. Gordon (ed.), *Michel Foucault, Power/Knowledge: Selected Interviews and Other Writings 1972-1977*. Brighton: Harvester Press, 1988, 166-82.
Foxhall, Katherine, 'Fever, Immigration and Quarantine in New South Wales, 1837-1840', *Social History of Medicine*, 24 (2011), 624-43.
Frandsen, Karl-Erik, *The Last Plague in the Baltic Region 1709-1713*. Copenhagen: Museum Tusculanum Press, 2010.
Frost, Robert I., *The Northern Wars 1558-1721*. Harlow: Pearson, 2000.
Fukuda, Mahito, *Netsu to Makoto ga areba*. Kyoto: Minerva, 2008.
Fukuyama, Francis, *The End of History and the Last Man*. New York: Avon Books, 1992.
Furber, Holden, *Rival Empires of Trade in the Orient, 1600-1800*. Minneapolis: University of Minnesota Press, 1976.
Gallagher, Nancy E., *Medicine and Power in Tunisia, 1780-1900*. New York and Cambridge: Cambridge University Press, 1983.
Gamsa, Mark, 'The Epidemics of Pneumonic Plague in Manchuria 1910-1911'. *Past and Present*, 190 (2006), 147-83.
Garza, Randal P., *Understanding Plague: The Medical and Imaginative Texts of Medieval Spain*. New York: Peter Lang, 2008.
Geggus, David, *Slavery, War, and Revolution: The British Occupation of Saint Dominique, 1793-1798*. Oxford: Clarendon Press, 1982.
Geyer, Martin H. and Paulmann, Johannes (eds), *The Mechanics of Internationalism: Culture, Society, and Politics from the 1840s to the First World War*. Oxford: Oxford University Press, 2001.
Gignilliat, John L., 'Pigs, Politics, and Protection: The European Boycott of American Pork, 1879-1891'. *Agricultural History*, 35 (1961), 3-12.
Gills, B.K. and Thompson, W.R. (eds), *Globalization and Global History*. London: Routledge, 2006.
Goodyear, James D., 'The Sugar Connection: A New Perspective on the History of Yellow Fever in West Africa'. *Bulletin of the History of Medicine*, 52 (1978), 5-21.
Gordon, Daniel, 'The City and the Plague in the Age of Enlightenment'. *Yale French Studies*, 92 (1997), 67-87.
Gottfried, Robert S., *The Black Death: Natural and Human Disaster in Medieval Europe*. London: Robert Hale, 1983.
Gottheil, Paul, 'Historical Development of Steamship Agreements and Conferences in the American Foreign Trade'. *Annals of the American Academy of Political and Social Science*, 55 (1914), 48-74.
Grafe, Regina, 'Turning Maritime History into Global History: Some Conclusions from the Impact of Globalization in Early Modern Spain', in M. Fusaro and A. Polomia (eds), *Research in Maritime History. No. 43: Maritime History as Global History*. St John's: International Maritime History Association, 2010, 249-66.
Gray, John, *False Dawn: The Delusions of Global Capitalism*. London: Granta, 2009.
Green, William A. *British Slave Emancipation: The Sugar Colonies and the Great Experiment 1830-1865*. Oxford: Clarendon Press, 1991 [1976].
Greger, Michael, 'Industrial Animal Agriculture's Role in the Emergence and Spread of Disease', in J. D'Silva and J. Webster (eds), *The Meat Crisis: Developing More Sustainable Production and Consumption*. London: Earthscan, 2010, 161-72.
Grob, Gerald N., *The Deadly Truth: A History of Disease in America*. Cambridge, Mass.: Harvard University Press, 2002.
Gummer, Benedict, *The Scourging Angel: The Black Death in the British Isles*. London: The Bodley Head, 2009.

Haakonssen, Lisabeth, *Medicine and Morals in the Enlightenment: John Gregory, Thomas Percival and Benjamin Rush*. Amsterdam and Atlanta: Rodopi Press, 1997.
Hamlin, Christopher, 'Politics and Germ Theories in Victorian Britain: The Metropolitan Water Commissions of 1867–9 and 1892–3', in R. MacLeod (ed.), *Government and Expertise: Specialists, Administrators and Professionals, 1860–1919*. Cambridge: Cambridge University Press, 1988), 110–27.
——, *Public Health and Social Justice in the Age of Chadwick: Britain, 1800–1854*. Cambridge: Cambridge University Press, 1998.
——, *Cholera: The Biography*. Oxford: Oxford University Press, 2009.
Hardy, Anne, 'Cholera, Quarantine, and the English Preventive System', *Medical History*, 37 (1993), 252–69.
Harley, C. Knick, 'Steers Afloat: The North Atlantic Meat Trade, Liner Predominance, and Freight Rates, 1870–1913'. *Journal of Economic History*, 68 (2008), 1028–58.
Harnetty, P., *Imperialism and Free Trade: Lancashire in the Mid-Nineteenth Century*. New York: Columbia University Press, 1972.
Harrison, Mark, 'Towards a Sanitary Utopia? Professional Visions and Public Health in India, 1880–1914'. *South Asia Research*, 10 (1990), 19–40.
——, *Public Health in British India: Anglo-Indian Preventive Medicine 1859–1914*. Cambridge: Cambridge University Press, 1994.
——, 'A Question of Locality: The Identity of Cholera in British India, 1860–1890', in D. Arnold (ed.), *Warm Climates and Western Medicine: The Emergence of Tropical Medicine, 1500–1900*. Amsterdam and Atlanta: Rodopi, 1996, 133–59.
——, *Climates and Constitutions: Health, Race, Environment and British Imperialism in India 1600–1850*. Delhi: Oxford University Press, 1999.
——, 'From Medical Astrology to Medical Astronomy: Sol-Lunar and Planetary Theories of Disease in British Medicine, c.1700–1850'. *British Journal for the History of Science*, 33 (2000), 25–48.
——, *Disease and the Modern World: 1500 to the Present Day*. Cambridge: Polity, 2004.
——, 'An "Important and Truly National Subject": The West Africa Service and the Health of the Royal Navy in the Mid-Nineteenth Century', in S. Archer and D. Haycock (eds), *Health and Medicine at Sea*. London: Boydell & Brewer, 2010, 108–27.
——, *Medicine in the Age of Commerce and Empire: Britain and its Tropical Colonies, 1660–1830*. Oxford: Oxford University Press, 2010.
Hatcher, John, *Population and the English Economy 1348–1530*. London: Macmillan, 1987.
Hazareesingh, Sandip, *The Colonial City and the Challenge of Modernity: Urban Hegemonies and Civic Contestations in Bombay (1900–1925)*. Hyderabad: Orient Longman, 2007.
Heaman, E.A., 'The Rise and Fall of Anticontagionism in France'. *Canadian Bulletin of the History of Medicine*, 12 (1995), 3–25.
Henderson, John, *Piety and Charity in Late Medieval Florence*. Oxford: Clarendon Press, 1994.
——, *The Renaissance Hospital: Healing the Body and Healing the Soul*. London and New Haven: Yale University Press, 2006.
Herlihy, David, *The Black Death and the Transformation of the West*. Cambridge, Mass.: Harvard University Press, 1997.
Hillemand, Bertrand, 'L' Épidémie de fièvre jaune de Saint-Nazaire en 1861'. *Histoire des sciences médicales*, 40 (2006), 23–36.
Hirst, Fabian L., *The Conquest of Plague: A Study of the Evolution of Epidemiology*. Oxford: Clarendon Press, 1953.
Hopkins, A.G. (ed.), *Globalization in World History*. London: Pimlico, 2002.
Horrox, Rosemary, *The Black Death*. Manchester: Manchester University Press, 1994.
Howard, Jr., William Travis, *Public Health Administration and the Natural History of Disease in Baltimore, Maryland 1797–1920*. Washington, DC: Carnegie Institution of Washington, 1924.

Howard-Jones, Norman, 'Origins of International Health Work'. *British Medical Journal*, 6 May 1950, 1032–37.
——, *The Scientific Background to the International Sanitary Conferences, 1851–1938*. Geneva: World Health Organization, 1975.
——, *The Pan American Health Organization: Origins and Evolution*. Geneva: WHO, 1981.
Howe, Anthony, 'Free Trade and Global Order: The Rise and Fall of a Victorian Vision', in Duncan Bell (ed.), *Victorian Visions of Global Order: Empire and International Relations in Nineteenth-Century Political Thought*. Cambridge: Cambridge University Press, 2007, 26–46.
Howell, Raymond, *The Royal Navy and the Slave Trade*. London: Croom Helm, 1987.
Hoy, Suellen and Nugent, Walter, 'Public Health or Protectionism? The German–American Pork War, 1880–1891'. *Bulletin of the History of Medicine*, 63 (1989), 198–224.
Huber, Valeska, 'The Unification of the Globe by Disease? The International Sanitary Conferences on Cholera, 1851–1894'. *Historical Journal*, 49 (2006), 453–76.
Hull, Terence H., 'Plague in Java', in N.G. Owen (ed.), *Death and Disease in Southeast Asia: Explorations in Social, Medical and Demographic History*. Singapore: Oxford University Press, 1987, 210–34.
Humphreys, Margaret, *Yellow Fever and the South*. Baltimore and London: Johns Hopkins University Press, 1992.
Hünniger, Dominick, 'Policing Epizootics: Legislation and Administration during Outbreaks of Cattle Plague in Eighteenth-Century Northern Germany as Continuous Crisis Management', in K. Brown and D. Gilfoyle (eds), *Healing the Herds: Disease, Livestock Economies, and the Globalization of Veterinary Medicine*. Athens, Ohio: Ohio University Press, 2010, 76–91.
Huppert, G., *After the Black Death*. Bloomington: Indiana University Press, 1986.
Iliffe, John, *The African AIDS Epidemic*. Oxford: James Currey, 2006.
Ingram, E. (ed.), *Two Views of British India: The Private Correspondence of Mr Dundas and Lord Wellesley: 1798–1801*. London: Adams & Dart, 1969.
Inikori, A., 'Africa and the Globalisation Process: West Africa 1450–1850'. *Journal of Global History*, 2 (2007), 63–86.
Iriye, Akira, *Cultural Internationalism and World Order*. Baltimore and London: Johns Hopkins University Press, 1997.
Javelosa, Josyline and Schmitz, Andrew, 'Costs and Benefits of a WTO Dispute: Philippine Bananas and the Australian Market'. *Estey Centre Journal of International Law and Trade Policy*, 7 (2006), 78–83.
Johnson, Ryan, '*Mantsemei*, Interpreters, and the Successful Eradication of Plague: The 1908 Plague Epidemic in Colonial Accra', in R. Johnson and A. Khalid (eds), *Public Health in the British Empire: Intermediaries, Subordinates, and the Practice of Public Health, 1850–1960*. London: Routledge, 2011, 135–53.
Jones, J.R., *The Anglo-Dutch Wars of the Seventeenth Century*. New York: Longman, 1996.
Jung-A, Song, 'South Korea Removes Ban on US Beef', *Financial Times*, 30 May 2008.
Kahn, Richard J. and Kahn, Patricia G., 'The *Medical Repository* – The First US Medical Journal (1797–1824)', *New England Medical Journal*, 337 (1997), 1926–30.
Kale, Madhavi, *Fragments of Empire: Capital, Slavery, and Indian Indentured Labor Migration in the British Caribbean*. Philadelphia: University of Pennsylvania Press, 1998.
Kaplan, Herbert H., *The First Partition of Poland*. New York and London: Columbia University Press, 1969.
Kastner, Justin, et al., 'Scientific Conviction amidst Scientific Controversy in the Transatlantic Livestock and Meat Trade'. *Endeavour*, 29 (2005), 78–83.
Keith, Robert C., *Baltimore Harbor*. Baltimore and London: Johns Hopkins University Press, 2005.
Kelly, Catherine, '"Not from the College, but through the Public and the Legislature": Charles Maclean and the Relocation of Medical Debate in the Early Nineteenth Century'. *Bulletin of the History of Medicine*, 82 (2008), 545–69.

Kelly, John, *The Great Mortality: An Intimate History of the Black Death*, London and New York: Fourth Estate, 2005.
Kickbusch, Ilona, 'The Development of International Health Policies – Accountability Intact?' *Social Science and Medicine*, 51 (2000), 979–89.
—— and de Leeuw, Evelyne, 'Global Public Health: Revisiting Healthy Public Policy at the Global Level'. *Health Promotion International*, 14 (1999), 285–8.
Kilbourne, Edwin D., 'A Virologist's Perspective on the 1918–19 Pandemic', in H. Phillips and D. Killingray (eds), *The Spanish Influenza Pandemic of 1918–19: New Perspectives*. London: Routledge, 2003, 29–38.
Kim, Jong-soo, 'Spaghetti Westerns Come to Seoul'. *Joong Ang Daily*, 3 July 2008, 10.
Kim, So-hyun, 'Expat, Korean Coalition plans Counter-Demonstration in Seoul'. *Korea Herald*, 4 July, 3.
Kim, Yon-se, 'President Appeals for End to Beef Row'. *Korean Times*, 26 June 2008, 1.
Kimball, Anna Marie, *Risky Trade: Infectious Disease in the Era of Global Trade*. Aldershot: Ashgate, 2006.
Kiple, Kenneth F., *The Caribbean Slave: A Biological History*. Cambridge: Cambridge University Press, 1984.
—— (ed.), *The African Exchange: Towards a Biological History of Black People*. Durham, NC: Duke University Press, 1987.
—— and Beck, S.V. (eds), *The Biological Consequences of European Expansion, 1450–1800*. Aldershot: Variorum, 1997.
Kissinger, Henry A., *A World Restored: Metternich, Castlereagh and the Problems of Peace 1812–22*. London: Weidenfeld & Nicolson, 1957.
Klein, H.S. and Engermann, S.L., 'A Note on Mortality in the French Slave Trade in the Eighteenth Century', in H.A. Gemery and J.S. Hogendorn (eds), *The Uncommon Market: Essays on the Economic History of the Atlantic Slave Trade*. New York: Academic Press, 1979.
Koolmees, Peter A., 'Epizootic Diseases in the Netherlands, 1713–2002: Veterinary Science, Agricultural Policy, and Public Response', in K. Brown and D. Gilfoyle (eds), *Healing the Herds: Disease, Livestock Economies, and the Globalization of Veterinary Medicine*. Athens, Ohio: Ohio University Press, 2010, 19–41.
Kraut, Alan M., *Silent Travelers: Germs, Genes, and the 'Immigrant Menace'*. New York: Basic Books, 1994.
——, 'Plagues and Prejudice: Nativism's Construction of Disease in Nineteenth- and Twentieth-Century New York City', in D. Rosner (ed.), *Hives of Sickness: Public Health and Epidemics in New York City*. New Brunswick: Rutgers University Press, 1995, 65–94.
Kudlick, Catherine J., *Cholera in Post-Revolutionary Paris: A Cultural History*. Berkeley: University of California Press, 1996.
Kuhnke, La Verne, *Lives at Risk: Public Health in Nineteenth-Century Egypt*. Berkeley: University of California Press, 1990.
Kunitz, Stephen J., *Disease and Social Diversity: The European Impact on the Health of Non-Europeans*. New York: Oxford University Press, 1994.
La Berge, Ann F., *Mission and Method: The Early Nineteenth-Century French Public Health Movement*. Cambridge: Cambridge University Press, 1992.
Ladurie, Emmanuel Le Roy, *The Mind and Method of the Historian*. Brighton: Harvester, 1981.
Lal, Deepak, *Reviving the Invisible Hand: The Case for Classical Liberalism in the Twenty-First Century*. Princeton, NJ: Princeton University Press, 2006.
Langford, Paul, *A Polite and Commercial People: England 1727–1783*. Oxford: Clarendon Press, 1989.
Law, Robin, *The Slave Coast of West Africa*. Oxford: Clarendon Press, 1991.
Lawrence, Christopher, 'Disciplining Disease: Scurvy, the Navy, and Imperial Expansion, 1750–1825', in D.P. Miller and P.H. Reill (eds), *Visions of Empire: Voyages, Botany, and Representations of Nature*. Cambridge: Cambridge University Press, 1996, 80–106.

Lee, Hee-ok, 'Spirit of the Times'. *Joong Ang Daily*, 4 July 2008, 10.
Leigh, Robert D., *Federal Health Administration in the United States*, New York: Harper, 1927.
Leung, Angela Ki Che, 'Diseases of the Premodern Period in China', in K.F. Kiple (ed.), *The Cambridge World History of Human Disease*. Cambridge: Cambridge University Press, 1993, 354–62.
Leung, A.K.C., 'The Evolution of the Idea of *Chuanran* Contagion in Imperial China', in A.K.C. Leung and C. Furth (eds), *Health and Hygiene in Chinese East Asia: Policies and Publics in the Long Twentieth Century*. Durham, NC and London: Duke University Press, 2010.
Leveen, E.P., *British Slave Trade Suppression Policies*. New York: Arno Press, 1977.
Lewis, Milton J., *The People's Health: Public Health in Australia, 1788–1950*. Westport, Conn.: Praeger, 2003.
——, 'Public Health in Australia from the Nineteenth to the Twenty-First Century', in M. Lewis and K. MacPherson (eds), *Public Health in Asia and the Pacific*. London: Routledge, 2008, 222–49.
Lian, Jun, *Zhungguo Gudai Yizheng Shilue*. Huhehaste: Nei Monggu Renmin, 1995.
Little, Lester K. (ed.), *Plague and the End of Antiquity: The Pandemic of 541–750*. Cambridge: Cambridge University Press, 2007.
Lloyd, Christopher, *The Navy and the Slave Trade*. London: Longmans, Green & Co., 1949.
Loh, Christine and Welker, Jennifer, 'SARS and the Hong Kong Community', in C. Low (ed.), *At the Epicentre: Hong Kong and the SARS Outbreak*. Hong Kong: Hong Kong University Press, 2004, 215–34.
Lovell, George W., 'Disease and Depopulation in Early Colonial Guatemala', in D.N. Cook and W.G. Lovell (eds), *The Secret Judgments of God: Native Peoples and Old World Disease in Colonial Spanish America*. Norman: University of Oklahoma Press, 1992, 51–85.
Lyons, F.S.L., *Internationalism in Europe 1815–1914*. Leiden: A.W. Sythoff, 1963.
Lyons, J.B., 'A Dublin Observer of the Lisbon Yellow Fever Epidémie'. *Vesalius*, 1 (1995), 8–12.
McCaa, Robert, 'Spanish and Nahuatl Views on Smallpox and Demographic Collapse in Mexico,' in R.I. Rotberg (ed.), *Health and Disease in Human History: A Journal of Interdisciplinary History Reader*. Cambridge, Mass.: MIT Press, 2000, 167–202.
McDonald, J.C., 'The History of Quarantine in Britain during the 19th Century'. *Bulletin of the History of Medicine*, 25 (1951), 22–44.
Machado, Manuel A., *Aftosa: A Historical Survey of Foot-and-Mouth Disease and Inter-American Relations*. Albany: State University of New York Press, 1969.
McKay, D. and Scott, H.M., *The Rise of the Great Powers 1645–1815*. London: Longman, 1983.
McKie, Robert, 'Swine Flu is Officially a Pandemic. But Don't Worry . . . Not Yet, Anyway'. *Observer*, 14 June 2009, 22.
McNeill, John R., 'The Ecological Basis of Warfare in the Caribbean, 1700–1804', in M. Utlee (ed.), *Adapting to Conditions: War and Society in the Eighteenth Century*. Tuscaloosa: University of Alabama Press, 1986, 26–42.
——, *Mosquito Empires: Ecology and War in the Greater Caribbean, 1620–1914*. Cambridge: Cambridge University Press, 2010.
—— and McNeill, William H., *The Human Web: A Bird's-eye View of World History*. New York: W.W. Norton, 2003.
McNeill, William H., *Plagues and Peoples*. New York: Monticello, 1976.
MacPherson, Kerrie L., *A Wilderness of Marshes: The Origins of Public Health in Shanghai, 1843–1893*. London: Oxford University Press, 1987.
——, 'Invisible Borders: Hong Kong, China and the Imperatives of Public Health', in M.L. Lewis and K.L. MacPherson (eds), *Public Health in Asia and the Pacific: Historical and Comparative Perspectives*. London: Routledge, 2008, 10–54.

Magee, Gary B. and Thompson, Andrew S., *Empire and Globalisation: Networks of People, Goods and Capital in the British World, c.1850–1914*. Cambridge: Cambridge University Press, 2010.

Maglen, Krista, 'A World Apart: Geography, Australian Quarantine, and the Mother Country'. *Journal of the History of Medicine and Allied Sciences*, 60 (2005), 196–217.

——, '"In this Miserable Spot Called Quarantine": The Healthy and Unhealthy in Nineteenth-Century Australian and Pacific Quarantine Stations'. *Science in Context*, 19 (2006), 317–36.

Mander, Jerry and Goldsmith, Edward (eds), *The Case Against the Global Economy and For a Turn towards the Local*. Berkeley: Sierra Club Books, 1996.

Manderson, Lenore, 'Wireless Wars in the Eastern Arena: Epidemiological Surveillance, Disease Prevention and the Work of the Eastern Bureau of the League of Nations Health Organization', in P. Weindling (ed.), *International Health Organizations and Movements 1918–1939*. Cambridge: Cambridge University Press, 1995, 109–33.

Mariani, Meredith T., *The Intersection of International Law, Agricultural Biochemistry, and Infectious Disease*. Leiden: Brill, 2007.

Markel, Howard, *Quarantine! East European Jewish Immigrants and the New York City Epidemics of 1892*. Baltimore: Johns Hopkins University Press, 1997.

Marks, Steven G., *Road to Power: The Trans-Siberian Railroad and the Colonization of Asian Russia, 1850–1917*. London: I.B. Tauris, 1991.

Marriot, Edward, *The Plague Race: A Tale of Fear, Science and Heroism*. London: Picador, 2002.

Mateos, Miguel Ángel Cuerya, *Puebla de los Ángeles en Tiempos de Una Peste Colonial: Una Mirada en Torno al Matlazahuatl de 1737*. Puebla: El Colegio de Michoacan, 1999.

Mayne, Alan, '"The Dreadful Scourge": Responses to Smallpox in Sydney and Melbourne', in R. MacLeod and M. Lewis (eds), *Disease, Medicine, and Empire: Perspectives on Western Medicine and the Experience of European Expansion*. London: Routledge, 1988, 219–41.

Mazlisch, B. and Buultjens, R. (eds), *Conceptualizing Global History*. Boulder, Col.: Westview Press, 1993.

Meade, Teresa, '"Civilizing Rio de Janeiro": The Public Health Campaign and the Riot of 1904'. *Journal of Social History*, 20 (1986), 301–22.

Meers, P.D., 'Yellow Fever in Swansea, 1865', *Journal of Hygiene*, 97 (1986), 185–91.

Mishra, Saurabh, 'Beyond the Bounds of Time? The Haj Pilgrimage from the Indian Subcontinent, 1865–1920', in B. Pati and M. Harrison (eds), *The Social History of Health and Medicine in Colonial India*. London: Routledge, 2009, 31–44.

——, *Pilgrimage, Politics, and Pestilence: The Haj from the Indian Subcontinent 1860–1920*. New Delhi: Oxford University Press, 2011.

Mohr, James C., *Plague and Fire: Battling Black Death and the 1900 Burning of Honolulu's Chinatown*. New York: Oxford University Press, 2005.

Moorehead, Alan, *Fatal Impact*. New York: Harper & Row, 1966.

Moote, A. Lloyd and Moote, Dorothy C., *The Great Plague: The Story of London's Most Deadly Year*. Baltimore and London: Johns Hopkins University Press, 2004.

Morelli, Giovanna et al., 'Yersinia pestis Genome Sequencing identifies Patterns of Global Phylogenetic Diversity'. *Nature Genetics*, 42 (2010), 1, 140–43.

Moreno, Leandro Ruiz, *La Peste Historica de 1871: Fiebre Amarilla en Buenos Aires y Corrientes*. Parana: Neuva Impressora, 1949.

Morgan, David, *The Mongols*. Oxford: Basil Blackwell, 1986.

Morton, R.S., *Venereal Diseases*. London: Penguin, 1974.

Mullet, Charles F., 'Politics, Economics and Medicine: Charles Maclean and Anticontagion in England'. *Osiris*, 10 (1952), 224–51.

Naphy, William G., *Plagues, Poisons and Potions: Plague-spreading Conspiracies in the Western Alps c.1530–1640*. Manchester: Manchester University Press, 2002.

—— and Spicer, Andrew, *The Black Death: A History of Plagues 1345–1730*. Stroud: Tempus, 2001.

Neill, Michael, *Issues of Death: Mortality and Identity in English Renaissance Tragedy*. Oxford: Clarendon Press, 1997.

Nicolson, Harold, *The Congress of Vienna: A Study in Allied Unity 1812–1822*. London: Constable & Co., 1946.

Norris, John, 'East or West? The Geographic Origin of the Black Death'. *Bulletin of the History of Medicine*, 51 (1977), 1–24.

Nozawa, Takuro, 'Impact of SARS: Business with China Remains Stable thus Far'. *Japan Watch*, 16 July 2003.

Nutton, Vivian, 'The Seeds of Disease: An Explanation of Contagion and Infection from the Greeks to the Renaissance'. *Medical History*, 27 (1983), 1–34.

——, 'Medicine in Medieval Western Europe', in L. I. Conrad, M. Neve, V. Nutton, R. Porter and A. Wear, *The Western Medical Traditions 800 BC to AD 1800*, Cambridge: Cambridge University Press, 1995, 139–206.

——, 'Did the Greeks Have a Word for It?, in L.I. Conrad and D. Wujastyk (eds), *Contagion: Perspectives from Pre-Modern Societies*. Aldershot: Ashgate, 2000, 137–62.

—— (ed.), *Pestilential Complexities: Understanding Medieval Plague, Medical History*, Supplement no. 27. London: Wellcome Centre for the History of Medicine at UCL, 2008.

Nye, Joseph S., *Power in the Global Information Age: From Realism to Globalization*. London: Routledge, 2004.

O'Brien, Patrick, 'Europe in the World Economy' in H. Bull and A. Watson (eds), *The Expansion of International Society*. Oxford: Clarendon Press, 1984, 43–60.

Oh, Byung-sang, 'Keep Church and State Separate'. *Joong Ang Daily*, 2 July 2008, 10.

Owen, R., *The Middle East in the World Economy 1800–1814*. London: I.B. Tauris, 1981.

Panzac, Daniel, *La Peste dans l'empire Ottoman 1700–1850*. Louvain: Peeters, 1985.

——, *Quarantaines et lazarets: l'empire et la peste*. Aix-en-Provence: Édisud, 1986.

——, *Commerce et navigation dans l'empire Ottoman au XVIIIe siècle*. Istanbul: Isis, 1996.

——, *Populations et santé dans l'empire Ottoman (XVLLe–XXe siècles)*. Istanbul: Isis, 1996.

Park, Sang-woo and Ser, Myo-ja, 'Despite Warning, Thousands Strike', *Joong Ang Daily*, 3 July 2008, 1.

Parker, Matthew, *Panama Fever: The Battle to Build the Canal*. London: Hutchinson, 2007.

Patterson, K. David, 'Yellow Fever Epidemics and Mortality in the United States, 1693–1905'. *Social Science and Medicine*, 34 (1992), 855–65.

Pauwelyn, Joost, 'The WTO Agreement on Sanitary and Phytosanitary (SPS) Measures as Applied in the First Three SPS Disputes'. *Journal of International Economic Law*, 4 (1999), 641–64.

Peard, Julyan G., *Race, Place, and Medicine: The Idea of the Tropics in Nineteenth-Century Brazilian Medicine*. Durham, NC and London: Duke University Press, 1999.

Pelling, Margaret, *Cholera, Fever and English Medicine 1825–1865*. Oxford: Oxford University Press, 1978.

——, *The Common Lot: Sickness, Medical Occupations, and the Urban Poor in Early Modern England*. London: Longman, 1998.

——, 'The Meaning of Contagion: Reproduction, Medicine and Metaphor', in A. Bashford and C. Hooker (eds), *Contagion: Historical and Cultural Studies*. London: Routledge, 2001, 15–38.

Pernick, Martin S., 'Politics, Parties and Pestilence: Epidemic Yellow Fever in Philadelphia and the Rise of the First Party System', in J. Walzer Leavitt and R.L. Numbers (eds), *Sickness and Health in America: Readings in the History of Medicine and Public Health*. Madison: University of Wisconsin Press, 1985.

Perrins, Robert J., 'Doctors, Disease and Development: Emergency Colonial Public Health in Southern Manchuria, 1905–1926', in M. Low (ed.), *Building a Modern Japan: Science, Technology, and Medicine in the Meiji Era and Beyond*. Basingstoke: Palgrave, 2005.

Pestana, Carla G., *The English Atlantic in an Age of Revolution 1640–1661*. Cambridge, Mass.: Harvard University Press, 2004.

Peterson, Susan, 'Epidemic Disease and National Security'. *Security Studies*, 12 (2002), 43–81.
Phillips, Howard and Killingray, David (eds), *The Spanish Influenza Pandemic of 1918–19: New Perspectives*. London: Routledge, 2003.
Phillipson, Coleman and Buxton, Noel, *The Question of the Bosphorus and Dardanelles*. London: Stevens & Hayes, 1917.
Phoofolo, Pule, 'Epidemics and Revolutions: The Rinderpest Epidemics in Late Nineteenth-Century Southern Africa'. *Past and Present*, 138 (1993), 112–43.
Pietschmann, Horst, (ed.), *Atlantic History: History of the Atlantic System 1580–1830*. Göttingen: Vandenhoeck & Ruprecht, 2002.
Pormann, Peter E. and Savage-Smith, Emilie, *Medieval Islamic Medicine*. Edinburgh: Edinburgh University Press, 2007.
Porter, Dorothy, *Health, Civilization and the State: A History of Public Health from Ancient to Modern Times*. London: Routledge, 1999.
Porter, Stephen, *The Great Plague*. Stroud: Sutton, 1999.
Postma, Johannes M., *The Dutch in the Atlantic Slave Trade*. Cambridge: Cambridge University Press, 1990.
Powell, J.H., *Bring Out Your Dead: The Great Plague of Yellow Fever in Philadelphia in 1793*. Philadelphia: University of Pennsylvania Press, 1949.
Price-Smith, Andrew T., *Contagion and Chaos: Disease, Ecology, and National Security in the Era of Globalization*. Cambridge, Mass.: MIT Press, 2009.
Prinzing, F., *Epidemics Resulting from Wars*. Oxford: Clarendon Press, 1916.
Pullan, Brian, 'Plague Perceptions and the Poor in Early Modern Italy', in T. Ranger and P. Slack (eds), *Epidemics and Ideas*. Cambridge: Cambridge University Press, 1992, 101–24.
Quataert, Donald, 'Population', in H. Inalcik and D. Quataert (eds), *An Economic and Social History of the Ottoman Empire*, vol. 2. Cambridge: Cambridge University Press, 1994, 777–97.
——, *The Ottoman Empire: 1700–1922*. Cambridge: Cambridge University Press, 2005.
Quétel, Claude, *History of Syphilis*. Baltimore: Johns Hopkins University Press, 1992.
Ramanna, Mridula, *Western Medicine and Public Health in Colonial Bombay*. Hyderabad: Orient Longman, 2002.
Rasor, E.L., *Reform in the Royal Navy: A Social History of the Lower Deck 1850 to 1880*. Hamden, Conn.: Archon Books, 1976.
Raychaudhuri, Tapan, Kumar, Dharma and Desai, Meghnad (eds), *The Cambridge Economic History of India*, vol. 2. Cambridge: Cambridge University Press, 1980.
Reynolds, L.A. and Tansey, E.M. (eds), *Foot and Mouth Disease: The 1967 Outbreak and its Aftermath. The Transcript of a Witness Seminar held by the Wellcome Trust Centre for the History of Medicine at UCL, on 11 December 2001*. London: The Wellcome Trust, 2003.
Riviere, W.E., 'Labour Shortage in the British West Indies after Emancipation'. *Journal of Caribbean History*, 4 (1972), 1–30.
Robertson, Robbie, *The Three Waves of Globalization*. London: Zed Books, 2003.
Roff, William R., 'Sanitation and Security: The Imperial Powers and the Nineteenth-Century Hajj', in R. Serjeant and R.L. Bidwell (eds), *Arabian Studies, VI*. Cambridge: Cambridge University Press, 1982, 143–60.
Rogaski, Ruth, *Hygienic Modernity: Meanings of Health and Disease in Treaty-Port China*. Berkeley: University of California Press, 2004.
Romano, Terrie M., 'The Cattle Plague of 1865 and the Reception of the "Germ Theory" in Mid-Victorian Britain', *Journal of the History of Medicine and Allied Sciences*, 52 (1997), 51–80.
Rosen, George, 'Cameralism and the Concept of Medical Police'. *Bulletin of the History of Medicine*, 27 (1953), 21–42.
——, 'The Fate of the Concept of Medical Police, 1780–1890'. *Centaurus*, 5 (1957), 97–113.
——, *A History of Public Health*. Baltimore: Johns Hopkins University Press, 1993 [1958].
Rosen, William, *Justinian's Flea: Plague, Empire and the Birth of Europe*. London: Viking, 2007.

Rosenberg, Charles E., *The Cholera Years: The United States in 1832, 1849, and 1866.* Chicago and London: Chicago University Press, 1968.
Rothenberg, Gunther E., 'The Austrian Sanitary Cordon and the Control of Bubonic Plague: 1710-1871'. *Journal of the History of Medicine and Allied Sciences*, 28 (1973), 15-23.
Ruggie, John G., *Constructing the World Polity: Essays on Internationalization.* London: Routledge, 1994.
Russell-Wood, A.J.R., *The Portuguese Empire, 1415-1808: A World on the Move.* Baltimore: Johns Hopkins University Press, 1992.
Sanderson, S.K., *Civilizations and World Systems: Studying World-Historical Change.* Walnut Creek, Calif.: AltaMira Press, 1995.
Scammell, G.V., *The World Encompassed: The First European Maritime Empires c.800-1650.* London: Methuen, 1981.
Schwartz S.B. (ed.), *Tropical Babylons: Sugar and the Making of the Atlantic World, 1450-1680.* Chapel Hill and London: University of North Carolina Press, 2004.
Sealey, Anne, 'Globalizing the 1926 International Sanitary Conference', *Journal of Global History*, 6 (2011), 431-55.
Ser, Myo-ja, 'Religious Groups Join Protests against Beef Deal'. *Joong Ang Daily*, 2 July 2008, 1.
———, 'Teachers' Union Joins Protests to Prevent Imports of US Beef'. *Joong Ang Daily*, 4 July 2008, 1.
Sharp, Walter R., 'The New World Health Organization'. *American Journal of International Law*, 41 (1947), 509-30.
Sheridan, Richard B., *Doctors and Slaves: A Medical and Demographic History of Slavery in the British West Indies, 1680-1834.* Cambridge and New York: Cambridge University Press, 1985.
Skinner, Quentin, *The Foundations of Political Thought. Volume I: The Renaissance.* Cambridge: Cambridge University Press, 1978.
———, *Visions of Politics. Volume II: Renaissance Virtues.* Cambridge: Cambridge University Press, 2002.
Slack, Paul, 'The Disappearance of Plague: An Alternative View'. *English Historical Review*, 34 (1981), 469-76.
———, *The Impact of Plague in Tudor and Stuart England.* Oxford: Clarendon Press, 1985.
———, 'Responses to Plague in Early Modern Europe: The Implications for Public Health', *Social Research*, 55 (1988), 433-53.
Smallman-Raynor, Matthew and Cliff, Andrew D., *War Epidemics: An Historical Geography of Infectious Diseases in Military Conflict and Civil Strife, 1850-2000.* Oxford: Oxford University Press, 2000.
Snowden, Frank M., *Naples in the Time of Cholera 1884-1911.* Cambridge: Cambridge University Press, 1995.
Snyder, Louis I., 'The American-German Pork Dispute, 1879-1891'. *Journal of Modern History*, 17 (1945), 16-28.
Spufford, Peter, *Power and Profit: The Merchant in Medieval Europe.* London: Thames & Hudson, 2002.
Stannard, David E., *American Holocaust: Columbus and the Conquest of the New World.* New York: Oxford University Press, 1992.
Stewart, Larry, 'The Edge of Utility: Slaves and Smallpox in the Early Eighteenth Century'. *Medical History*, 29 (1985), 54-70.
Stiglitz, Joseph E., *Globalization and its Discontents.* London: Penguin, 2002.
Stolberg-Wernigerode, Otto zu, *Germany and the United States of America during the Era of Bismarck.* Reading, Penn.: Henry Janssen Foundation, 1937.
Striffler, Steve and Moberg, Mark (eds), *Banana Wars: Power, Production, and History in the Americas.* Durham, NC: Duke University Press, 2003.
Subramanyam, Sanjay, *The Political Economy of Commerce: Southern India 1500-1650.* Cambridge: Cambridge University Press, 1990.

Sugihara, K. (ed.), *Japan, China and the Growth of the Asian International Economy 1850-1949*. Oxford: Oxford University Press.

Sussman, George D., 'Was the Black Death in India and China?', *Bulletin of the History of Medicine*, 85 (2011), 319-55.

Sutphen, Molly, 'Striving to be Separate? Civilian and Military Doctors in Cape Town during the Anglo-Boer War', in R. Cooter, M. Harrison and S. Sturdy (eds), *War, Medicine and Modernity*. Stroud: Sutton, 1998, 48-64.

Sutter, Paul S., 'Nature's Agents or Agents of Empire? Entomological Workers and Environmental Change during the Construction of the Panama Canal'. *Isis*, 98 (2007), 724-54.

Swanson, M.W., 'The Sanitation Syndrome: Bubonic Plague and Native Policy in the Cape Colony, 1900-1909'. *Journal of African History*, 18 (1977), 387-410.

Takeda, Junko T., *Between Crown and Commerce: Marseille and the Early Modern Mediterranean*. Baltimore: Johns Hopkins University Press, 2011.

Taylor, A.J.P., *The Struggle for Mastery in Europe 1848-1918*. Oxford: Clarendon Press, 1954.

Taylor, Alan M. and Williamson, Jeffery G., 'Convergence in the Age of Mass Migration', *European Review of Economic History*, 1 (1997), 27-63.

Thomas, Hugh, *The Slave Trade: The History of the Atlantic Slave Trade 1440-1870*. London: Picador, 1997.

Tomlinson, B.R., *The Economy of Modern India, 1860-1970*. Cambridge: Cambridge University Press, 1993.

Traynor, Ian, 'EU Official Accused of "Alarmism" for Telling Travellers to Avoid Americas'. *Guardian*, 28 April 2009, 4.

Tuchman, Barbara W., *A Distant Mirror: The Calamitous Fourteenth Century*. London: Macmillan, 1978.

Tuckman, Jo and Booth, Robert, 'Four-year-old Could Hold Key Clue in Mexico's Search for Ground Zero'. *Guardian*, 28 April 2009, 4-5.

Twigg, Graham, *The Black Death: A Biological Reappraisal*. New York: Schocken, 1984.

Twitchett, Denis, 'Population and Pestilence in T'ang China', in *Studia Sino-Mongolica*. Wiesbaden: Franz Steiner Verlag, 1979, 35-68.

Vagts, Alfred, *Deutschland und die Vereinigten Staaten in der Weltpolitik*. New York: Macmillan, 1935.

Wallerstein, Immanuel, *The Modern World System: Capitalist Agriculture and the Origins of the European World-Economy in the Sixteenth Century*. New York: Academic Press, 1974.

Wallis, Patrick, 'A Dreadful Heritage: Interpreting Epidemic Disease at Eyam, 1666-2000'. *History Workshop Journal*, 61 (2006), 31-56.

Walløe, Lars, 'Medieval and Modern Bubonic Plague: Some Clinical Continuities', in Nutton (ed.), *Pestilential Complexities*, 59-73.

Walton, Look Lai, *Indentured Labor, Caribbean Sugar: Chinese and Indian Migrants to the British West Indies, 1838-1918*. Baltimore and London: Johns Hopkins University Press, 1993.

Washbrook, David, 'India in the Early Modern World Economy: Modes of Production, Reproduction and Exchange'. *Journal of Global History*, 2 (2007), 87-111.

Washer, Peter, *Emerging Infectious Diseases and Society*. Basingstoke: Palgrave, 2010.

Watts, Sheldon, *Epidemics and History: Disease, Power and Imperialism*. New Haven: Yale University Press, 1997.

——, 'From Rapid Change to Stasis: Official Responses to Cholera in British-Ruled India and Egypt: 1860-c.1921'. *Journal of World History*, 12 (2001), 349-51.

Webster, C.K., *Palmerston, Metternich and the European System 1830-1841*. London: The British Academy, 1934.

——, *The Congress of Vienna 1814-1815*. London: Thames & Hudson, 1963.

Weindling, Paul (ed.), *International Health Organizations and Movements 1918-1939*. Cambridge: Cambridge University Press, 1995.

Weisz, George, *The Medical Mandarins: The French Academy of Medicine in the Nineteenth and Early Twentieth Centuries*. Oxford: Oxford University Press, 1995.
Whitlock, Ralph, *The Great Cattle Plague: An Account of the Foot-and-Mouth Epidemic of 1967–8*. London: John Baker, 1968.
Whitmore, T.M., *Disease and Death in Early Colonial Mexico: Simulating Amerindian Depopulation*. Boulder, Col.: Westview Press, 1991.
Wilkinson, Henry C., *Bermuda from Sail to Steam: A History of the Island from 1784 to 1901*. London: Oxford University Press, 1973.
Wilkinson, Lise, *Animals & Disease: An Introduction to the History of Comparative Medicine*. Cambridge: Cambridge University Press, 1992.
Williams, R.C., *On Guard against Disease from Without: The United States Public Health Service, 1798–1950*. Washington, DC: Commissioned Officers Association of the United States Public Health Service, 1951.
Williamson, Jeffery G., 'The Evolution of Global Labour Markets since 1830: Background Evidence and Hypotheses'. *Explorations in Economic History*, 32 (1995), 141–96.
——, 'Globalization, Convergence, and History'. *Journal of Economic History*, 56 (1996), 277–306.
Wilson, Adrian, 'On the History of Disease-Concepts: The Case of Pleurisy'. *History of Science*, 38 (2000), 271–319.
Wilson, F.P., *The Plague in Shakespeare's London*. Oxford: Clarendon Press, 1927.
Woods, Abigail, *A Manufactured Plague? The History of Foot and Mouth Disease in Britain*. London: Earthscan, 2004.
Worboys, Michael, 'Germ Theories of Disease and British Veterinary Medicine, 1860–1890'. *Medical History*, 35 (1991), 308–27.
Wright, Ronald, *Stolen Continents: The Americas through Indian Eyes since 1492*. Boston: Houghton Mifflin, 1992.
Yen-Fen, Tseng and Chia-Ling, Wu, 'Governing Germs from Outside and Within Borders', in A.K.C. Leung and C. Furth (eds), *Health and Hygiene in Chinese Asia*. Durham, NC and London: Duke University Press, 2010, 225–72.
Zamoyski, Adam, *Rites of Peace: The Fall of Napoleon & the Congress of Vienna*. London: Harper Press, 2007.
Zessin, K.-H., 'Emerging Diseases: A Global and Biological Perspective'. *Journal of Veterinary Science*, 53 (2006), 7–10.
Zuckerman, Arnold, 'Plague and Contagionism in Eighteenth-Century England: The Role of Richard Mead'. *Bulletin of the History of Medicine*, 78 (2004), 273–308.
Zylberman, Patrick, 'Civilizing the State: Borders, Weak States and International Health in Modern Europe', in A. Bashford (ed.), *Medicine at the Border: Disease, Globalization and Security, 1850 to the Present*. Basingstoke: Palgrave Macmillan, 2006.

Unpublished theses

Foxhall, Katherine, 'Disease at Sea: Convicts, Emigrants, Ships and the Ocean in the Voyage to Australia, c.1830–1860', University of Warwick Ph.D. thesis, 2008.
Stark, James F., 'Industrial Illness in Cultural Context: *La maladie de Bradford* in Local, National and Global Settings, 1878–1919'. University of Leeds Ph.D. thesis, 2011.
Taylor, Sean P., '"We Live in the Midst of Death": Yellow Fever, Moral Economy and Public Health in Philadelphia, 1793–1805', Northern Illinois University Ph.D. thesis, 2001.
Towner, Katrina E., 'A History of Port Health in Southampton, 1825 to 1919'. University of Southampton Ph.D. thesis, 2008.
Varlik, Nükhet, 'Disease and Empire: A History of Plague Epidemics in the Early Ottoman Empire (1453–1600)', University of Chicago Ph.D. thesis, 2008.

Web-based articles

Agnote, Dario, 'Calls to scrap aerial spraying of bananas in Philippines', 16 June 2010, www.dirtybananas.org.

Andrews, Bill, '500 quarantines in new SARS outbreak', *Edinburgh Evening News*, 26 April 2004, http://news.scotsman.com/print.cfm?id=470902004.

Asamoa-Baah, Anarfi, 'Can new infectious diseases be stopped? Lessons from SARS and avian influenza', *OECD Observer*, no. 243, May 2004, http://www.oecdobserver.org.news.printpage.php

BBC News, 'UK BSE Timeline', http://news.bbc.co.uk/1/hi/uk/218676.stm.

Beech, Hannah, 'The Quarantine blues', *Time Asia*, http://www.time.com/time/asia/magazine/article/0,13673,501030519-451009,00.html.

Bell, David, Jenkins, Philip and Hall, Julie, 'World Health Organization Global Conference on Severe Acute Respiratory Syndrome', CDC Emerging Infectious Diseases, http://www.cdc.gov/ncidod?EID/vol9no9/03-0559.htm.

Bennett, Cortlan, 'Alarm as China's wild animal trade is blamed for "new case of SARS"', 3 January 2004, http://www.telegraph.co.uk/news/worldnews/asia/china/1450911.

Bicknell, Tom, 'Japan grabs Philippine banana supplies', http://www.fruitnet.com/content.aspx?cid=4190&ttid=17.

Biotechnology Working Party, 'First cases of BSE in USA and Canada: risk assessment of ruminant materials originating from USA and Canada', European Medicines Agency, London, 21 July 2004, www.emea.europa.eu/docs/en_GB/document. . ./WC500003697.pdf.

Bowcott, Owen, 'Swine flu could kill 65,000 in UK, warns chief medical officer', 16 July 2009, http://www.guardian.co.uk/world/2009/jul/16/swine-flu-pandemic-warning-helpline.

Browning, Philip, 'Asian currencies: how SARS could cause a trade war', *Herald Tribune*, 15 May 2003, http://www.iht.com/articles/2003/05/15/edbow_ed3_.php.

Cogan, James, 'South Korean government turns to repression to curb protests', 3 July 2008, http://www.wsws.org/articles/2008/jul2008/skor-j03.shtml.

Compassion in World Farming, 'The role of the intensive poultry production industry in the spread of avian influenza', February 2007, http://www.apeiresearch.net/document_file/document_2007076104415-1.pdf.

De Vegara, Evergisto, 'La epidemia de fiebre amarilla de 1871 en Buenos Aires', http://www.ieela.com.ar.

Durcot, Christian, Arnold, Mark, de Koeijer, Aline, Heim, Dagmar and Calvas, Didier, 'Review on the epidemiology and dynamics of BSE epidemics', *Veterinary Research*, 39 (2008), http://www.vetres.org.

Fajardo, Sarah C., 'Philippines–Australia agricultural trade dispute: the case of quarantine on Philippine fruit exports', International Gender and Trade Network, Monthly Bulletin, June 2002, vol. 2, www.genderandtrade.net/Regions/Asia.html.

Fidler, David, 'The swine flu outbreak and international law', 13, 27 April 2009, http://www.asil.org/insights090427.cfm.

——, 'International law and the *E. coli* outbreaks in Europe', *Insights*, 15, 6 June 2011, http:www.asil/org/pdfs/insight110606.pdf.

Gallagher, Peter, 'Bringing quarantine barriers to account', *Australian Financial Review*, 10 April 2003, http://www.inquit.com/article/228/bringing-quarantine-barriers-to-account.

Gratzer, David, 'SARS 101', *National Review*, 19 May 2003, http://www.manhattan-institute.org/cgi-bin/apMI/print.cgi.

Greger, Michael, 'CDC confirms ties to virus first discovered in US pig factories', *The Humane Society of the United States*, 28 August 2009, http://www.humane.society.org/news/2009/04/swine_flu_virus_origin_1998_042909.html.

Haines, Lester, 'Bird flu pandemic inevitable, says WHO', *Science*, 8 September 2005, http://www.theregister.co.uk/2005/09/08/bird_flu_pandemic/print.html.

Harcourt, Tim, 'The economic effects of SARS: what do we know so far?', http://www.austrade.gov.au/Default.aspx?PrintFriendly=True&ArticleID=6039.

Katz, Amanda and Wahlert, Sarah, 'SARS prompts crackdown on wildlife trade', http://www.buzzle.com/editorials/5-26-2003-40813.asp.

Kazmi, S.M.A., 'SARS gone, effect lingers on Indo-China annual trade', http://www.indianexpress.com/oldstory.php?storyid=27793.

Kingsland, Sally, Submission to the Senate Select Committee on the Free Trade Agreement, 30 April 2004, http://www.aph.gov.au/senate_freetrade/submissions/sublist.html.

Knight, John, 'Underarm bowling and Australia–New Zealand Trade', 18 July 2005, http://www.australianreview.net/digest/2005/07/knight.html.

Landos, John, 'Where is the SPS agreement going?', http://www.apec.org.au/docs/landos.pdf.

Lee, Jong-Wha and McKibbin, Warwick J., 'Globalization and disease: the case of SARS', *Brookings*, 20 May 2003, http://www.brookings.edu/papers/2003/0520development_lee.aspx?p=1.

Levitt, Tom, 'Asian factory farming boom spreading animal diseases like avian influenza', *Ecologist*, 11 February 2011, http:www.theecologist.org.

Lewis, Peter, 'Banana industry anxiously awaits import decision', 23 June 2002, http://www.abc.net.au/landline/stories/s586538.html.

Limberger, Michael, 'City, government and public services in Antwerp, 1500–1800'. https://lirias.hubrussel.be.

McDermott, John and Grace, Delia, 'Agriculture-associated diseases: adapting agriculture to improve human health', International Livestock Research Institute Policy Brief, February 2011, http://.www.ilri.org.

McGreal, Chris, Carrell, Severin and Wintour, Patrick, 'Swineflu – Global Threat Raised'. *Guardian*, 28 April 2009. http://www.guardian.co.uk/world/2009/apr/27/swine-flu-race-to-contain-outbreak, accessed 18/04/2012.

MacKenzie, Debora, 'Pork industry is blurring the science of swine flu', *New Scientist*, 30 April 2009, http://www.newscientist.com/blogs/shortsharpscience/2009/04/why-the-pork-industry-hates-th.html.

Martin, Daniel, 'Sir Liam Donaldson quits the NHS ... but critics say resignation is two years too late', 16 December 2009, http://www.dailymail.co.uk/article-1236179/Sir-Liam-Donaldson-quits-NHS—critics-say-resignation-years-late-html.

Miller, Brad, 'Aerial spraying issue turns seesaw court battle', 29 November 2007, http://ipsnews.net/news.asp?idnews=40264

Mugas, Neil V., 'Laguna hospital placed in SARS quarantine', *Manila Times*, 7 January 2004, http://www.manilatimes.net/national/2004/jan/07/yehey/top_stories/20040107top5.html.

Nawal, Allan, Tupas, Jeffrey M. and Badilla, Joselle, 'Banana firms to contest aerial spraying ban at CA', *Inquirer*, http://services.inquirer.net/print/print.php?article_id=20070924-90391.

O'Sullivan, Terrence, 'Globalization, SARS and other catastrophic disease risks: the little known international security threat', http://allacademic.com/meta/p73646_index.html.

Palmer, Daniel, 'Agreement to import bananas from the Philippines leads to grower fears and IGA ban', *Australian Food News*, 5 March 2009, http://www.ausfoodnews.com.au/2009/03/05.

Pena, Aurelio A., 'Australia's farmers scared of losing market to Philippine bananas', *Philpress*, http://goldelyonn.wordpress.com/2009/06/11.

Pew Commission on Industrial Farm Animal Production, 'Putting meat on the table: industrial farm animal production in America' (2008), www.pewtrusts.org.

Porter, Andrew, 'Disease pandemic "inevitable" in Britain warns House of Lords', *Daily Telegraph*, 20 July 2008, http://www.telegraph.co.uk/health/2437908.

Purdey, Mark, 'An explanation of mad cow disease', http://www.ourcivilisation.com/madcow/madcow.htm.

Rosenthal, Elisabeth, 'Bird flu is linked to global trade in poultry', *International Herald Tribune*, 12 February 2007, http://www.iht.com/bin/print.php?id=4568849.

'Short history of the OIE', http://www.oie.int/Eng/OIE/en_histoire.html.

Simpson, Natasha, 'Philippines takes Aust to WTO over banana ban', 15 July 2003, http://www.abc.net.au/am/content/2003/s902217.html.

Smith, Harvey and Lattimore, Ralph, 'The search for non tariff barriers: fire blight of apples', Department of Economics and Marketing Discussion Paper no. 44 (1997),

International Trade Policy Research Centre, Department of Economics and Marketing, Lincoln University, Canterbury, http://researcharchive.lincoln.ac.nz/dspace/bitstream/ .../1/cd_dp_44.pdf.
Sudworth, John, 'Political price paid in beef row', BBC News, 5 June 2008, http://news.bbc.co.uk/1/hi/world/asia-pacific/7436914.stm.
——. 'S Korea head admits beef failings', 7 July 2008, http://news.bbc.co.uk/1/hi/world/asia-pacific/7492562.stm.
Teel, Geena, 'Ottawa threatens trade action if pork ban isn't lifted', *National Post*, 5 May 2009, http://www.nationalpost.com/m/story.html?id=1564824.
Thornton, Philip, 'WTO warns SARS Could Halt Revival of International Trade'. *Independent*, 24 April 2003, www.independent.co.uk/news/business/news/wto-warns-sars-could-halt-revival-of-international-trade-595505.html, accessed 24/03/2009.
Tyson, Carl N., 'Texas fever', http://digital.library.okstate.edu/encyclopedia/entries/T/TE022.html.
Vaile, Mark, 'Quarantine is helping Australia reach the world's markets', 29 August 2001, http://www.maff.gov.au/releases/01/01237wtj.html.
Vasilyeva, Nataliya, 'US: Lifting pork imports could boost Russia's WTO talks', http:blog.targana.com/n/us-lifting-pork-imports-could-boost-russias-wto-talks-73287/.
Workman, Daniel, 'Swine flu infects pork trade policies in China', 5 May 2009, http://world-trade-organization.suite101.com/article.cfm/swine_fluinfects_pork_trade.

Websites
American Government Archive, www.america.gov.
American Phytopathological Society, www.apsnet.org.
American Society of International Law, www.asil.org.
Animal Health Alliance, www.america.gov.
Area Development, www.areadevelopment.com/newsitems/4-27-2009.
Asia Pacific Economic Cooperation (APEC), www.apec.org/apec.
Asia Pulse News, www.highbeam.com.
Australia Ministry for Agriculture, Fisheries and Forestry, www.maff.gov.au.
Australian Ministries for Foreign Affairs and Trade, www.foreignminister.gov.au.
BBC News, www.news.bbc.uk.
Bernama, www.bernama.com.
Biotechnology Working Party, www.emea.europa.eu.
Bridges Weekly Trade News Digest, www.ictsd.org/weekly/03-07-17/story5.html.
California Farm Bureau Federation, www.cfbf.com.
CBC, www.cbc.ca.
CBS, www.cbsnews.com.
Centers for Disease Control, www.cdc.gov.
China, www.china.org.cn.
China Post, www.chinapost.com.
China View, www.news.xinhuanet.com.
Chosun, www.english.chosun.com.
CNN, www.cnn.com.
Daily Mail, www.dailymail.co.uk.
Daily Telegraph, www.telegraph.co.uk.
Dawn, www.dawn.com.
Department of Food, Agriculture and Trade, Australia, www.dfat.gov.au.
Durian Post, www.durianpost.wordpress.com.
Ecologist, www.theecologist.org.
English DBW, www.english.dbw.cn.
English Peoples Daily, english.peoplesdaily.com.cn.
European Union, www.eubusiness.com.
Food and Agriculture Organization, www.fao.org.
Fox News, www.foxnews.com.

Grain, www.grain.org.
Guardian, www.guardian.co.uk
Herald Sun, www.news.com.au/heraldsun.
Hindu, www.theHindu.com.
Horticulture and Home Pest News, www.ipm.iastate.edu.
Hypocrites, www.hypocrites.com.
Infoweb, www.infoweb.newsbank.com.
Inquirer, www.inq7.net.
Institute for Animal Health, www.iah.bbsrc.ac.uk.
International Herald Tribune, www.iht.com.
International Livestock Research Institute, www.ilri.og.
International Socialism, www.isj.org.uk.
Medical News Today, www.medicalnewstoday.com
National Marketplace News, www.ams.usda.gov.
National Pork Producers Council www.nppc.org.
National Post, www.nationalpost.com.
New Zealand Herald, www.nzherald.co.nz.
New Zealand Ministry of Foreign Affairs, www.mfat.govt.nz.
Northeast News, www.northeastnews.ca.
Official Gazette of the Republic of the Philippines, www.gov/ph.
OIE, www.oie.int.
Oxford Dictionary of National Biography, www.oxforddnb.com.
People's Daily Online, www.english.peoplesdaily.com.cn.
Pew Trusts, www.pewtrusts.org.
Resist, www.lists.resist.ca.
Reuters, www.reuters.com.
Sydney Morning Herald, www.smh.com.au
Third World Network, www.twnside.org.
United States Department of Agriculture, www.ers.usda.gov.
United States Embassy, Tokyo, www.tokyo.usembassy.gov.
University of Oregon, www.iea.uoregon.edu.
WattAgnet, www.wattagnet.com.
World Health Organization, www.who.int.
World Trade Organization, www.wto.org.

Index

Page numbers with 'n' are notes.

Aberdeen, Lord 74, 76, 91, 93, 102
Academy of Medicine (France) 56, 68, 219
Accra 197–8
Addington, A.H. 90
Aden 77
   cholera 145–6, 150–4, 158, 163–5, 169
   plague 154
Adorno, Marquis Botta 43–4
Aedes aegypti mosquito 20
Africa
   cattle plague 221
   and HIV/AIDS 258
   plague 192–3
   slave trade 17–21, 26, 117
   South Africa 234
Ahmed I 6
AIDS 258
airlines 260, 267
Albert (steam ship) 82, 91
Ali, Muhammad 68–70
Almeida, Dr 92
Alvarado, Ignacio 125
Americas 16–23
   see also Canada; Mexico; United States of America
Amsterdam 15, 24–7, 123
Andrew, Dr 203–4
Angel Island, quarantine station 131
Anglo-Boer War 192
animal diseases 211, 248
   BSE 249–56
   cattle plague 212, 213, 214–21

   foot-and-mouth disease 222, 226, 229–32
   H1N1 (swine flu) 269–75
   H5N1 (avian flu) 266–9, 275
   lung sickness 213–14, 224–5
   SARS 264
   Texas fever 222–4
anthrax 6, 163, 242
anticontagionists 50, 58–9, 60, 69
Antilles 19
APEC *see* Asia-Pacific Economic Cooperation
Argentina
   cattle plague 221
   export treaty with USA 229
   OIE 228
   plague 192
   yellow fever 115–17
Asia-Pacific Economic Cooperation (APEC) 264
Asian 'Tigers' 247, 257–8
Astrology, as cause of disease 10
Atrato, RMS 114
Australia
   free-trade agreement with US 245–6
   and New Zealand fruit ban 236, 237, 239–41
   and Philippines dispute 241–6
   plague 182, 194–6
   and quarantine 240–1, 242
Austria-Hungary
   cattle plague 215
   cholera 139

Phylloxera vastatrix 233, 234
plague
 and Britain 165–6
 and Persian Gulf 164–5
 sanitary measures 36–7, 40, 172–3
  cordons 76
  and international agreement on 71–9
  and US pork 225
Ayre, Joseph 71

bananas 241–5
Barbados 19, 20, 21, 103
barracoons 18
Basra (Bussorah) 154, 159
Ibn Battuta 5
Bayley, E.C. 158
Belgium 225, 227, 268
Bell, Charles 217
Benjamin, Georges C. 263
Bernard, Sidney 83–4
Berne Agreement (1878) 233–4
bills of health/certificates 39, 40–1, 52, 146, 227
biosecurity 269, 275
Biosecurity Australia 243, 244
bioterrorism 263
Biraben, J.-N. 7
birds, wild 269
 see also H5N1
Black Death
 ix-x 7
 see also plague
black sigatoka disease 241
Blair, Dr 111
Blane, Sir Gilbert 57–8, 66
Boa Vista (Cape Verde Island) 80, 82, 90, 94, 97, 100, 102–6, 109
Board of Agriculture (Britain) 219–20
Bombay, Government of, and cholera 149–53
Bombay Gazette 180, 181–2
Bombay, Government of 146, 157, 179–85
Borromeo, Carlo 13
Boston, United Fruit Company 133
bovine pleuropneumonia 213–14
Bovine Spongiform Encephalopathy (BSE) 249–56
Bowerbank, Dr 104
Bowring, John 71, 96
Boyle, Ross 241
Bradford (West Yorkshire) 163

Brazil
 ban on US grain 237
 cattle plague 221
 and international agreement of sanitary measures 134
 plague 174–9, 182, 192
 yellow fever 108, 115, 117–18
Bright, John 96
Britain
 African colonies 196–8
 and Argentinean meat 229–32
 and Austro-Hungary 165–6
 BSE 249–50
 cotton industry 68–9
 H1N1 270, 271
 and outbreak of plague in Mesopotamia 167–70
 and Persia 159–61
 sanitary measures
  animal disease 213, 214, 216–19, 224
  and cholera 148
  debate on quarantine 96–9, 106
  international agreement on 71–9
  plague 26–8, 30–2, 33–6, 41–3, 51–2
  yellow fever 107, 110, 111–15, 118–19
 SARS 257
 and Suez Canal 157
 textile industry 163
 and the Third International Sanitary Conference 142
 Treaty of London 72–3
 yellow fever, *Eclair* incident 83–106
British India Steam Navigation Company 147, 155–6, 158, 164, 166–7
British Medical Journal 190
Bryden, James Lumsdaine 148
BSE *see* Bovine Spongiform Encephalopathy
buboes 6, 7, 189
Buchanan, George 113–14
Buchanan, William 132
Buckle, Captain 90
Budd, William, iv
Buenos Aires (Argentina) 115–17
Buenos Aires Standard (newspaper) 116
burials 166
Burnett, Sir William 86–8, 88–9, 90, 98
Bush, George W. 260–1
business community, involvement in sanitary measures 121–2, 264
Bustillo, Col. J.M. 116

Caffrey, Donelson 127-8
Caldwell, Charles 56
California Farm Bureau Federation 245
Calmette, A. 189
Calvert, Henry 168
Canada
  BSE 250, 251
  H1N1 270
  SARS 257, 262
Cape Verde islands 95
capital flight 264
carazzo 6
Castaldi, Dr 164
cattle diseases
  bovine pleuropneumonia 213-14
  BSE 249-56
  cattle plague 212, 213, 214-21
  lung sickness 213-14, 224-5
  Texas fever 222-4
Cattle Diseases Act 1866 (Britain) 219
Cattle Diseases Act 1866 (India) 220
cattle plague 212, 213, 214-21, 227
Cecil, Robert Arthur Talbot Gascoyne- *see* Salisbury, Lord
Central America 17
certificates/bills of health 39, 40-1, 52, 146, 227
Chadwick, Edwin 100
Chanovitz, Steve 274
Chappell, Greg 239
Chataud, Captain 28
Chervin, Nicolas 65
China
  and capitalism 247
  cholera 203
  H1N1 271-2
  Manchurian epidemic of plague 201-5
  plague 2-3, 5, 11, 154, 174-9
  SARS 257, 258-62, 259
  and US 237, 248, 274-5
Chinese Eastern Railway (CER) 201
Chisholm, Colin 58, 59-60
cholera 66-7, 139-53, 171, 314n.19
  discussed at Paris conference (1851) 78-9
  epidemiology 139-40, 142, 147, 148, 172
  Shah's visit to Baghdad 160-1
  and the War of the Triple Alliance 115
  see also individual countries
Chung Woon-chan 253
Church 56, 216-18, 254
City of Cork, SS 188

civil unrest
  Bombay 181
  Korea 254-5
  Portugal 190
*v*CJD *see* Creutzfeldt-Jakob Disease, variant
Clement XI 212
climate
  and plague 46-7
  and potato blight 233
  and yellow fever 87, 101, 108, 113, 117
Clot, Antoine Barthélémey (Clot-Bey) 69
coast fever 81, 89, 92
Cobden, Richard 96
coffee 118
Coffey, Mr 83, 84
Colbert, Jean-Baptiste 29
Colombia 124, 130, 131, 132
Colombian Exchange 17
Colorado beetle (*Leptinotarsa decemlineata*) 234
Colucci Pasha 168
Columbus, Christopher 16
Colvill, William 160, 165, 167
Compassion in world Farming 269
Conference of American Republics 198-9
conferences 72
  Emerging Infections Network 2003 (Thailand) 264
  International Conference of American States 134-5
  International Sanitary Conference 1851 (Paris) 78-9, 101-2, 140
  International Sanitary Conference 1866 (Constantinople) 141-4
  International Sanitary Conference 1874 (Vienna) 151
  International Sanitary Conference 1892 (Venice) 172-3, 182
  International Sanitary Conference 1893 (Dresden) 172-3
  International Sanitary Conference 1897 (Venice) 184-5, 186
  International Sanitary Conference 1903 (Paris) 197
  International Sanitary Conference 1911 (Mukden) 203
  International Sanitary Conference 1926 (Paris) 209
  International Sanitary Conference 1938 (Paris) 210
  Pan American sanitary conference (1902) 134

veterinary surgeons 1863 (Hamburg) 215
veterinary surgeons 1865 (Vienna) 215
Washington Convention (1905) 134
Congress of Vienna 72
Conseil Sanitaire Maritime et Quarentine 210
Constantinople 3
　Council of Health 74, 159, 165
　International Sanitary Conference (1866) 141–4
　plague 73
Consular Commission of Health (Egypt) 70
'contagion' 8, 9, 10, 65
　and 'infection' 64
　theories of 45–9
　see also *individual diseases*
contagionists 51, 56–7, 58, 61, 217
Contagious Diseases (Animals) Acts (Britain) 234
containment of disease 265
Cooper, Anthony Ashley *see* Shaftesbury, Lord
cordons 46
　land 37, 40
　　Accra 197
　　Austro-Hungary 36–7, 76, 165
　　and cholera 143
　　discussed in Paris conference (1851) 79
　　France 65, 65–6
　　Marseilles 29–30
　　Persian Gulf area 162–3
　　Portugal 190
　　Spain 189
　　Venetian 37
Corfu 63
Corn Laws 84–5, 96
Corps of Engineers (US) 130
cotton industry/goods 6, 44, 47, 68, 149, 180, 181–2
councils of health 8, 14–15
Country Landowners' Association 231
Cowdrey, Jonathan 60
Creutzfeldt-Jakob Disease, variant (*v*CJD) 249–56
cricket test match incident 239
Crimean War 79
Cuba 119, 129, 136, 199
Cullen, William 53
culling of livestock 231, 267, 268, 269
Cuningham, J.M. 147–9, 154, 156, 158, 169, 170–1

Dauntless, HMS 103
Day of Humiliation 216–17
de Lesseps, Ferdinand 130
de Macazaga, Don Jorje 33
de Mendonia, Diego 32–3
de Rochefort, Charles 19
de Ségur Dupeyron, P. 70–1
Defoe, Daniel 8
Demerara 111
Dempster, Dr 104
dengue fever 258
Denmark 25, 27, 38, 216
Derby, Lord (Edward Henry Stanley, 15th Earl of Derby) 165–6, 168
Destructive Insects Act (Britain) 234
Devèze, Jean 65
Dickson, Sir J. 162
diet 21, 92, 97, 233, 245
diphtheria 124
diplomacy 38, 75
disinfection
　cattle plague 212
　plague 39, 78, 180, 187, 190, 202, 205
　yellow fever 121, 123, 164
Dispute Settlement Board (DSB) 238–9, 242
dithane 244
Donaldson, Sir Liam 271
Dresden, International Sanitary Conferences 172–3
Dubois, Cardinal 30
Dubrovnik 8, 14
'dwarf bunts' 237
dysentery 18

East India Company 42, 58, 60, 61
East Indies 25
Ebola 258
Eclair, HMS 80–106
E. coli, x
Ecuador 132, 134, 137, 198
Egypt
　plague 4–5, 166–70, 185–8
　sanitary measures 68–70, 73
　cholera 146, 186
　plague 157, 166–73, 185–8
　Suez Canal 153–8, 210
　war with Turkey 72
Eliot, Edward Granville *see* St Germans, Lord
Emerging Infections Network 264
environment, and disease 20–1

Environmental Protection Agency (EPA) 244
Epidemic Diseases Act 1897 (Bombay) 182–3
'epidemic fever' 59
Erasmus 13
Estcourt, Commander Walter 82–3, 85, 89–91
European Community/Union 230, 256, 268, 270
evacuation 198
evasion of quarantine 55

face masks 266
FAO *see* Food and Agriculture Organization
Federal Horticultural Board (US) 235–6
Federalist (newspaper) 55
Ferro, Paskal Joseph 37
Finlay, Carlos 129
'fire-blight' 236–40
First Plague Pandemic 1
Fitch, Asa 233
fleas 2, 4, 180, 184, 193
Food and Agriculture Organization (FAO) 230, 237, 279, 280
foot-and-mouth disease 222, 226, 229–32
Formosa (Taiwan) 207, 208, 235, 262–3
France
  BSE 250
  Phylloxera vastatrix 233–4, 234
  plague 3
  sanitary measures
    and international agreement of 70–9
    plague 12, 26–31, 44–5, 68, 182, 190
    and US pork 225
    and yellow fever 107, 109, 110
    and Treaty of London 72–3
  yellow fever 111–12
Frank, John Peter 36
free trade 81
  agreement with USA and Australia 245–6
  and Pork War 225–6
  regulation on trade in plants 234–5
  and SARS 256–65
  and the WTO 247
  and yellow fever outbreak 109
Free Trade Agreement 245–6
freedom, quarantine as anti- 51, 52
fumigation 38–9
  damage by 29, 133, 208
  Formosa 208
  and Korea 208

methods of 39, 42, 144, 208
used to damage goods 77
wool and Suez Canal 164
for yellow fever 134, 138
fungicides 241, 243–4

Galen 9
Gallagher, Peter 242
Gallway, Captain 93
Gamgee, John 215, 216, 218–19, 223–4
Garfield, James 125
Gazeta Medica da Bahia 117–18
General Agreement on Tariff and Trade (GATT) (1947) 236–7
General Board of Health (Britain), and the *Eclair* incident 99–101
Genoa 8
Georgia (US) 22, 120
'germ governance' 265, 266
Germany
  and *Phylloxera vastatrix* 233, 234
  and SARS 257
  and Suez 172
  and US pork 225–6
Gibraltar 31–2, 34–6, 45, 86, 93, 107
Girard, Stephan 65
Glasgow 192
Glass, Rear Admiral Henry 132
globalization 247–8, 263
  and SARS 257, 261
Gloucester Chronicle 85
Gloucester Journal 85
God, as cause of plague 9–12
  see also Church
Goethals, Col. G.W. 130
Good Neighbor Policy (US) 229
Gorgas, Col. William 130, 137
governance, and disease 117, 120, 129
Granville, A.B. 63
grapes/vines 233–4, 245
Great Depression 228
Great Irish Famine 232–3
Great Northern War 24, 27
Great Powers 72–3, 203, 210–11
Greece 143, 225
Green, John 62
Gretzer, David 265
Growler (ship) 83, 90, 93
Guayaquil (Ecuador) 137
Guiteras, Juan 136
Guyon, Dr 109–10

H1N1 (swine flu) 269–75
H5N1 (avian influenza) 266–9, 275
haemorrhagic fevers 258
Hamburg 65, 67, 192, 215
Hammett, Bashaw 35
Hankey (ship) 59
Hankin, E.H. 201
Harcourt, Tim 261
Haridwar (India) 147
Harris, James Howard *see* Malmesbury, Lord
Harte, Mr 83
Harvey, Alexander 101
Hawaiian Islands 132, 193, 199, 200
Hayes, Rutherford B. 124
Hecla, HMS 113
Heiser, Victor 204–6
Hejaz 144, 146, 193
Herald Tribune 260
Heschl, Richard 225
HIV/AIDS 258
Hobhouse, John Cam 62
Holland *see* Netherlands
Holles, Thomas Pelham *see* Newcastle, Lord
Hong Kong
  and H5N1 267
  plague 175–9, 181
  and SARS 259–60, 262
Hong Kong Retail Management Association 259
Hot Zone diseases 258
Howard, John 51–2, 243
Howell, Arthur 156
Hu Jintao 259
Hume, Joseph 96
'humours', four 9
Huxham, John 47

immigration 176, 198–9, 204, 232
incubation periods
  bovine pleuropneumonia (lung sickness) 214
  cholera 143
  plague 194
  SARS 259
  Texas fever 222
  yellow fever 104, 119
India
  cattle plague 220, 221
  cholera 144–53
  cotton industry 68
  and plague 5–7, 167, 170–3
  sanitary measures 143, 182–3
    plant diseases 235
  SARS 258
  see also Bombay
India Medical Service 183
'infection', and contagion 64
Infectious Diseases Ordinance 197–8
infectiousness 9–10
influenza 17, 248, 263
  avian (H5N1) 266–9, 275
  swine (H1N1) 269–75
Ingram, Dale 47, 56
inoculation
  cattle plague 218, 221
  plague 183, 197
  smallpox 19
international agreement on sanitary measures 71–6, 102, 124–5, 206–10, 226
International Conference of American States 134–5
international health organizations 198
International Health Regulations (IHRs) 238, 265, 272
International Plant Protection Convention (IPPC) 237, 238
International Sanitary Conferences 171
  Constantinople (1866) 141–4
  Dresden (1893) 172–3
  Mukden (1911) 203
  Paris (1851) 78–9, 101–2, 140
  Paris (1903) 197
  Paris (1926) 209
  Paris (1938) 210
  Venice (1892) 172–3, 182
  Venice (1897) 184–5, 186
  Vienna (1874) 151
International Telegraphic Union 141
International Union of Weights and Measures 141
Islam
  and belief of cause of plague 10–11
  Shi'a Muslims 169
  terrorism 242, 243, 263
  tibb ('Islamic' medicine) 162
Isles of Scilly 88
Isthmian Canal Commission (ICC) 130, 131
Italy
  Phylloxera vastatrix 233, 234
  quarantine 13
  sanitary measures 12, 172–3
  see also Messina; Milan; Palermo; Pisa; Venice; Verona

Jahangir 6
Jamaica 111

James, Surgeon-Major 177
James, Sydney Price 137
Jameson, Horatio 65
Janibeg 3
Japan 5, 193, 204–5, 207–8
   ban on US beef 251
   ban on US fruit 237
   ban on US pork 273
Jefferson, Thomas 54, 55
Jenkinson, Robert Banks *see* Liverpool, Lord
Jews 7, 8, 29, 39
Johnson, James 66
Jong-Wha Lee 261
Jorge, Ricardo 188
journals, medical 59–60, 117–18
J.P. Morgan 259

Kaffa (Crimea) 3
Kenny, Peter 92
Kidderminster (Worcestershire) 163
Kim Kwang-il 254
King, Gilbert 98–9, 100–1
Kitasato Shibasaburo 177
Koch, Robert 172, 221
Korea 5, 208, 235
   see also South Korea
Korean Confederation of Trade Unions 254–5
Korean Teachers' and Education Workers' Union 255
Kroomen 82

La Democracia (newspaper) 116
'la maladie de Siam' 22
La Plata, RMS 103
Ladurie, Emmanuel Le Roy 211
Lancet 85, 101, 111, 114
Lancisi, Giovanni Maria 212–13
land cordons 40, 143
   Accra 197
   Austro-Hungary 36–7, 76, 165
   discussed at 1851 Paris conference 79
   France 65–6
   Marseilles 29–30
   Persian Gulf area 162–3
   Portugal 190
   Spain 189
lazarettos 8, 14, 15, 51
   Egypt 70
   France 110
   US 54, 55
   see also quarantine
League of Nations 206–7, 210, 227, 228

Leclainche, E. 227
Lee Myung-bak 251–6
Leptinotarsa decemlineata (Colorado beetle) 234–5
Levant 4–5, 8, 60, 71
Levant Company 42, 52, 61–2
Lisbon (Portugal) 108, 110
Liverpool, Lord (Robert Banks Jenkinson, 2nd Earl of Liverpool) 61
Livorno (Tuscany) 14, 43
Louisiana (US) 119
Lowson, James 177, 182
'lung sickness' 213–14, 224–5
Lyons, sanitary measures 12

Ma Zhaoxu 273
McCall, William 217
MacGregor, J. 74–5
McGrigor, James (later Sir James) 57, 58
Mackenzie, Mordach 48
McKibbin, Warwick J. 261
Maclean, Charles 60–2, 66
Mclure, Mr 83
McNeill, William H. 2
McWilliam, Dr 97–8, 100
'mad cow disease' *see* BSE
Madison, James 55
Mahmut II 74
mail, and quarantine 169
malaria 18, 87, 91, 99, 130
Malmesbury, Lord (James Howard Harris, 3rd Earl of Malmesbury) 102
Manchurian Plague Prevention Service (MPPS) 203
Manila (Philippines) 178, 193, 207, 242, 243
Manson, Sir Patrick 137
Marine Hospital Service (MHS) (US) 126–7, 131, 132
Marseilles (France) 3, 14, 192
   sanitary measures 8, 28–30, 39–41, 182
Martin, Samuel 63–4
Martinique (Lesser Antilles) 19
Mead, Richard 21, 48
measles 17, 113, 124, 153
Meat Inspection Acts 1906–7 (US) 226
Mecca 149–50
   cholera 73, 139–41, 145, 147
   plague 154
media 264, 265
medical inspection
   plague 167–8, 178–9, 183, 185, 189–90, 203, 207
   yellow fever 115, 121, 123

medical periodicals 59–60, 66, 85, 101, 190
Medical Repository 59–60
Medico-Chirurgical Review 66, 101
Mélier, François 109, 112, 114
merchandise, as carrier of plague 4, 8–10, 15–16, 38–9, 44, 47
merchants
  power of 15–16, 67
  self-interests of 36–45, 50
  subscribers to medical journals 59
Mesopotamia 159, 160
  plague 154, 161, 163, 164, 166–70
Messina (Sicily) 3
Messrs Gray, Dawes & Co. 155
metaphor, seed and soil 54
Methuen Treaty (1704) 31
Metternich, Prince Clemens Lothar Wenzel 73, 74–5, 89
Mexico 198, 270, 271
Milan (Italy) 4
Miller, Edward 59
Milroy, Gavin 73, 101, 104
Mississippi (US) 119, 120
Mitchill, Samuel Latham 59
mohair 163
Mongolia/Mongols 193
Monroe Doctrine (US) 135
Montenegro, José Verdes 189–90
Morris, Henry 41
mosquitoes 258
  and yellow fever 20, 129, 130, 138
MPPS (Manchurian Plague Prevention Service) 203
Muhammad bin Tughluq 5
Mukden, International Sanitary Conference 203
Murchison, Charles 119
Mustafa Bey 70

NAFTA (North American Free Trade Agreement) 256
Naples 14, 92–4, 192
Napoleon III 141
Nasir al-Din Shah 160
National Board of Health (US) 121
National Farmers' Union 231
National Review 265
Native Passenger Ships Act 1858 (India) 143, 145
Naval Intelligencer 85
Navigation Act (Britain) 26
Netherlands 15, 24–7, 123
  bovine pleuropneumonia 214
  and the Levant Company 62
  sanitary measures 15, 24–7

New Orleans (Louisiana) 108, 119, 122
New Scientist 274
New World 16–23
New Zealand 228
  fire-blight 236, 237, 239–41
  Phylloxera vastatrix 234
Newcastle, Lord (Thomas Pelham Holles, 1st Duke of Newcastle upon Tyne) 33, 34, 35
Nixon, J.P. 166–7
North West Fruit Growers' Association 235
notification of disease 40, 125–6, 134, 137, 196
  Pan American Sanitary Code (1924) 208
  phylloxera 233–4
  plague 184, 192
  and telegraphy 150, 206–7, 209
Notter, J. Lane 185

Office International des Épizooties (OIE) 227–32
Office Internationale d'Hygiène Publique 198, 206, 210
O'Halloran, Thomas 65
O'Hara, James *see* Tyrawley, Lord
opium trade 179
Oporto (Portugal) 188–92
Ordinances of Pistoia 8
Orinoco, RMS 110
Ottoman Empire 11, 72–4
  cholera 73–4, 141–2
  and Persia 159–66
  sanitary measures 35–6
    against cholera 144–5
    against plague 182
  see also Egypt

Pacific Mail Steamship Company 131–2
Pacific Steam Navigation Company 131–2
Packington, Sir John 103
Paiba, Mr 59
Palermo (Sicily) 14
Palmerston, Lord (Henry John Temple, 3rd Viscount Palmerston) 73, 74, 102
Pan American Bureau of Health 134
Pan American Commercial Conference (1931) 229
Pan American Sanitary Bureau 136
Pan American Sanitary Code (1924) 208
Pan American sanitary conference (1902) 134

Panama Canal 130–8, 199
pandemic disease
  H1N1 270
  SARS 265
  see also H1N1; plague; yellow fever
Papon, J.P. 40
Paraguay 115–16, 116, 192
Parana, RMS 103
parasites 223, 233
Paris
  1851 International Sanitary Conference 78–9, 101–2, 140
  1903 International Sanitary Conference 197
  1921 veterinary conference 226–7
  1926 International Sanitary Conference 208
  1938 International Sanitary Conference 210
Pariset, Étienne 65–6
Parisian Institut Pasteur 189
Parsis 181
'patente brute/nette/soupçonnée/touchée' 40–1
Penang (Malaysia) 193
Peninsular and Oriental Company 146–7
Penny, Dr 177
People's Conference Against Mad Cow Disease (Korea) 255
Persia 6, 159–66
Persian Gulf 159–66
Peru 233, 234
peste/pestis 7
Philadelphia (US), yellow fever 52, 53–6
Philippines 132, 193, 204
  animal disease 220
  plant disease 235, 241–6
Phylloxera vastatrix 233–4
Phytophthora infestans 233
Pierce, Claude 132
pigs see H1N1; pork
pilgrims/pilgrimage 73, 144–5, 149–50
  cholera outbreak at Mecca 140–1, 147
  third pandemic of plague 193
Pisa 8, 14
Pitt, William (Pitt the Elder) 52
Pitt, William (Pitt the Younger) 52
plague 71
  Arabian Sea area 153–8
  causes of 4, 7, 180, 184, 202
    contemporary views of 7–11, 38, 46–7, 56, 177, 186–7
  compared to yellow fever 21–2
  origins of 1–3
  Persian Gulf area 161, 163

  pneumonic 202
  refusal to use term 28
  Suez Canal area 153–8
  see also individual countries
Plague of Justinian, xi 1
Plague of San Carlo 12
Plagues and Peoples (McNeill) 2
plant diseases/pests 211, 232–46
  black sigatoka 241
  Colorado beetle 234
  'dwarf bunts' 237
  'fire-blight' 236–40
  grapevine diseases 233–4
  phylloxera 233–4
  potato blight 232–3, 233
  regulation and sanitary measures 233–46
Plant Protection Act 1912 (Korea) 235
Plant Quarantine Act 1912 (US) 235
Playfair, Lyon 217–18
pneumonic plague 17, 202
Poland, Great Northern War 27
population
  collapse 1, 2
  flight 181, 183
pork 225–6, 272–5
Port Quarantine Law 1922 (Japan) 207
Portugal
  and the Eclair incident 94–5
  Phylloxera vastatrix 233, 234
  plague 188–92
  sanitary measures 27, 31–4
    and International Sanitary Conference 78
  yellow fever 108, 110
Potato Famine 232–3
preparedness, pandemic 265, 267
Prideaux, W.F. 164
privateers 43
protectionism 228–9, 265
  after GATT 237
  and H5N1 266–75
  quarantine as 235
  and SARS 264
Prussia 37, 67, 213, 216
Public Health Act 1848 (Britain) 99
public health workers 258
Puerto Rico 132
Pym, Sir William 75, 86, 88–9, 93, 97–9, 110

qi 11
quarantine 8, 11, 12–16, 45–6, 94, 275
  for cholera 143–4
  evasion penalties 55

profitability of 14
reform 51–2, 51–67
and SARS 260, 261, 266
used against business rivals 13–16, 79, 235
see also *individual countries*
Quarantine Act 1710 (Britain) 27–8
Quarantine Act 1721 (Britain) 30
Quarantine Act 1753 (Britain) 51
Quarantine Act 1780 (Britain) 44
Quarantine Act 1825 (Britain) 83, 102, 121, 168
Quarantine Act 1870 (India) 147, 149
Quarantine Act 1893 (US) 131

railways 121, 131, 140–1, 201, 215
Ransom, Dr 205
rats 4, 184, 193, 195–6
  extermination 207, 208
  flea (*X. cheopis*) 180
Ravenscroft, E.W. 157
recession 248, 258, 272, 274
Republic of Ragussa (Dubrovnik) 8
Richardson, Dr 85–6
rinderpest *see* cattle plague
Rio de Janeiro (Brazil) 108, 109, 110, 117–18
risk management 201, 209
River Medway, lazaretto 52
Robinson, Sir William 176
Rochester, Bishop of 9
Rockefeller Foundation 137, 138
rodents 2, 4, 6, 184
  see also rats
Rogers, Dr 84
Roh Moo-hyun 252
Romania 193
Roosevelt Corollary 135
Roosevelt, Franklin D. 228–9
Roosevelt, Theodore 130, 135
Royal College of Physicians 61, 62
Royal Mail Company 101, 110
Rush, Benjamin 52–3, 55–6, 59
Russell, Alexander 52
Russell, Patrick 52
Russia
  and ban on US pork 273
  cattle plague 215
  cholera 139
  sanitary measures 38–9
    and international agreement of 78–9
  third pandemic 193

St Christopher (island) 19
St Domingue (French Caribbean colony) 53, 54
St Germans, Lord (Edward Granville Eliot, 3rd Earl of St Germans) 102
St Thomas (Caribbean island) 110, 114–15
Salimberi, A.-T. 189
Salisbury, Lord (Robert Arthur Talbot Gascoyne-Cecil, 3rd Marquess of Salisbury) 156, 170
Samoilovich (Samoilowitz), Danilo 38
Sanitary, Maritime and Quarantine Board of Egypt 185, 186–7
sanitary measures/precautions xii–xiii, 12
  bills of health 40–1, 52, 146
  culling of livestock 231, 267, 268, 269
  effect of *Eclair* incident on 102–6
  evacuation 198
  and H5N1 267–9
  as instruments of statecraft 24–49
  international agreement on 71–9, 124–5, 206–10, 211–12
  Ordinances of Pistoia 8
  rat extermination 207
  and SARS 260–2, 266
  segregation 177, 181, 183, 195, 196
  see also cordons; disinfection; fumigation; *individual countries*; medical inspection; quarantine
Sanitary and Phyto-Sanitary (SPS) Regulations 238
sanitary reform 22, 117–18, 120
  Bombay 180–1
  Hong Kong 176–7
  Portugal 191
Sardinia 89
SARS (Severe Acute Respiratory Syndrome) 256–66
Savannah (Georgia) 120
scarlet fever 150, 194
Schudel, Alex 268
Second Plague Pandemic 2
secularization 12
seed and soil metaphor 54
segregation 177, 181, 183, 195, 196
Sénac, Jean Baptiste 46
September 11 attacks 242, 263, 265
Seven Years War 42–3, 212
Seville 15
Shadwell, A. 191
Shaftesbury, Lord (Anthony Ashley Cooper, 7th Earl of Shaftesbury) 100

Shanghai 178, 205
Shi'a Muslims 169
Siam (Thailand) 220
Siberian marmots (tarbagans) 202
Silesia 63
silk routes 3
Simon, Sir John 113–14
Simpson, Captain 100–1
Simpson, William 178–9, 197
Singapore 260, 262
slave trade 17–21, 26, 117
    abolition of 105
    and the *Eclair* 81, 91
    revolt in St Domingue 53
Slovenia 257
smallpox 17, 18, 124, 171, 314n.19
Smith, Elihu Hubbard 59
Smith, John 62
Smith, Thomas Southwood 100
Smithfield Foods 269, 272
smuggling 26, 29, 213, 215
Snow, John 142
Snow, P.C.H. 181
Soane, John 52
South Africa 234
South America 18–19
South American Steamship Company 131
South Korea
    H5N1 266
    and US beef 250–6
    and US pork 273
South Manchuria Railway Company 201
Southwell, Sir Robert 27
Southwood Committee 249
sovereignty 263
Spain
    and H1N1 270
    Phylloxera vastatrix 234
    plague 3
    sanitary measures 15, 31–6, 189, 192, 225
    SARS 257
    yellow fever 65
Spanish fever *see* Texas fever
Special Plague Service (Egypt) 188
splenic fever *see* Texas fever
Spooner, John C. 127
SPS Agreement 238, 239, 241, 246, 248, 272, 275
SPS (Sanitary and Phyto-Sanitary Regulations) 238, 245
Stangate (Standgate) Creek (Kent) 83–4
Stanley, Edward Henry *see* Derby, Lord
Start, Mr 33

steamships 87, 111, 115, 123, 140–1
Straits Convention 73
Suez Canal area 136, 146–53, 167, 172–3, 182, 210
sugar plantations 20–1
Suggestions for the Prevention and Mitigation of Epidemic and Pestilential Diseases (Maclean) 61
Sullana, Ihtizadel 161–2
Sutton, Jim 239
Swansea (Wales) 113
Sweden 25, 27
swine flu 269–75
Switzerland 233, 234
Sydenham, Thomas 46
syphilis 17
System of Complete Medical Police, A (Frank) 36

Taiwan (Formosa) *see* Formosa (Taiwan)
Tamar, RMS 109
Tangier 35
tarbagans (Siberian marmots) 202
Tariff Act 1879 (Germany) 225
ta'un 6, 7
telegraphic service 132, 133
    and disease notification 150, 209
Temple, Henry John *see* Palmerston, Lord
Tennessee (US) 119, 120
terrorism 242, 263
Texas fever 222–4
textiles, trade in 6
Thailand 220
    and H5N1 267, 268
    and SARS 257, 264
Thiers, (Louis) Adolphe 73
Tholozan, J.D. 161–2, 164
Thompson, John Ashburton 196
Thomson, W. Taylor 162–3
tibb ('Islamic' medicine) 162
Tibet 193
ticks 223
Tigris and Euphrates Company 161
time zones, agreement of international 141
The Times 84, 91
Timoni, Dr 47
tourism, and SARS 259–60, 261
Trans-Caucasian railway 141
Trans-Siberian Railway 201
transportation
    airlines 260, 267
    and animal disease 221–2
    railways 121, 131, 140–1, 186, 201, 215
    steam vessels 107, 123
    see also Eclair

Treaty of London (1840) 72
Treaty of Utrecht (1713) 31
Trichinella spiralis 225
Trieste 192
Triple Alliance (1882) 172
Tripoli, and plague 60, 154
'tropicalists' 117
Tunisia 37–8, 63, 70
Turkey 72–4, 225
Tuzuk-i-Jahangiri (Jahangir) 6
Tyne, RMS 110
typhus, ix 6, 17
Tyrawley, Lord (James O'Hara, 2nd Baron Tyrawley) 33, 34

Understanding on Rules and Procedures Governing the Settlement of Disputes (WTO) 240
United Fruit Company of Boston 133, 138
United Kingdom *see* Britain
United Nations 230
   Food and Agriculture Organization (FAO) 230, 237, 279, 280
United Provinces *see* Netherlands
United States of America
   animal diseases
      bovine pleuropneumonia 214
      BSE 250–1, 253
      cattle, Texas fever 222–4
      Pork War 225–6
   business involvement in sanitary measures 121–2
   and China 248, 274–5
   cholera 139
   H1N1 270, 272
   H5N1 267–8
   and intervention in South America 135, 137
   and the OIE 227, 228, 232
   and the Panama Canal 130–8
   plague 182, 192, 198–201
   plant diseases 234–5
      and bananas from Philippines 244
      and free-trade agreement with Australia 245–6
      fruit export to Japan 237
   SARS 257, 260–1
   yellow fever 20
      sanitary measures 52–6, 63–5, 107, 108, 119–29
Universal Postal Union 141
Uruguay 115, 116, 228

Vassiliou, Androulla 270
Venice
   Council of Health 8
   International Sanitary Conference (1892) 172–3, 182
   International Sanitary Conference (1897) 184–5, 186
   sanitary measures 12, 16, 37, 43
   war with Tunisia 37–8
ventilation 91, 99, 111, 217
Verona (Italy) 13
Vienna 151, 215
Vietnam
   H5N1 266–7
   SARS 257
Virginia (US) 120
Vivian, C. 170

waba 6, 7
Wacha, Dinshaw Edulji 181
Wade, John 58
Wallachia 38
war
   Anglo-Boer War 192
   caused by sanitary measures 37–8
   Crimean 79
   and disease 71
   Prussia and Denmark 216
   Seven Years 42–3
   Turkey and Egypt 72
   USA and Spain 129
   War of the Austrian Succession 212
   War of the Spanish Succession 31, 212
   War of Triple Alliance 115
Warren, Henry 21–2
Washington Convention (1905) 134
Watson, Mr 104
Watts, Sheldon 148
Webster, Noah 59
Wellington, Arthur Wellesley, 1st Duke of 57
Wen Jiabao 259
West Indies 25, 62–3, 104
White, F. Norman 207
Wilbin, John 101
Wilby, Dr 109
winter wheat 237
Woodhull, Alfred 120
wool trade 163–4
World Bank 260, 265, 278
World Health Organization (WHO) x, 210, 230, 238, 265
   and H1N1 270, 272, 274
   and H5N1 267
   and SARS 257, 258–9

World Organization for Animal Health (WOAH) 232, 253, 268, 272–4
World Trade Organization (WTO) 237–46, 263
  and ban on pork 272–5
  and ban on US beef 251
Worsley, Henry 32
Wu Liande, Dr 202
Wyman, Walter 127–9

X. cheopis (rat flea) 180

Yap, Arthur 243
yellow fever 19–23, 20, 52–9, 57, 99, 107, 258
  cause of 112, 113, 114, 115, 117
  contagionist view 58
  Eclair incident 80–9, 94–9, 103
  incubation period 119
  Panama Canal 130–8
  sanitary measures 115–29
  symptoms 95
  tropical origins 71, 108–15
Yersinia pestis 2, 7

Zeeland 26–7
Zhong Nanshan, Dr 257